AF572609

Environmental Technology in the Oil Industry

Environmental Technology in the Oil Industry

Edited by

S.T. Orszulik
Research & Technology Manager
Autotype International Ltd
Wantage

BLACKIE ACADEMIC & PROFESSIONAL
An Imprint of Chapman & Hall
London · Weinheim · New York · Tokyo · Melbourne · Madras

Published by Blackie Academic & Professional, an imprint of Chapman & Hall, 2–6 Boundary Row, London SE1 8HN, UK

Chapman & Hall, 2–6 Boundary Row, London SE1 8HN, UK

Chapman & Hall GmbH, Pappelallee 3, 69469 Weinheim, Germany

Chapman & Hall USA, Fourth Floor, 115 Fifth Avenue, New York NY 10003, USA

Chapman & Hall Japan, ITP-Japan, Kyowa Building, 3F, 2-2-1 Hirakawacho, Chiyoda-ku, Tokyo 102, Japan

DA Book (Aust.) Pty Ltd, 648 Whitehorse Road, Mitcham 3132, Victoria, Australia

Chapman & Hall India, R. Seshadri, 32 Second Main Road, CIT East, Madras 600 035, India

---

First edition 1997

Typeset in 10/12pt Times by Cambrian Typesetters, Frimley, Surrey

Printed in Great Britain by Hartnolls, Bodmin, Cornwall

ISBN 0 7514 0347 4

A Catalogue record for this book is available from the British Library

Library of Congress Catalog Card Number 96–86560

∞ Printed on acid-free text paper, manufactured in accordance with ANSI/ NISO Z39.48-1992 (Permanence of Paper)

Contents

Contributors

A. Ahnell	BP International Ltd, Chertsey Road, Sunbury-on-Thames, Middlesex TW16 7LN, UK
H. Amiry	BP International Ltd, Chertsey Road, Sunbury-on-Thames, Middlesex TW16 7LN, UK
C.I. Betton	Group Environment Department, Burmah Castrol Trading Ltd, Burmah Castrol House, Pipers Way, Swindon, Wilts SN3 1RE, UK
P. Calow	Department of Animal and Plant Sciences, University of Sheffield, Sheffield S10 2TN, UK
D. Caudle	Sound Environmental Solutions Inc., 11111 Katy Freeway, Suite 1004, Houston, TX 77079, USA
T. Coley	Petroleum Quality Consultant, 29 Hanson Road, Abingdon, Oxon OX14 1YL, UK
M.D. Day	Mobil Exploration & Production US Inc., New Orleans Division, 1250 Poydras Building, New Orleans, LA 70113-1892, USA
A.B. Doyle	Sound Environmental Solutions Inc., 11111 Katy Freeway, Suite 1004, Houston, TX 77079, USA
F.V. Jones	Sound Environmental Solutions Inc., 11111 Katy Freeway, Suite 1004, Houston, TX 77079, USA
M.H. Marks	Mobil Exploration & Production US Inc., New Orleans Division, 1250 Poydras Building, New Orleans, LA 70113-1892, USA
E. Martin	CONCAWE, Madouplein 1, B-1030 Brussels, Belgium
H. O'Leary	BP International Ltd, Chertsey Road, Sunbury-on-Thames, Middlesex TW16 7LN, UK
D.M. Ong	Department of Law, University of Essex, Wivenhoe Park, Colchester, Essex CO4 3SQ, UK
S.S.R. Pappworth	Sound Environmental Solutions Inc., 11111 Katy Freeway, Suite 1004, Houston, TX 77079, USA

G. Peet Consultant, Friends of the Earth Representative at the International Maritime Organization, Heemraadssyngel 193, 3023 CB Rotterdam, The Netherlands

S.C. Rapson RSK Environment Limited, 172 Chester Road, Helsby, Cheshire WA6 0AR, UK

A.A. Ryder RSK Environment Limited, 172 Chester Road, Helsby, Cheshire WA6 0AR, UK

H. Sutherland BP International Ltd, Chertsey Road, Sunbury-on-Thames, Middlesex TW16 7LN, UK

A.K. Wojtanowicz Department of Petroleum Engineering, Louisiana State University, Baton Rouge, LA 70803-6417, USA

1 Introduction

A. AHNELL and H. O'LEARY

1.1 Environmental technology

Perhaps the place to start this book is with definitions of the two key words [1]:

- **Technology** – the scientific study and practical application of the industrial arts, applied sciences, etc., or the method for handling a specific technical problem.
- **Environmental** – all the conditions, circumstances and influences surrounding and affecting the development of an organism or group of organisms.

Environmental technology is the scientific study or the application of methods to understand and handle problems which influence our surroundings and, in the case of this book, the surroundings around oil industry facilities and where oil products are used. Traditionally the phrase has meant the application of additional treatment processes added on to industrial processes to treat air, water and waste before discharge to the environment. Increasingly the phrase has a new meaning where the concept is to create cleaner process technology and move towards sustainability.

1.2 The beginning

As we begin our discussion of environmental technology, it is important to take a few moments to remember how we became so involved with this substance, oil. Regardless of our opinions about its use, oil is, and has been, the key resource in the twentieth century. From humble beginnings as a medicine and a lamp oil, oil has become the energy of choice for transport and many other applications and the feedstock for a major class of the material used today, plastic.

It is in some ways ironic that oil, initially the cheap fuel for lighting that improved many peoples' lives, next the enabler of affordable motorized personal transport and later the solution to the air pollution problems caused by coal, has become one of the chief environmental concerns of the late twentieth century. Often the fuel of choice because of price and convenience, oil was once also the 'environmentally friendly' choice. Long before the 1950s, London suffered from 'pea souper' fogs caused by stagnant air patterns and emissions from open coal fires which resulted in serious respiratory problems. These fogs caused hospitals to fill with sufferers of respiratory ailments. As a result, 'smokeless zones' were enacted and coal gas and then oil became the heating fuels of choice.

It can truly now be said we exist in a Hydrocarbon Society [2], the paradox being that we want the mobility and convenient energy oil provides but we also want a clean environment. In recognizing the need for oil, we also need to ensure that the environment is respected.

1.3 The environmental effects of the oil industry

What kind of impact does the oil industry have? One way to begin to assess this aspect is to look at the emissions, in terms of both their effect and the quantity. Although emissions data for industry worldwide are not available, some companies are now publishing their data. The data in this chapter are from BP's *New Horizons* annual HSE report, which is published as part of a policy to improve communication of the company's HSE performance [3].

1.3.1 Effect of emissions

(a) Volatile organic compounds (VOCs). The principal effect of VOCs is their local ambient ozone-forming potential in combination with nitrogen oxides and sunlight. Ozone can affect the respiratory system in humans and affect plant growth. Some VOCs, such as benzene and 1,3-butadiene, are also potentially hazardous to health if high concentrations occur. Methane can be considered separately from other VOCs as its main impact is its global warming potential, which is second only to that of carbon dioxide.

(b) Sulphur oxides (SO_x). Sulphur oxides lead to acid rain. This may corrode buildings and increase the acidity of poorly buffered soil, thereby affecting productivity. Lakes and rivers can also be affected in such a way that higher life forms, such as fish, cannot survive. Acid rain can also lead to forest damage such as defoliation.

(c) Nitrogen oxides (NO_x). Along with VOCs and sunlight, NO_x can combine to increase ambient ozone and cause photochemical smog, particularly where there is no air dispersion. NO_x can also cause acid rain. Inhalation of NO and NO_2 can affect the respiratory system directly.

(d) Carbon oxides (CO_2/CO). Carbon dioxide is the predominant greenhouse gas which could bring about global climate change. Carbon monoxide increases the lifetime of VOCs by atmospheric chemistry and can also produce ozone in its own right, although slowly. Inhalation of carbon monoxide can inhibit the transportation of oxygen around the body, but this is usually reversible.

(e) Hydrocarbons in water. Oil is a naturally occurring, biodegradable material, although the process is slow. There are different effects, from lowering the oxygen level in water due to biodegradation, to the gross contamination caused by oil spills. Contaminants in the soil can leach into groundwater and thereby pollute a potable source. Some aromatic hydrocarbon components are toxic to aquatic life.

1.3.2 *Quantity of emissions*

As can be seen in Tables 1.1 and 1.2, emissions from the oil industry represent a very small percentage of the actual production, 0.1% from

Table 1.1 Emissions from BP exploration and production activities (tonnes)

	1990	1991	1992	1993	1994
Emissions to air:					
VOCs				20 080	19 528
Methane				17 416	16 942
Sulphur oxide (SO_x)				1 592	1 588
Nitrogen oxide (NO_x)				23 885	25 821
Carbon monoxide (CO)				14 396	13 656
Particulates				392	398
Total emissions to air				77 761	77 933
Discharges to water:					
Oil in produced water	588	616	888	995	1 293
Oil on muds and cuttings	4 827	2 988	848	806	1 285
Total oil discharges to water	5 415	3 604	1 736	1 801	2 578
Total on-site disposal				17 415	0
Total off-site disposal				11 146	43 036
Total number of spills	349	370	229	272	306
Total tonnes spilled	257	287	61	142	268
Total emissions and discharges				108 264	123 803
Total production				111 674 000	120 253 000
Emissions as a percentage of production				0.10%	0.10%

Table 1.2 Emissions from BP refining and marketing activities (tonnes)

	1990	1991	1992	1993	1994
Emissions from oil refining					
Emissions to air:					
Hydrocarbons	101 133	98 400	98 590	85 690	82 450
Sulphur oxide (SO_x)	106 910	96 840	99 035	112 105	107 121
Nitrogen oxide (NO_x)	34 730	33 730	35 130	31 695	25 685
Carbon monoxide (CO)	19 180	24 600	11 100	17 510	11 040
Particulates	6 730	6 945	7 615	6 565	5 316
Total emissions to air	268 683	260 515	251 470	253 565	231 612
Discharges to water:					
Oils and greases	491	497	514	658	199
Sulphides	149	146	149	54	16
Ammonia	714	828	785	480	304
Phenols	205	214	172	97	15
Total suspended solids	1 384	1 126	1 157	1 178	775
Total discharges to water	2 943	2 811	2 777	2 467	1 309
Chemical oxygen demand (COD)	7 672	8 804	7 370	6 785	4 449
On-site and off-site disposal					
On-site	14 330	11 885	25 077	21 817	20 151
Off-site	198 443	59 465	99 473	76 774	111 506
Total on- and off-site discharges	212 773	71 350	124 550	98 591	131 651
Total refining emissions and discharges	484 399	334 676	378 797	354 623	364 578
Total refining throughput	81 603 000	81 042 000	84 996 000	87 069 000	82 263 000
Emissions as a percentage of production	0.59%	0.41%	0.45%	0.41%	0.44%
Emissions from oil marketing					
VOC emissions to air	97 040	97 060	94 070	90 050	84 150
Total gasoline sales	26 900 000	27 600 000	29 000 000	31 000 000	26 500 000
Emissions as a percentage of throughput	0.36%	0.35%	0.32%	0.29%	0.32%

upstream exploration and production and 0.5% from downstream processes of marketing and refining. In all sectors most of these emissions (60%+) are atmospheric, one third are solid waste and only 1–2% are discharges to water. Figures 1.1 and 1.2 show this diagrammatically. The actual compositions of emissions vary by sector.

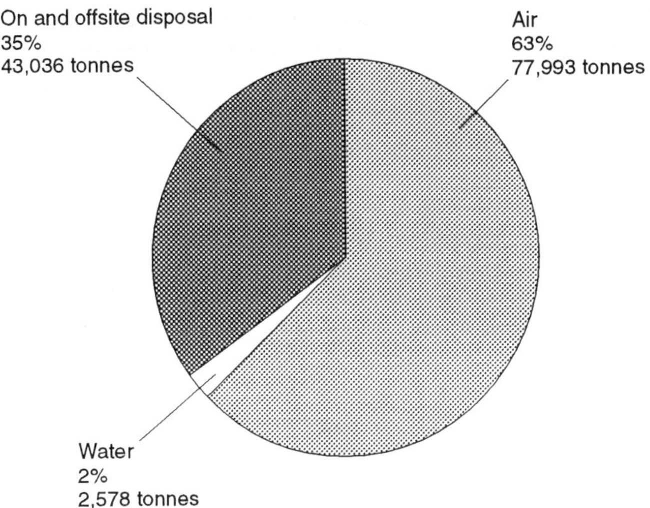

Figure 1.1 Where emissions go – exploration and production 1994.

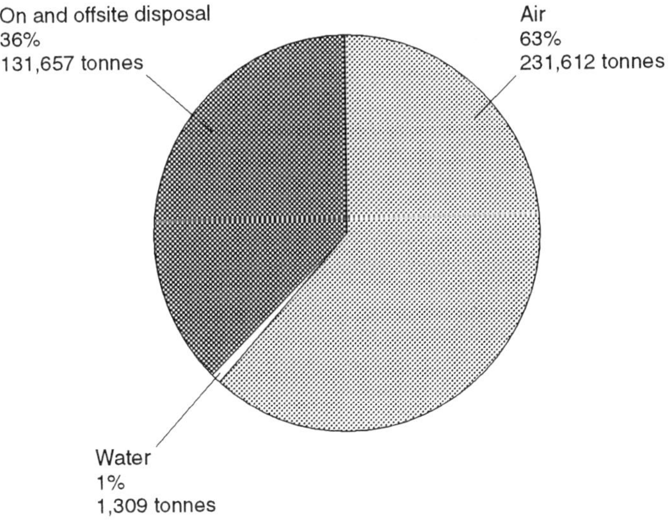

Figure 1.2 Where refinery emissions go – 1994.

(a) Exploration and production (E&P). The majority of the air emissions in E&P arise from the use of fuel or from controlled flaring and venting, which are necessary for safe operation. Almost half of the emissions are hydrocarbons, consisting predominantly of methane (Fig. 1.3). The remaining emissions, principally NO_x, SO_x and CO, are produced during

fuel combustion. CO_2 is not included in this data set because its impact is much lower on a per tonne basis and its impact on climate is still uncertain.

Oil contamination of the sea from exploration and production activities arises nowadays primarily from the discharge of 'produced water', which itself comes from the reservoir and is cleaned typically to 30 ppm before discharge. Historically some oil was discharged in oil-based drilling 'muds', but this has greatly diminished recently. Oil spills to the sea from exploration and production typically account for less than 5% of the total

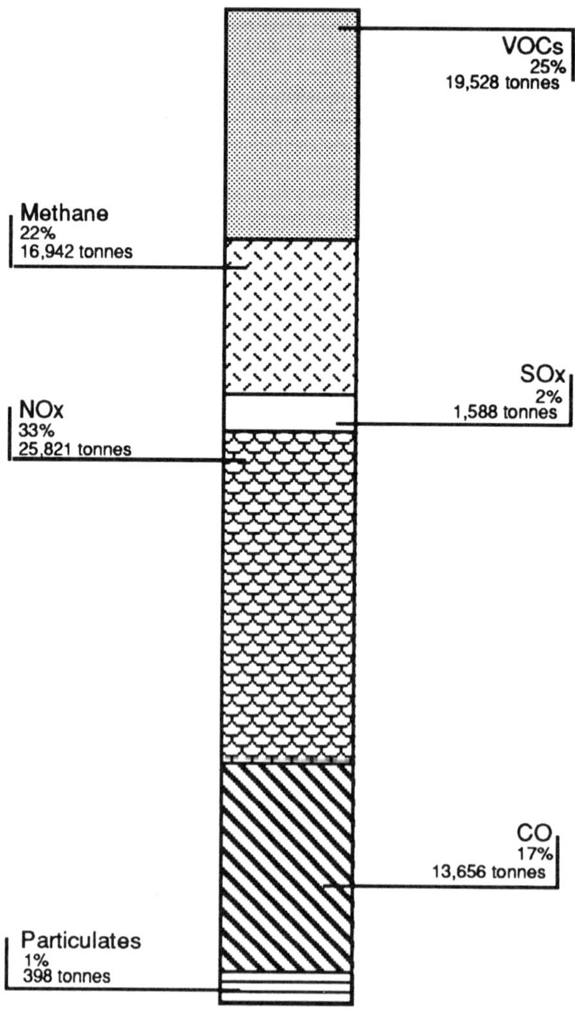

Figure 1.3 Exploration and production air emission components.

oil discharged from all sources. For example, natural seepage in the North Sea is four times as great as oil spills from E&P activities.

Table 1.1 provides the exploration and production data in tabular form.

(b) Marketing and refining. The compositions of the air emissions are shown in Fig. 1.4. The emissions are due both to the losses to the atmosphere of hydrocarbons (36%) and to the combustion products, which occur in the refining process as the chemical composition of the oil is modified to meet the product demand. Also note the downward trends in hydrocarbons emissions over time in Fig. 1.5. As the hydrocarbons lost to the air represent lost intermediates or products, there can be opportunities

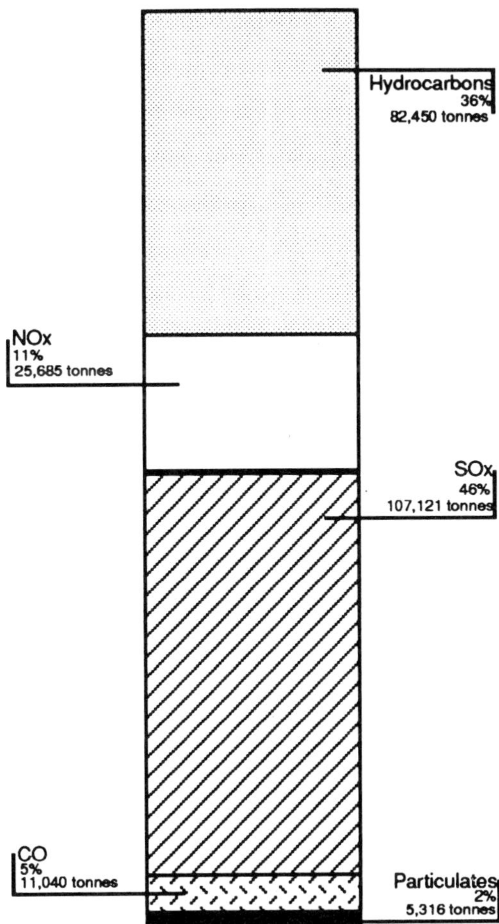

Figure 1.4 Components of refinery air emission – 1994.

to improve process efficiency, albeit small. For example, complete elimination of the hydrocarbon air emissions changes the efficiency by only 0.2%.

Water discharges in refining are increasingly due to process water, as storm water and cooling water are handled separately and recycled wherever practicable. Here efforts to reduce the amount of water used and to improve the treatment of the water prior to discharge are contributing to further reductions in the amount of contaminants discharged (Fig. 1.6). Solid waste represents the disposal of process materials such as tank sludges and soil from the closure of impoundments. There is an overall decreasing trend as refineries become more able to avoid production of these wastes and treat them in increasingly sophisticated ways before disposal.

The environmental emissions from the final part of the oil industry chain, that of distributing and selling the final product to the consumer, are essentially all VOC emissions occurring during transfer of the product. With a gasoline, about 0.3% is lost, but this amount is steadily reducing as additional vapour recovery equipment is being installed.

Table 1.2 provides the refinery and marketing data in tabular form.

1.4 Oil industry response

Pollution can be seen as a waste product and environmental management has become a major part of the oil industry. Historically, environmental

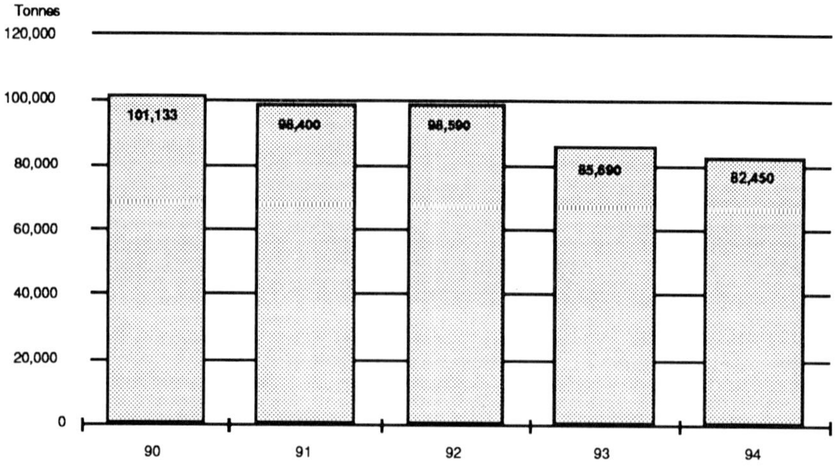

Figure 1.5 Refinery emissions to air.

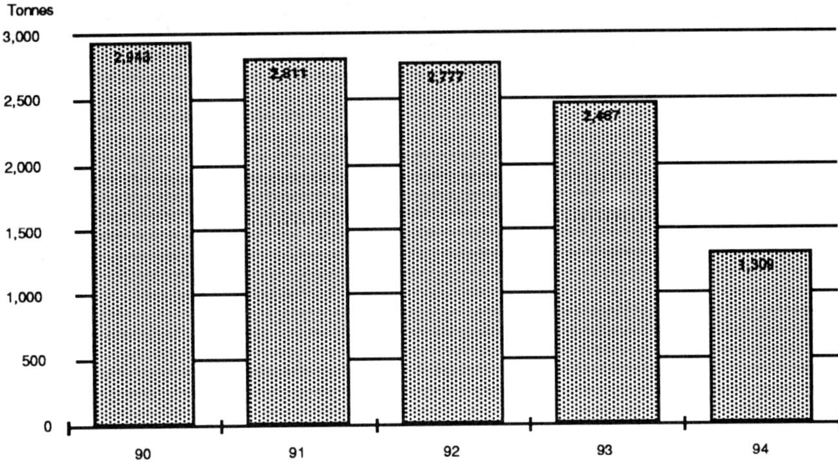

Figure 1.6 Refinery discharges to water.

management has been predominantly 'end-of-pipe' pollution control but over the last 10 years the focus has been shifting towards pollution prevention. Obviously in this introduction it is only possible to skim the surface of these areas and subsequent chapters will go into much greater detail. All pollution control techniques are very dependent on plant and process specifics.

1.4.1 Pollution control

(a) Production

Produced water. Historically, efforts have been concentrated on the separation of oil and water and the key technologies are separators, hydrocyclones and produced water reinjection.

Drilling mud. Traditionally, oil-based muds have been used. The main types of pollution control technologies are substitution by biodegradable synthetic muds and water-based muds, treatment of drill cuttings, e.g. solvent extraction and thermal treatment process, and reinjection of the ground-up cuttings into an impermeable formation. Also, ship to shore for waste treatment and disposal can be an option.

Air. Reduction of venting and flaring together with improved operational procedures and leakage minimization are some of the most cost-effective technologies applied in refineries. Purge substitution or management, flare gas recovery, compression and reuse are other control measures.

(b) Refining

Wastewater. The main pollution controls are source segregation and effluent treatment facilities. Treatment facilities can include gravity separation, e.g. APIs, plate interceptors; advanced treatment, e.g. flocculation, filtration; and biological treatment, e.g. biofilters, activated sludge. About 85% of refineries in Western Europe apply all these methods [4].

Air. The two major groups of air pollutants are VOCs and combustion products. Fugitive emissions which are responsible for the majority of VOC emissions can be reduced by improved maintenance and inspection regimes, by effective operating procedures, by improved seals on tanks and valves and by implementing vapour recovery systems. Combustion products can be reduced by improving energy efficiency, by process modifications such as low NO_x burners or dry low NO_x systems, and end-of-pipe systems such as flue gas desulphurization, e.g. Claus plants.

Waste. Sludge handling, waste minimization, recycling, management systems and regeneration (e.g. catalysts) are involved. Disposal methods include recycling, reuse and alternative fuel use, incineration (with or without energy recovery), landfill and land farming [4] (see Fig. 1.7).

(c) Marketing

Air. The key control systems are reduction of vapour pressure of the fuel, on-board vehicle carbon canisters, specially designed filling nozzles, hoses and lines to transfer vapour from vehicle tanks to service station tanks.

Groundwater. The main forms of pollution control are overfill protection, e.g. high-level alarms, and inventory control for surface water run-off, a three chamber interceptor being used. For new installations, pollution control may include secondary containment where required, e.g. double-bottomed tanks and second sleeves on piping, corrosion-resistant tanks and piping, fibre-glass underground storage tanks and closed drainage systems.

(d) Transport

Spill prevention. On sea-going tankers, double-skin vessels are being used and commissioned and procedures are continually improving. Also,

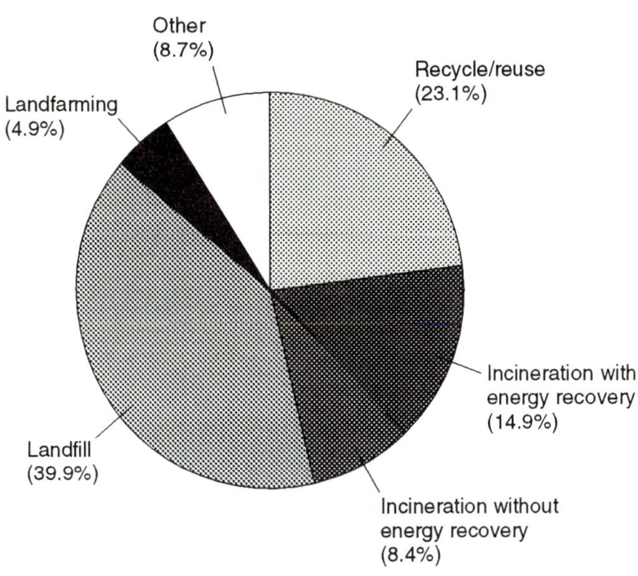

Figure 1.7 Refinery waste disposal methods [4].

ballast is segregated to avoid discharge of oily ballast water. On road tankers, bottom loading has been implemented. In loading and unloading areas, impermeable surfaces are used to prevent spills reaching underlying groundwater.

Vapour recovery. Some of the main sources of VOCs come from tanker loading and unloading; the major control technologies are closed-loop systems and vapour recovery units, liquid absorption (usually kerosine), liquefaction by refrigerated cooling and membrane systems.

1.4.2 Pollution prevention

Pollution is a wasted resource, incurring raw material costs, disposal costs, expensive treatment and increased liability from environmental risk. The oil industry has been aware for many years that it makes both environmental and commercial sense to prevent and minimize pollution wherever possible.

The basic concepts of pollution prevention or waste minimization are to identify all sources of waste (where waste includes all pollutant emissions: atmospheric, aqueous and solid discharges to all media), quantify these losses and evaluate opportunities to reduce the waste such as reduce at source, reuse or recycle (see Fig. 1.8) [4]. Examples of where these concepts have been applied are shown in Table 1.3.

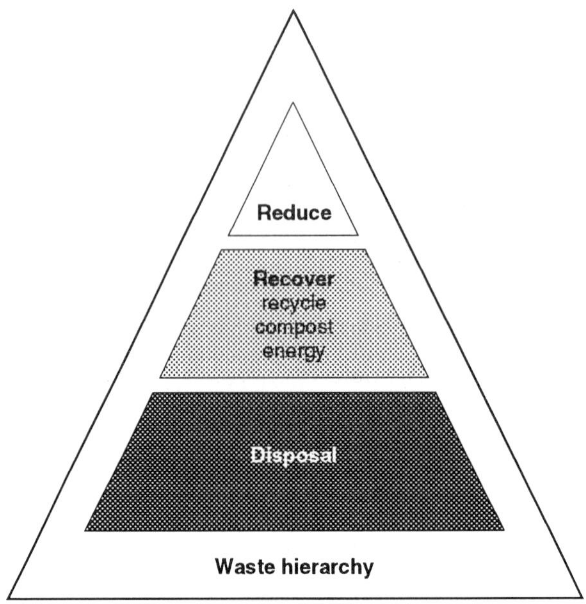

Figure 1.8 Waste hierarchy.

1.5 Oil industry future: design for the environment

The most effective way forward for environmental technology is to design-in environmental considerations, in much the same way as mechanical strength and solvent and catalyst characteristics are. There are two ways for the industry to design for the environment, that is, within facility design and within the product specification.

1.5.1 Design out the production problems

A new drilling technique – extended reach drilling (ERD), sometimes called 'horizontal drilling' – has allowed the development of reservoirs in environmentally sensitive areas, by keeping the drilling and production facilities away from the most sensitive locations, such as at Poole Harbour in Dorset, UK. This type of drilling also allows greater production from minimum facilities, which is both cost effective and environmentally beneficial.

Operators in Alaska, BP and Arco, have moved away from using surface reserve pits for muds and cuttings (a large-volume but low-toxicity waste stream) and have developed downhole injection techniques for the disposal of waste muds and cuttings to eliminate the need for surface

Table 1.3 Pollution prevention

Reduced emissions	Zero discharge to sea of drilling waste by annular reinjection Drilling wastes are the combination of drilling muds and cuttings from wells. By grinding and injecting these wastes into the impermeable layers of rock formation where they came from there is: ● no contamination of the environment; ● energy efficiency – no transportation of waste; ● cost saving in transportation and disposal charges
Reduced waste	Minimization of liquid effluent Surveys of refineries have been able to identify an average of 30% reduction in effluent flow and to reduce future capital expenditure on end-of-pipe treatment
Reduced emissions	Flare reduction scheme Flaring from BP's North Sea operations have been reduced by over 20% without additional cost by target setting, reporting, optimization and improving awareness and cooperation between onshore and offshore expertise to ensure the best solutions
Substitution	Lubricant substitution Replacement of a listed toxic catalyst lubricant with limestone, which is non-toxic, has resulted in: ● zero toxic emissions from this source; ● savings in raw material costs; ● reduced particulate emissions
Recycling	Recycling refinery oily waste By reducing the water content of the solid waste and blending with fuel to use as cement kiln fuel, it was possible to: ● reduce solid waste to landfill; ● save disposal costs; ● remove existing waste handling treatment; ● obtain approval from regulatory agencies and local community

discharge into reserve pits. In addition to the benefit of zero discharge of drilling wastes, the surface area of a well pad can be significantly reduced by as much as 70%.

1.5.2 Adjust the product

The major products of the oil industry are fuel – oil, diesel, gas and aviation fuel – and chemicals for plastics and solvents. The most immediate environmental concern concerning oil industry products is fuel oil used for transportation and its effect on air quality.

Clean fuel is generally thought to be the fuel which produces the most environmentally benign emissions, ultimately just water and CO_2, or even zero emissions. In considering what is the most environmentally benign fuel, it is important to consider the life-cycle of the fuel, including the energy (and hence combustion emissions) required to produce the fuel.

In the USA, the extreme air quality problems experienced in some areas such as Los Angeles have led to the extensive use of 'reformulated gasoline' which contains oxygenates to aid complete combustion and reduce hydrocarbons and CO, although this can lead to an increase in NO_x emissions. Oxygenates can also be used in diesel fuel, such as the 20% addition of rape seed methyl ester (RSME). This reduces hydrocarbon and CO emissions, although NO_x emissions are slightly increased and the hydrocarbon content of the particulates is under study [5].

Worldwide, there has also been a thorough and on-going evaluation of alternative fuels ranging from natural gas to solar power. At present, with the exception of niche markets such as city centres and electric cars, there is no economic alternative to hydrocarbon-powered vehicles. There is also no best or easy answer to an ideal hydrocarbon source, although research is always on-going.

Consider the case of an unleaded fuel additive, methyl *tert*-butyl ether (MTBE), which was originally developed as part of a research effort to find a commercial use for a waste by-product. It was found that MTBE was an efficient octane enhancer that could be used as an alternative to the lead-based additives widely used in gasolines. Lead is harmful to health and can also poison catalysts, particularly the platinum catalysts in car exhausts. Unleaded gasoline contains between 10 and 15% MTBE. This is a case where a decision in favour of the environment opened up a new and interesting business area.

There are many factors involved in the effects that transport can have on air quality, not only fuel but also vehicle inspection and maintenance programmes, advanced vehicle technologies and non-technical measures such as traffic management and early car scrappage. All factors need to be assessed in determining the most effective way forward [6]. Obviously, progress and product adjustment need an integrated approach.

The European Union has reported on its study 'Assessing Air Quality for European Auto/Oil'. The aim of the study was to consider all factors affecting air quality arising from mobile sources. The initial findings are that urban air quality should significantly improve as a consequence of already agreed measures and that CO and benzene should meet required air quality targets. Catalyst-equipped cars seem to have been particularly effective in reducing benzene emissions. However, NO_x is not predicted to be reduced and requires further measures. The report also indicates that the extent of exceeding NO_x air quality targets varies with cities and that differing measures would be required. For example, alternative fuel options would probably not be needed in 80% of cases but may well be required in some areas such as Athens.

Although more analysis is required, initial analysis indicates that a significant improvement in ozone air quality between 1990 and 2010 is anticipated as a consequence of already agreed measures. Also, further measures on transport are unlikely to produce significant improvements, however stringent the measures are. Even elimination of all vehicle emissions would do little for peak ozone levels. As a result, the European Commission suggests that an 'integrated' rather than 'transport-targeted' approach is necessary in targeting residual ozone problems in Europe.

1.6 Summary

Oil is integral to our society and is likely to continue to be so. The oil industry does produce emissions to the environment but these emissions are continually being minimized by the application of improved 'end-of-pipe' technology, improved design of facilities and modifications of the product specification. Further chapters in this book will deal with all these issues in much more detail.

In summarizing the issues of environmental technology and the oil issues, this chapter ends with a quotation from a former chairman of Chevron in 1987 [7]: 'Today we are in a very real sense a society of environmentalists. We all want clean water and pure air . . . Most people in industry, like most people in general, place a value on a wholesome environment . . . Time has come for industry to move beyond compliance'.

References

1. Nuefeldt, V. (ed.) (1988) *Webster's New World Dictionary*, 3rd College edn, Simon and Schuster, New York.
2. Yergin, D. (1991) *The Prize*, Simon and Schuster, New York.
3. *New Horizons, Health, Safety, and Environmental Report 1994* (1995), British Petroleum, London.
4. Concawe (1995) *Oil Refineries Waste Survey – Disposal Methods, Quantities and Costs, 1993 Survey*. CONCAWE Report No. 1/95.
5. CONCAWE (1995) *Alternative Fuels in the Automotive Market*. CONCAWE Report No. 2/95.
6. CONCAWE (1995) Assessing Air Quality for European Auto/Oil and European Auto/Oil: Initial Findings, *CONCAWE Review*, **4**(2), 6–7 and 10–11.
7. Smart, B. (ed.) (1992) *Beyond Compliance, a New Industry View of the Environment*, World Resources Institute, Washington, DC.

2 International legal developments in environmental protection: implications for the oil industry

D.M. ONG

2.1 Introduction

This chapter will address various international, regional, comparative and domestic legal issues that have arisen regarding several different aspects of the oil industry. In keeping with the 'cradle to grave' approach adopted by this book and implicit in the layout of the chapters within it, this one will begin by discussing the legal effect of the recent entry into force of the 1982 UN Convention on the Law of the Sea on the present regime governing the offshore development and maritime transport aspects of the industry.

First, this chapter will examine the jurisdictional powers for the overall protection of the marine environment that are incidental to the exploitation rights over offshore petroleum resources in continental shelf areas and exclusive economic zones allocated to coastal states by the new Law of the Sea Convention. Next, the regulation of offshore installations in these areas, including their pollution controls and the vexed question of their disposal (at sea or otherwise), will be discussed.

The chapter will then proceed to describe the latest developments in the international and national environmental laws relating to the maritime oil transport industry. These run across a whole gamut of issues ranging from proposed navigational restrictions on oil carrying and other hazardous cargoes, to the development of operational discharge standards, oil spill contingency measures and liability and compensation for oil spill damage.

Last but certainly not least, this chapter considers the legal implications of the entry into force of the 1992 UNEP Framework Convention on (Global) Climate Change for the oil industry as a whole, especially in respect of measures for carbon dioxide reduction through the imposition of carbon taxes and the possibilities for joint implementation in the achievement of emission targets between developed and developing countries. Other key legal developments arising from the Climate Change Convention, such as the need to curb emissions of volatile organic compounds, which contribute to low-level ozone, a cause of smog in cities, will also be examined.

Throughout this chapter the emphasis will be upon the new or more stringent obligations and higher standards required by a variety of legislative sources, whether at the international, regional or domestic level, and the consequent need for new or improved technologies within the oil industry in order to comply with these. An important question, which needs to be asked in relation to this perceived drive towards improved environmental performance by the oil industry as a whole, is the extent to which these new technical developments may be seen to vindicate the now widely held public perception of the industry, and in particular the major oil companies, as being both greedy and wasteful in their general business practices. To the extent that this perception has been more or less successfully mitigated by the oil industry, a related question arising from this, which will also be examined, concerns the capacity (or otherwise) of the oil industry to be more pro-active in its response to new or continuing environmental issues.

Before beginning the sectoral analysis of the various international legal developments for environmental protection which affect the oil industry, it may be worth considering certain theoretical explanations for the fact that international environmental law generally, and marine environmental law in particular, have currently become a growth industry. One such explanation derives from the application of economic theory to the development of property rights. This analysis suggests that a primary function of property rights is to guide incentives in order to achieve a greater internalization of externalities. Furthermore, property rights develop to internalize externalities when the gains from internalization become greater than the costs of internalization (Demsetz, 1967). This explanation was initially developed to explain the process of extension of private property rights over land, but Pearson argues that in the case of environmental resources, the same economic forces result in the extension of governmental regulations limiting the use of such resources. Furthermore, he suggests that these same economic forces are now at work in the area of international environmental resources generally and ocean resources in particular (Pearson, 1975, p. 18). Current and recent multilateral efforts to regulate ocean resources use can, therefore, be similarly explained. Ocean resources have traditionally been viewed as common property and subjected to a regime of free access. Over time, technological improvements, new markets, increases in demand and growing congestion costs soon combine to result in increasing the gains from internalizing externalities. At that point the prevailing property rights system gives way to more efficient patterns (Pearson, 1975, p. 19).

Another explanation which invokes a wealth appropriation motive can be added to the above efficiency argument. This suggests that the process of drawing away from a regime characterized by high seas freedoms and free access to high seas resources which is now under way, and its replacement by an extension of coastal states' rights over those resources as well as restrictions on the use of the oceans for waste disposal, are also motivated by the wish to bring the newly found wealth of the oceans under national control (Pearson, 1975, p. 22). Thus, the extension of national sovereign rights over these resources (such as fish and hydrocarbons), and the implicit or explicit limitation of other, more traditional rights (such as navigation and pollution), reflects both the attempts to appropriate new wealth and to mitigate the costs of externalities associated with ocean resource use (Pearson, 1975, pp. 22–3).

2.2 The entry into force of the 1982 UN Convention on the Law of the Sea: implications for marine environmental protection in the oil industry

Among the many international legal developments concerning environmental protection that have emerged in the years since the so-called Earth Summit in 1992 [also known as the UN Conference on Environment and Development (UNCED) in Rio de Janeiro], the entry into force of the 1982 UN Convention on the Law of the Sea on 14 November 1994, one year after the deposit of the 60th instrument of its ratification at the UN, as required by its art. 308, is arguably the 'main event'. As one writer has put it, 'during the decade following the Stockholm Conference (on the Human Environment), the environmental provisions of the United Nations Convention on the Law of the Sea (UNCLOS) constituted the single most important step forward toward the progressive development of international environmental law' (Nanda, 1995a, p. 257). The Convention is the only global agreement that provides comprehensive coverage of all aspects of the various uses, abuses and resources of the world's oceans (Broadus and Vartanov, 1994, p. 223). The same commentator also noted recently that the coming into force of the 1982 UN Convention on the Law of the Sea 'represents an important step forward in international environmental law, for it raises to binding treaty status the ideals of Principle 21 of the Stockholm Declaration, and strives to balance environmental protection and resource management with the requirements of free navigation' (Nanda, 1995b, p. 657). As Birnie and Boyle (1992, pp. 252–3) point out, 'The Convention thus attempts for the first time to provide a global framework for the rational exploitation and conservation of the sea's resources and the protection of the environment, which can be seen as a system for sustainable development, and as a model for the evolution of international environmental law'.

It is, therefore, certainly a Convention which will have great impact on the oil industry, both for the foreseeable future and in global terms. Indeed, it is already possible to chart the influence of the new law of the sea treaty on several different sectors of the marine or maritime aspects of the oil industry. These fields include the applicable principles and rules for the apportionment of the Continental Shelf and its hydrocarbon resources (not in themselves of environmental import but, as we shall see, raising several important jurisdictional issues with environmental implications); the regulation of the offshore installations utilized to extract the above-mentioned resources, as well as the eventual disposal (at sea or otherwise) of these structures once the extraction is completed; and the code of conduct for the maritime transportation of the crude products of such extraction.

One lesson in international environmental negotiation that has been learned already from the negotiation process of the 1982 Convention itself relates to the fact that this was the first intergovernmental negotiating conference in which many newly independent, developing countries actively participated. Their participation was significant in two ways: first, their arrival on the international stage further upset the already uneasy balance that had hitherto obtained between the two superpowers and their allies in the immediate aftermath of World War II and the institution of the United Nations system. The Group of 77 – a group of developing states which now number more than 120 – constituted a majority whenever a one state–one vote system was utilized to resolve stalemates occurring during the negotiation process. Second, and perhaps even more important, these states were generally united under the ideological flag of their demand for a New International Economic Order (NIEO), which basically called for an end to the perceived exploitative practices of the rich, developed (formerly colonial) countries which disadvantaged poor, developing (ex-colonial) countries, the restructuring of global economic and trade relations to reflect the above goal, as well as the redistribution of wealth through transfer of technology and resources. The developing countries' participation and stance during the law of the sea treaty negotiations was a portent of things to come, especially in respect of international negotiations on environmental issues. This was the case, for example, in the UNCED negotiations, where the above tension was manifested in the efforts of developed and developing countries, respectively, to promote environmental versus developmental priorities. The increasingly democratic nature, and consequent uncertainty, of multilateral environmental negotiations at the international level are thus an important factor for the oil industry to take into account when evaluating the evolving legal regime to control global climate change in the aftermath of the Rio 'Earth Summit' (Sebenius, 1993, p. 193). This point will be considered further in section 2.5, which is devoted to the impact on the oil industry of international policy on climate change.

As noted above, the 1982 Convention represents the first major undertaking among states to protect the world's oceans in their entirety against all potentially polluting maritime activities, as opposed to the largely piecemeal, regional and specific activity-related international law-making processes that previously characterized developments in this field of international environmental law. This general legal obligation of states to protect and preserve the marine environment extends throughout all maritime zones, from internal waters and coastal ports to the high seas (Broadus and Vartanov, 1994, p. 226). Part XII of the Convention, which is devoted to the protection and preservation of the marine environment, is an important advance on earlier and other conventions relating to various aspects of marine pollution since it formulates the obligation of environmental protection in terms which are comprehensive to all sources of marine pollution. As noted below, it applies to pollution from ships, land-based sources, sea-bed operations, dumping and the atmosphere, and provides the framework for the series of treaties, both global and regional, which have been negotiated, or remain to be so, on each of these topics (Birnie and Boyle, 1992, p. 255).

Part XII, while not providing specific rules and standards regarding the activities it aims to regulate, nevertheless lays down the broad legal framework within which all law making on the marine environment must now take place. Article 192 sets the tone for the whole of this Part by providing, for the first time in any global convention on the law of the sea, a general legal obligation upon all states to protect and preserve the marine environment. Indeed, since the obligation to protect and preserve laid down in art. 192 is given implicit priority over the more traditional (and much abused) sovereign right of states to exploit their natural resources laid down in art. 193, it is arguable that these provisions represent a stronger statement for the sustainable development of the world's natural resources than the more widely known principles enunciated in the 1972 Stockholm and 1992 Rio Declarations (Birnie and Boyle, 1992, p. 255). Thus, as Birnie and Boyle again note, these articles on the marine environment represent the culmination of a process of international law making which has effected a number of fundamental changes in the international law of the sea. Of these, perhaps the most important is that pollution in the form of dumping of wastes or discharges can no longer be regarded as an implicit freedom of the seas; rather, the diligent control of all sources in order to prevent pollution is now a matter of comprehensive legal obligation affecting the marine environment as a whole, and not simply to safeguard the interests of other states in being protected from pollution damage (Birnie and Boyle, 1992, p. 253). Article 194 elaborates the content of this obligation significantly to show that its coverage extends not only to states and their marine environment, but extends also to cover the marine environment as a whole, including the high seas areas of the world's oceans (Birnie and Boyle, 1992, p. 255). Moreover, the 'marine environment' for this purpose includes 'rare and fragile ecosystems as well as the habitat of depleted, threatened or endangered species and other forms of marine life' (art. 194.5), and is thus not confined to the protection of economic interests, private property or the human use of the sea (Birnie and Boyle, 1992, p. 255).

As noted above, however, Part XII is strong in laying down a comprehensive framework for the taking and enforcing of measures on all the major sources of pollution but weak in indicating precisely when a violation occurs and what consequences flow from this as far as liability is concerned (Birnie, 1993, p. 15). Other writers also point out that Part XII does not contain concrete marine pollution standards, nor does it purport to substitute for special agreements. Rather, its main objectives are to delimit states' competence pertaining to the establishment of concrete national and international rules and standards to prevent, reduce and control pollution of the marine environment, and to ensure that its parties apply and implement these (Broadus and Vartanov, 1994, p. 226).

At its most basic, bilateral level, the 1982 Convention provides coastal states with a formal recognition of their right, and indeed duty, under international law to protect the marine environment in the large areas of sea-bed and superjacent waters that are now within their sovereign and jurisdictional scope, if not actual territorial domain. These areas are the Continental Shelf and the Exclusive Economic Zone. For example, while the Convention provides the coastal state with sovereign rights over the Continental Shelf, for the purpose of exploring and exploiting its natural resources (art. 77), nevertheless, the coastal state is obligated to adopt law and regulations to prevent, reduce and control pollution of the marine environment arising from or in connection with sea-bed activities subject to its jurisdiction (art. 208.1). Furthermore the coastal state has to ensure that such laws and regulations shall be no less effective than international rules, standards and recommended practices and procedures (art. 208.3). As has been noted, this obligation to implement minimum international standards for the safety of operations concerning the exploration and exploitation of the sea-bed within national jurisdiction is a new one, imposed upon states parties by the Convention (Nanda, 1995a, p. 260).

2.3 The regulation of offshore petroleum installations for the protection of the marine environment

This particular sector of the oil industry is characterized both by a distinct lack of instruments for environmental protection at the global level and, where regional and national regulation has in fact taken place, by the continuing lack of uniformity and complementarity of such controls. This is true both in the case of the regulation of pollution from such activities and also in relation to the more vexed question of the disposal of such offshore installations at sea (or otherwise) at the end of their usefulness to the industry. As may be seen, the environmental issues raised by the sea-bed activities of offshore petroleum installations fall into two categories. These cover, first, operational discharges resulting from the day-to-day running of the platform and, second, the decommissioning and consequent partial or complete removal of the offshore installation, once it has come to the end of its productive life. These two categories are covered by different sectoral regimes under international law, roughly commensurate with the type of polluting activity that is being undertaken, albeit from an offshore installation rather than a sea-borne vessel.

In the first category, the requirement for state measures to control pollution from installations and devices used in natural resource extraction operations is provided for under art. 194.3(c) of Part XII of the 1982 UN Convention on the Law of the Sea, along with various other sources of marine pollution. In contrast, art. 208 is directed specifically at coastal states and not at states generally, and requires that pollution from such sources shall be subject to coastal state measures designed to minimize such pollution to the fullest extent possible. In relation to pollution from sea-bed activities, art. 208 further provides that coastal states must adopt laws and regulations to prevent, reduce and control pollution of the marine environment arising from or in connection with sea-bed activities subject to their jurisdiction. There is no explicit requirement, as is the case with the regulation of vessel-source operational discharge, for example, to give effect to generally accepted international rules and standards on this issue, with the sole exception of an exhortation that states should try to harmonize their policies in this connection at the regional level (art. 208.4). This omission is a tacit reminder that to date there has not been a single international convention setting down global standards for pollution from sea-bed activities. Moreover, states are obliged to establish global and regional rules and standards on this issue (art. 208.5). In lieu of international regulation in this field, the provisions in Part XII of the 1982 UNCLOS have merely paved the way for the possible future introduction of international rules and standards for such pollution by requiring that nationally promulgated measures shall be 'no less effective' than whatever future international rules and standards on this issue are finally agreed (art. 208.3), whatever this may mean.

In this sense, the views expressed in section 2.2, regarding the generalized rather than the particular legal effects of the 1982 UNCLOS, are supported by the above analysis. The Convention does not contain any specific rules, procedures and standards by which states should be guided or which states should adopt, when trying to implement their duties in relation to pollution from sea-bed activities. The Convention is only concerned with assigning and allocating responsibilities between states (either generally or exclusively, in this case coastal states only) for designating the promulgating rules and standards. With respect to certain sources of pollution, such as dumping and vessel-source pollution, states are obliged to act through the competent international organization generally accepted to be the IMO in order to establish international rules and standards. However, this is not the case in relation to the present source of pollution, i.e. from sea-bed activities. Offshore petroleum operations are, by reasons of geology, restricted to the confines of the continental shelf as defined by art. 76 of the Convention (i.e. including the shelf, slope and part of the rise) and will therefore always fall and be carried out under the jurisdiction of an individual coastal state. Hence, the rules and standards in accordance with which these operations should be conducted in order to prevent damage to the marine environment are generally only national rules and standards, designed and adopted as the case may be by the coastal state (Taverne, 1994, p. 117).

Therefore, with respect to the first category, as one leading publicist in international environmental law matters concluded a few years ago, in a survey of existing international instruments, there are few specific, mandatory international (as opposed to national) requirements or measures for the protection of the marine environment from pollution from offshore oil and gas activities (Birnie, 1990, p. 219). Of these instruments, 'it is apparent that very few indeed apply directly, still less appropriately, to offshore installations and the activities thereon' (Birnie, 1990, p. 203). The nearest thing to a globally applicable body of laws, providing the opportunity for the setting of international standards for polluting activities, is the non-legally binding, and carefully so-entitled, 1982 UNEP Guidelines Concerning the Environment Related to Offshore Mining and Drilling Within the Limits of National Jurisdiction, which incorporated the conclusions of a study by a UNEP Group of Experts on Environmental Law on the same issue (Birnie, 1990, pp. 208–12). As its title suggests, these UNEP Guidelines are recommendatory only, and are not subject to the usual procedures of signature and ratification that are formally required in the case of treaties to indicate the consent of participating states to the obligations contained therein. The UNEP Guidelines have, however, formed the basis of a Protocol to the UNEP Regional Seas Programme's Kuwait Convention for the protection of the marine environment in the Arabian/Persian Gulf, and the IMO-promulgated advisory guidelines for the disposal at sea of offshore installations in 1989. This non-binding but globally applicable instrument will be discussed later.

The lack of even a framework-type binding international agreement within this sector of the oil industry, setting down globally verifiable operational discharge standards for offshore installations, is a cause for much environmentalist concern. As a direct result of such lacunae, the practice of domestic states regarding such potential sources of pollution is extremely varied both in their promulgation of measures and, more important, their enforcement. In a case study of two oil and gas platforms off the Malaysian coastline, for example, it was found that neither project had been subjected to environmental impact assessment (EIA) reporting requirements in its planning stages, thus creating possible future problems of interpretation regarding the incidence and sources of environmental degradation in the area within its immediate vicinity (Low, 1990, p. 234). At the other extreme, local community opposition to proposed offshore oil exploration off the northern Californian coastline in the USA has been so successful that no amount of argument by the prospective oil companies, government scientists or federal bureaucrats appears to be convincing enough to dilute the strength of this opposition (Freudenberg and Gramling, 1994). In another national case example, that of the UK, it was recently reported that the UK Government's international nature watchdog, the Joint Nature Conservation Committee (JNCC), had advised the UK Department of Trade and Industry to ban oil companies completely from exploring in certain new areas, and to place restrictions on drilling and seismic surveys during certain parts of the year on almost all the other new exploration blocks off the west coast of Scotland in an attempt to prevent seabird losses (Brown, 1995).

In the second category, the problem of full or partial removal of the offshore installations at the end of their production lives is covered in general terms by provisions in both the 1982 Convention and its precursor in this respect, the 1958 Geneva Convention on the Continental Shelf. Following their removal, the further question of their disposal at sea may be considered generally to fall within the provisions on ocean dumping, contained in art. 210 of Part XII of the 1982 Convention, as well as various global and regional instruments on the same topic, depending on which of these the coastal state concerned has signed and ratified. The proposed alternative of disposal at sea of such installations at the end of their shelf lives is, however, steeped in controversy, and has been the focus of much media and public attention recently in the case of the *Brent Spar* debacle. These questions are dealt with later in this section.

As noted by several commentators, the *Brent Spar* incident amply demonstrated the power of public opinion in modifying, changing and eventually even reversing both big business and even governmental policy and practice, especially when it is mobilized by committed special-interest pressure groups (such as Greenpeace) on a single burning issue (e.g. Lascelles, 1995). This was certainly the case in the Shell company's climbdown over its plans for the deep-sea disposal through dumping of the disused *Brent Spar* platform, even though most of the available information and opinion on this issue would suggest that this is the most environmentally effective option, at least from the point of view of threats to individual and public health, and the onshore environment in the vicinity of any disposal site on land. It is notable that it was not only opinion in the UK that was aroused but also in the western European region.

An interesting conclusion that may be arrived at in the aftermath of *Brent Spar* is that, once aroused, public opinion may be sufficient even to overturn decisions that are based on sound science and law. However, as noted below, the international regime governing waste disposal at sea through dumping is not well equipped to handle the specific question of disposal of disused oil platforms or installations. Therefore, would the situation have arisen if the legal regime for the disposal of oil platforms had been stricter? *Brent Spar* appears to show us that laws and conventions have little meaning in particular incidents which are, or are turned into, crisis situations by media publicity and politically unpopular decision making. In the longer term, the affair raises profound questions concerning the place that environmental considerations should occupy in business decisions. It will also affect government thinking on the environment, not just in the UK but also in Europe and in much of the industrialized world. Will it always be the case from now on that the oil industry has to take actions which are seen to be for the public good, as well as for the industry? If the *Brent Spar* incident does in the end set a precedent for a more politically sensitive approach to controversial business issues, rather than one based solely on the observance of existing environmental regulatory requirements, the result is bound to create more uncertainty for companies and the higher costs that go with this greater uncertainty.

The interesting question arising from the above point is how much the general public, as taxpayers and consumers, is willing to pay for the establishment of higher, more costly environmental standards for industry generally, and the oil industry in particular. In relation to *Brent Spar*, for example, it has been estimated by Shell that it now faces a bill of £46 million to dismantle the disused platform on land, compared with the £11 million it incurred under the original scheme to tow it out into the Atlantic and sink it. Under present UK tax rules, taxpayers are expected to pay for roughly 55% of the cost of decommissioning of oil production platforms in the UK sector of the North Sea, because oil companies are entitled to offset against tax up to this amount of the cost of dismantling the rigs. Thus, if Shell's experience leads to more onshore disposals and the UK government does not change the law, taxpayers could end up paying hundreds of millions of pounds extra to subsidize abandonment programmes (Reguly, 1995).

On the other hand, there are already some suggestions that the oil industry is working hard to avoid the public relations fiasco that characterized its approach to the *Brent Spar* incident. Among the proposals are more direct contact between the oil industry and environmental pressure groups and more frequent contact between the industry and the European Commission and European Parliament (Corzine, 1995). In this context, however, it is important that the oil industry does not make the same mistake as American oil companies did in California by assuming that merely providing for better communication of information will help to dispel any lingering feelings that the general public might have against the oil and gas industry, whether it be concerning exploration, production or the disposal of installations involved in these activities (Freudenberg and Gramling, 1994).

With respect to the question of the whole or partial removal of offshore installations, it is interesting to note that unlike nearly all other areas of the oil industry, which have seen an unprecedented rise in recent decades of environmentally associated regulation, the public international law regime on this issue has been rendered more rather than less ambiguous with the entry into force of the 1982 Convention and its replacement for its parties of the regime instituted by 1958 Geneva Convention on the Continental Shelf. Again, this is in keeping with some of the more critical reviews of the 1982 Convention which suggested that the generalized style of language utilized to paper over disagreement on controversial issues during the negotiations would 'come home to roost' when the Convention came into force and would inevitably become subject to intense legal scrutiny as to the precise obligations that UNCLOS purports to assign to coastal and other states. As two commentators have recently noted, the key questions in the case of the abandonment of offshore structures are whether the installations have to be removed, and if not, whether they may be left wholly or partly in place (Jones and Saunders, 1994, p. 240). To these questions, however, must also be added the further important one of the legality of disposal at sea after the whole or partial removal of these installations.

Article 5.5 of the 1958 Geneva Convention on the Continental Shelf provides that initially due notice must be given of the construction of any such installations and that permanent means for giving warning of their presence must be maintained. Furthermore, any installations which are abandoned or disused must be entirely removed, although the ultimate means of their disposal is not specified, even in general terms. As the two writers above note, however, there has been for some time now evidence of a tendency among directly involved states to avoid entire removal as a mandatory policy and instead to try and establish standards and criteria for partial removal which might meet with general acceptance (Jones and Saunders, 1994, p. 24). The UK government, for example, has argued that when complying with its international obligations under the 1958 Convention, the relevant provisions must be interpreted in a manner consistent with customary international law and furthermore that it supports a generally held view that art. 5.5 of the 1958 Convention does not now reflect customary international law. According to the UK, this provision also must be read in the context of art. 5.1 of the 1958 Convention, which requires only that there must be no unjustifiable interference with legitimate uses of the sea, such as navigation, fishing and scientific research. There is some support for this view in that it is commonly accepted that there is not enough evidence of a clear practice on the part of non-ratifying states for entire removal of offshore structures to support the view that the 1958 Convention as a whole has come to be accepted as representing customary international law. Moreover, the UK also provides other arguments in support of its view, to the effect that the total removal obligation is no longer binding upon it, such as claiming that (1) this requirement has ceased to have legal effect since it has fallen into desuetude, (2) the provision was established to overcome problems of navigational safety which through improved technology and experience have shown themselves not to require the total removal of structures and (3) the circumstances which gave rise to the requirement for total removal have changed fundamentally (Jones and Saunders, 1994, p. 241).

There is also a widely held view that the rule of customary international law governing this issue of the abandonment of offshore structures at sea is now reflected in the principles set out in the 1982 UN Convention on the Law of the Sea, rather than the 1958 Convention, especially now that the 1982 Convention is in force (Jones and Saunders, 1994, p. 241). As mentioned earlier, the provisions under the 1982 Convention are much more ambiguous in their phraseology, thus allowing some support for the argument against mandatory total removal of abandoned structures. Article 60.3 of the 1982 UNCLOS states that any installation or structure that is disused shall be removed to ensure safety of navigation, taking into account any generally accepted international standards established in this regard by the competent international organization. Such removal shall also have due regard to fishing, the protection of the marine environment and the rights and duties of other states. Appropriate publicity should be given to the depth, position and dimensions of any installations or structures not entirely removed. This article seems to hold to a basic rule of removal but which is capable of being modified if agreed international standards permit.

Now that the 1982 Convention is in force, once the UK and other major offshore oil producing states ratify this treaty, the possibility of only partial rather than full removal of abandoned installations, subject to regard for any international standards, may lead to the development of a new rule of customary international law releasing states from the total removal rule obtaining under the 1958 Convention (Jones and Saunders, 1994, p. 241). Article 60.3 therefore provides a starting point for a new practice on abandonment. As has been noted, this article appears to permit, in certain circumstances only, partial rather than full removal of offshore installations and, moreover, appears to grant coastal states a measure of discretion in the methods utilized for the abandonment of these structures once they are decommissioned insofar as they are silent on this aspect. These methods may involve leaving the facilities in place; partial removal, followed by toppling of the facilities; whole or partial removal of the facilities and subsequent dumping thereof in deep water; and whole or partial removal followed by transfer onshore for dismantling and landfill disposal. Article 60.3 also refers to standards set by the competent international organization, which in the case of the 1982 UNCLOS is generally accepted, when the reference is to 'the' and not 'a' competent body, to be the International Maritime Organization (IMO). In this respect, the IMO adopted 'Guidelines and Standards for the Removal of Offshore Installations and Structures on the Continental Shelf and in the Exclusive Economic Zone' (henceforth, IMO Guidelines) in October 1989, which laid down detailed standards relating to the removal of different categories of offshore structures in different circumstances. For example, they require the complete removal of (1) all existing structures in less than 75 m of water and weighing less than 4000 tonnes in air (exclusive of any deck or superstructure which would anyway need to be removed) and (2) of new structures put in place after 1 January 1998 in less than 100 m depth and weighing less than 4000 tonnes. In cases of partial removal or toppling, there must be an unobstructed water column over any remains of 55 m. However, these Guidelines apply only the generally accepted rule of removal, which is itself regarded as subject to the possible exception of allowing partial removal or even non-removal, where it does not interfere with other uses of the sea. The coastal state, having jurisdiction over the structure, may decide whether they should remain in whole or in part, by reference to such criteria as safety, environmental effects, effects on other uses of the sea, risk of shift, excessive costs, technical feasibility, risks of injury to personnel engaged in removal and determination of any possible new use or other reasonable justification for allowing the installation or structure, or parts of it, to remain. Furthermore, the coastal state can apply these criteria on a case-by-case basis (Jones and Saunders, 1994, pp. 241–2 and 248–9).

In this context, it is important to note the different positions adopted by various groups of states during the negotiation of these IMO Guidelines in 1988–89. Many states, especially developing countries, opposed the possibility of leaving any offshore structures *in situ* on the Continental Shelf. States with major oil industries, however, were only in favour of partially removing the huge concrete platforms of certain structures, although they were willing to agree to remove these totally in shallower waters. As with international negotiations on most issues, the final outcome in the Guidelines reflects a compromise position between these two standpoints.

As mentioned earlier, the further question of whole or partial removal for the purpose of deep ocean dumping may be considered under the global and, where applicable, regional regimes governing dumping generally, owing to the lack of a specific body of rules relating to the dumping of these structures alone. In this context, both the 1972 London Convention on the Prevention of Marine Pollution by Dumping, also known as the London Convention (which came into force in 1977), and the 1972 Oslo Convention for the Prevention of Marine Pollution by Dumping in the North Atlantic region (to be replaced soon, when it comes into force, by the 1992 Paris Convention, which amalgamates both this convention and the 1974 Paris Convention on Land-based Sources of Marine Pollution) need to be discussed. The London Convention parties have agreed that abandonment on site and toppling of oil platforms should be considered to be dumping and thus falling within the provisions of this Convention. The state party proposing disposal will therefore be required to apply for a permit for dumping, which can only be granted according to the criteria set down in the London Convention (Jones and Saunders, 1994, p. 242).

The parties to the regional Oslo Convention have drawn up further specific guidelines regarding abandonment in waters within its geographical scope, in the North Atlantic. While the IMO Guidelines apply to all cases of proposed abandonment, the Oslo guidelines only apply to disposal at sea of fixed structures (although the text also deals with disposal *in situ*). The technical criteria for disposal at sea are not necessarily mandatory and a contracting party could issue a permit in appropriate cases even if only some of the criteria are met. The Oslo Convention currently applies stringent criteria governing the depth of waters (2000 m) and distance from the nearest landfall (150 nautical miles; ca. 275 km) for deep-water dumping (Jones and Saunders, 1994, p. 242).

The existence of slightly more international guidance, standards and criteria (as opposed to international conventions or treaties) on this question of full or partial removal and consequent disposal at sea of these structures has not reduced the potential for controversy over the merits (or otherwise) of deep-ocean or sea-bed dumping of wastes, which is really the broad environmental issue here.

2.4 Developments in the international legal regime governing the environmental effects of the maritime oil transport industry

In stark contrast to the nascent and almost non-existent state of global (as opposed to regional) environmental regulation of offshore oil and gas platforms (section 2.3), and the still emerging international legal regime for the mitigation of global climate change (section 2.5), the system of international legal rules governing the control of vessel-source marine oil pollution is now both well established and multifarious in terms of the number of different international instruments covering various aspects of vessel-source marine oil pollution. As noted by one commentator, this was the first form of marine pollution to arouse the attention of the world community and, therefore, the first to receive concerted political and legal treatment at the international level (Rémond-Gouilloud, 1981, p. 196). These international instruments range, for example, from international conventions between states governing deliberate, operational oil discharges from vessels (1973/78 MARPOL Convention) and high seas intervention in the case of accidental oil spills from vessels in distress (1967 Intervention Convention), to voluntary international agreements between industrial participants (as opposed to states) to provide for compensation in the event of damages sustained from such accidental spillage on the basis of strict liability, subject to previously defined limits of compensation (TOVALOP, CRISTAL and PLATO).

Similarly, unlike the former – and almost non-existent – regime for the control of pollution from offshore installations and the latter emerging regime for curbing greenhouse gas emissions, the regime in issue in this section has developed through a well established international organization with authority originally over shipping issues, primarily those concerning maritime safety, although it has subsequently greatly developed its role in the protection of the marine environment from vessel-source pollution. This is the International Maritime Organization (IMO), a specialized agency of the United Nations. The IMO Assembly is composed of all member states, meeting every other year in plenary sessions. In 1977, the Assembly revised its constituent instrument in order to give it constitutional authority to expand its role beyond purely consultative and advisory functions. The Assembly's most important international law-making function is the issuing of recommendations to member state governments for the adoption of regulations and guidelines concerning maritime safety and marine pollution control. Also, by providing states with an on-going diplomatic forum for the discussion of such issues, the IMO [previously known as the Inter-Governmental Maritime Consultative Organization (IMCO)] has, since the 1970s, facilitated the transformation of these concerns into global agreements and has established effective equipment standards that have removed many practical and legal barriers that impeded enforcement of earlier international agreements (Mitchell, 1993, p. 184). A great many IMO recommendations stem from authority granted to IMO under multilateral conventions dealing with these matters, complementing the Assembly's powers under the IMO Convention itself.

Again, as in the international negotiation process for the 1982 UN Convention on the Law of the Sea noted above, the arrival of many newly independent developing states within the ranks of the IMO during the 1970s made a substantial difference to the international regulation of this issue area. Many of these countries had little domestic experience of, or concern for, marine oil pollution but supported strong controls because they expected few direct costs from such regulation and hoped that pollution controls would help establish jurisdictional precedents, especially in support of increased coastal state jurisdiction, during the then ongoing Third UN Conference on the Law of the Sea (UNCLOS III) negotiations, which culminated in the 1982 Convention. Developed states lacking strong oil and shipping interests, such as Canada, Australia and New Zealand, also took environmentalist positions. Only this combination of factors allowed environmentally concerned governments to overcome the resistance of the oil and shipping interests, and the maritime governments that supported them, in order to obtain agreement on the expensive but effective equipment requirements necessary to reduce intentional discharges. By 1973, these developments provided the votes needed to counter the power of maritime states and industry and to adopt international regulation that began to require real national and industrial policy responses. While the shift did not occur overnight, the international political bargaining process at the IMO was no longer weighted exclusively in favour of shipping interests (Mitchell, 1993, pp. 186 and 193). This has enabled the IMO not only to set international standards for ocean shipping, but also to attempt to ensure the efficacy of these standards through flag, coastal and port state compliance and enforcement (Kirgis, 1995, p. 715). Having finally achieved stringent regulations on paper, especially in respect of deliberate vessel-source marine oil pollution through the 1973/78 MARPOL treaty, the IMO has in recent years sought to redirect its focus to compliance. Progress on this front has been limited, however, by the fact that many developing states have opened shipping registers, even though they can offer little control over the ships registered under their flag, as outlined below.

This is not to suggest, however, that the present international legal regime is a wholly successful one. As one commentator has noted recently, 'During the last two decades, marine oil pollution has become an emotional and evocative issue as an increasing number of major oil spills have wreaked havoc on aquatic and bird life and on the fishing and tourism industries of coastal states' (Ellis, 1995, p. 31). She cites United Nations estimates that some 600 000 tons, approximately 4.2 million barrels, of oil spillages and seepages occur annually as a consequence of normal oceanic shipping activities. In this respect, it is important to point out that although dramatic accidental oil spills generate intense publicity, the amount of oil finding its way into the sea through such accidental oil spills is only about 12.5% of the total. Indeed, as Mitchell notes, the intentional discharge of oil during tanker operations has consistently overshadowed accidents as the major source of ship-related oil pollution that soils beaches and oils sea-birds (Mitchell, 1993, p. 183). In fact, most marine oil pollution originates either from land-based sources or through deliberate discharges by ships. Somewhat more controversially, Ellis also suggests that the use of supertankers has increased the hazard inherent in transporting oil by sea because of the far greater quantity of oil that is spilled when an accident occurs (Ellis, 1995, pp. 31–2).

Despite the great number of international conventions on this issue of accidental and/or deliberate vessel-source marine oil pollution which have received broad acceptance and support in the international community, to date these conventions have proven to be inadequate for controlling this source of pollution. As Mitchell notes, for example, nations have sought international regulations to address intentional vessel-source oil pollution for more than six decades, yet it is only in the last decade and a half that oil entering the ocean from tanker operations has begun to decrease (Mitchell, 1993, p. 183).

Two reasons have been suggested as the main causes for this failure. The first and more important reason, especially in the short term, is that the concept of the national sovereignty of states over their registered vessels which fly their flags serves to prevent the successful enforcement of the current system. The primarily flag state jurisdiction over shipping is notoriously weak in this respect. Pollution controls are not easily applied on vessels of many different 'nationalities' that move almost continuously around the world, in and out of every kind of maritime jurisdictional regime, many of which seldom, if ever, return to ports or waters of their states of nationality. Even in the presence of strong evidence that an unlawful discharge has occurred, it is often easy for the suspected offender to avoid detection by an investigating port or coastal state and thus evade proceedings against it. The difficulties inherent in applying normal judicial procedures against vessels are also a function of the economic structure of the world shipping industry. Until recently, up to two-thirds of the world's merchant shipping was registered in a small number of states, many of which were more interested in protecting the commercial interests of the ships flying their flag than in protecting the marine environment. Furthermore, many of these 'flag states' were, and many still are, also 'flags of convenience' states, that is, they maintain open registry systems, permitting the registration of almost any vessel, regardless of its true nationality or, more importantly for these purposes, the seaworthiness of the vessels. Such states are often reluctant to implement any treaty-based restraints upon their registered shipping, or even to sign and ratify vessel-source pollution control treaties (Rémond-Gouilloud, 1981, p. 196). Since the flag state itself must first consent to and then enforce any conventional obligations and take enforcement action against any breach by ships registered in that state, the national interests of the flag state in its shipping industry may prevent it from enforcing its jurisdiction over its own vessels (Ellis, 1995, p. 33). As Mitchell notes further, 'even when domestic calls for action (in relation to the oil pollution problem) have been loud, a government's international support for strong measures depended on the level of opposition from domestic oil and shipping concerns' (Mitchell, 1993, p. 192).

Oil companies also initially resisted any regulation, but have supported international negotiation to avert the competitive disadvantage inherent in unilateral regulation of the problem by coastal states, the marine environments of which may be damaged by their activities. Even when enough states with domestic oil pollution concerns felt that the necessity to take some initial international steps had arrived, they could only convince oil and shipping interests and the states representing them in the international arena that they should at least accept some no-cost pollution controls in order to avoid a patchwork of unilateral legislation. However, when efforts to impose real costs on the industry began to emerge, the industry quickly showed its power both to ignore existing and proposed regulations and to demand less costly rules (Mitchell, 1993, p. 192).

The second, longer term and more systemic-based, reason that international conventions have failed is because environmental protection generally has little value within current economic models. Mainstream economic thought has, until recently, either ignored the environment or had considerable difficulty in placing a 'value' on it. Environmental protection was externalized by economists who found it either too insignificant or too difficult to include as a cost associated with human activity (Ellis, 1995, p. 32). Again therefore, it can be noted that the driving force behind most international environmental regulation in this area has been economic in its origin, rather than due to any increasing industrial awareness of the environmental consequences of their activities on the part of the oil companies. As noted earlier in the Introduction (section 2.1), it is only through progressive 'internalization' of environmental costs within the overall industry cost–benefit analysis that incentives to clean up its activities and reduce its pollution output become evident. The economic theory requiring the internalization of environmental externalities has manifested itself in law as the 'polluter-pays' principle (Ellis, 1995, p. 57). Applying this principle, the costs of pollution, whether damaging local, regional or global environments, should be borne by the producer or company that causes the pollution, rather than falling on the general community, whether local or international, to be paid through reduced environmental quality or increased taxation in order to mitigate the environmentally degrading effects of such pollution (Ellis, 1995, citing Gaines, 1991, pp. 468–9 and 487). However, many commentators still contend that this is an economic principle for the distribution of costs rather than a legal one, although it does appear in some treaties now. For example, it is recognized in the Preamble of the 1990 Oil Pollution Preparedness, Response and Co-operation (OPRC) Convention (see below). At the regional level, it is included in the 1992 Paris Convention for the Protection of the Marine Environment of the North East Atlantic as a general principle to be applied, along with the 'precautionary' principle, under art. 2(2) of the Convention, which is yet to enter into force. Ultimately, however, the consumer does pay, rather than the polluting company, through the transferring by the latter of its costs by increasing the price of its product, even though insurance may absorb some of the burden.

Even the implementation of the polluter-pays principle has not been without its problems. This is mainly due to its assumption that everything, including the environment, has a market price attached to it (Ellis, 1995, p. 57). Although this process does provide the environment with a notional market value, the quantification of this value is predicated upon how much environmental damage is actually seen to affect the economic interests of humanity or groups of individuals, rather than by how much it affects the global ecosystem. Oil pollution of the high seas, for example, is not considered 'damage', as defined in the 1969 Civil Liability Convention, as no individual property interests are injured. Later Protocols to this Convention are less rigid but they have only just entered into force, at the end of May 1996. Although the pollution exists, ecosystems have been damaged and a delicate ecological balance adversely affected, neither the existing system nor the polluter-pays concept provides a realistic or workable solution (Ellis, 1995, p. 58).

In particular, the value of any damage to the ecological balance of the oceans *per se* due to marine oil pollution is not susceptible to quantification utilizing the usual methods. While it would be preferable to focus on the overall depletion of the ecosystem and biodiversity as a 'value' in itself, the incorporation of such an ecosystem value approach within the usual economic cost–benefit analysis would be problematic. To deal effectively with such global externalities requires a degree of international co-operation that has been noticeably lacking in this sector of human activity. Current international conventions provide a weak and ineffectual regime to enforce the compliance of shipping safety and environmental standards. In contrast, they provide a well structured process for victims of oil pollution to gain compensation, albeit liability under them being limited. This too can be regarded as a reinforcement of the market-oriented economy in that it is easier and less costly to try to quantify and compensate for damage to property and economic interests suffered by individuals (subject to limitation of liability) than it is to prevent such damage from occurring in the first place (Ellis, 1995, p. 58). The scale of some recent spillages, however, as in the *Amoco Cadiz* and *Exxon Valdez*, not to mention the *Braer* and now *Sea Empress* groundings, have called this system of strict but limited liability into question.

In the following pages a short description is given of the present regime governing accidental and deliberate vessel-source oil pollution. This analysis will also show the evolution and eventual inclusion of the polluter-pays principle within the developing legal regime. An interesting point in relation to this development is that the most progressive system of strict liability and compensation for environmental damage caused by marine oil pollution was set up by the oil and shipping industries themselves by way of voluntary agreements in order to supplement the negotiating efforts towards a conventional regime, and in return for definitively set limits for the amounts to be compensated. As Ellis has noted above, given our current theories of economics, the polluter-pays principle provides the best means to ensure that all the costs of oil production, transportation, refinement and consumption, including damage caused by it to the environment, are incorporated into its price. However, the problem of initiating and enforcing the polluter-pays principle at the international level highlights the inadequacy of the current international legal structure.

The overall lack of success of this regime is further evidenced by the newly evolving trend involving increasing attempts by coastal states in many different regions of the world to legislate unilaterally for jurisdiction to enforce higher discharge standards and even navigational restrictions over errant vessels within their national maritime zones. This expansive jurisdictional trend is both *quantitative*, in terms of incorporating both the exclusive economic zone (EEZ), as well as the territorial waters of coastal states, and *qualitative*, in terms of providing for more stringent rules and higher standards regarding deliberate or accidental vessel-source marine oil pollution. The discussion in this section will be rounded up with an examination of a few examples of this trend in state practice and the obvious threat it poses for the continued freedom of navigation currently enjoyed by the maritime sector of the oil transport industry throughout the world's oceans.

2.4.1 *The Law of the Sea Convention*

Again, however, it is important to note the impact that the 1982 UN Convention on the Law of the Sea has had on this issue area. Indeed, Ellis argues that the Convention is perhaps the most far-reaching attempt to date to prevent and regulate marine pollution, as the obligations of coastal, port and flag states are all considered within its provisions (Ellis, 1995, p. 42). In this issue area, as in other areas of maritime activity, states have also attempted to establish a broad framework of rules under the Convention, as part of its 'umbrella' style. At the time when the negotiations under the Third UN Conference on the Law of the Sea (UNCLOS III) began, there was a dearth of conventional and customary international law concerning the prevention and control of pollution in the marine environment in general, and vessel-source pollution in particular. The provisions drafted under UNCLOS III and now in force in the 1982 Convention have gone a long way towards filling this void, both in and of themselves and also in reflecting and giving impetus to the particular treaties and other agreements discussed below (Schneider, 1981, p. 210). However, as noted in the Introduction, although the 1982 UNCLOS now provides a comprehensive mechanism to prevent and regulate marine pollution, its wording has been criticized as being ambiguous and relying too heavily on generalized formulations (Ellis, 1995, p. 42, citing Dzidzornu and Tsamenyi, 1991, p. 281).

The standard-setting articles dealing with operational and accidental vessel-source pollution, as well as dumping, were especially controversial during their negotiation and are extremely complicated in their wording within the text of the Convention. In skeletal form, the regime has been described in this manner: first, in the territorial sea, coastal states may exercise their sovereignty to establish anti-pollution laws and regulations, [arts 21.1(f) and 211.4] provided that they 'shall not apply to the design, construction, manning or equipment of foreign ships unless they are giving effect to generally accepted international rules and standards' (art. 21.2). Second, in the exclusive economic zone, coastal states can legislate to prevent dumping (art. 210) and may establish certain other laws and regulations giving effect to generally accepted international rules and standards for ship-generated pollution (art. 211.5). Third, there are certain supplementary provisions for the adoption of special mandatory measures for prevention of vessel-source pollution within particular, well defined 'special areas' of the exclusive economic zones of coastal states due to their oceanographic and ecological conditions, as well as the protection or utilization of their resources. These measures, which may include navigational restrictions as well as discharge standards, can only be enacted by the coastal states after consultation with other states, through the appropriate international organization, which in this case would be IMO (art. 211.6) (Schneider, 1981, p. 210).

In the area of enforcement competence, the 1982 UN Convention introduced several important jurisdictional innovations, especially related to port and coastal state enforcement. While traditionally regarded as the sole province of flag states, new enforcement powers are now recognized to lie with port and coastal states. Port states are empowered to undertake certain enforcement procedures in respect of vessel-discharge violations – even those occurring outside their internal waters, territorial seas and exclusive economic zones (art. 218). Also, under certain clearly delineated circumstances, coastal states will be able to cause proceedings and other measures to be taken in respect of violations of national law and regulations, not only in their territorial sea but also in their EEZs, where these give effect to international rules and standards (art. 220). Enforcement powers are, however, subject to certain highly detailed safeguards to make sure that freedom of navigation is maintained alongside efforts to ensure environmental protection (Part XII, Section 7, arts 223–233). These safeguards include a basic provision requiring non-discrimination against foreign vessels (art. 227), flag state pre-emption rights over enforcement of violations (requiring, at the request of the flag state, suspension of proceedings that may have been initiated by other states) (art. 228), establishment of monetary penalties only for violations beyond internal waters, except in cases of 'wilful and serious' acts of pollution in the territorial sea (art. 230), and a special article on safeguards with respect to straits used for international navigation (art. 233).

The 1982 Convention has therefore been criticized as being dominated by and catering to maritime commercial interests and the states in which they are located (Ellis, 1995, p. 42, citing Stephenson, 1992, p. 133). Despite meaningful concessions to both port and coastal states, the regime of enforcement by the 1982 UNCLOS continues to recognize the flag state as the principal repository of jurisdiction over its own vessels (Ellis, 1995, p. 42, citing Collins, 1987, p. 288). Although there have been moves to improve the monitoring of compliance by flag states of international discharge standards with the establishment of a new IMO body, the Flag State Compliance Sub-Committee, it is difficult to see how much this will achieve when each flag state's sovereignty continues to be jealously guarded. In an area of such economic and political importance, it is difficult to believe that this new Sub-Committee will be given sufficient authority to influence the dominant *laissez-faire* trend (Ellis, 1995, p. 42). Indeed, it has been noted that practice by the major maritime states since the signing of the 1982 UNCLOS confirms that coastal state economic and environment interests will remain in second place to the interests of flag states (Ellis, 1995, p. 42, citing Dzidzornu and Tsamenyi, 1991, p. 287).

2.4.2 The Intervention Convention

When the *Torrey Canyon*, the third largest oil tanker in the world at the time, went aground in 1967, two conventions had already been adopted concerning marine pollution. One, the 1962 Convention on Nuclear Powered Ships was inapplicable, whilst the other, the 1954 Convention for the Prevention of Pollution of the Seas by Oil, was ineffective owing to the unwillingness of state parties to prosecute the owners of vessels flying their flag for violations under the Convention. Furthermore, observers unanimously agreed that finding evidence of a prohibited oil discharge was almost impossible in most cases (Rémond-Gouilloud, 1981, p. 198). The difficulty of establishing a causal link between the perpetrator of any polluting action and the damage to the environment resulting from such action is exacerbated by the relative paucity of general rules of customary international law for the protection of the environment. As one leading practitioner noted, under customary international law, there are no rules or standards related to the protection of the environment as such (Brownlie, 1974, p. 1, cited in Ellis, 1995, p. 38). The question is whether this is still wholly true after the Stockholm and now Rio Declarations, and what has been state practice in the past 24 years or so.

All efforts before 1967 can be regarded as belonging to a 'pre-historic' era of marine pollution control. Previous to that date, harbours, oyster beds and some vulnerable fishing grounds were the only marine areas subject to adequate protection under antiquated national legislation. As to international controls, the real work started in 1967, when the British and French governments, alarmed by the considerable damage they had suffered from the *Torrey Canyon* oil spill, convened an inter-governmental conference specifically in order to address this issue of intervention in the high seas in the case of a pollution emergency. Since then much of the legal development that has taken place in the field of marine pollution control has, often very conspicuously, been accelerated, facilitated or even initiated as a direct result of other comparably serious pollution incidents that have threatened to pose a serious hazard or cause major damage (Rémond-Gouilloud, 1981, pp. 198–9).

The 1967 conference culminated in the first of a number of international conventions to emerge after the *Torrey Canyon* incident, namely the 1969 International Convention Relating to Intervention on the High Seas in Cases of Oil Pollution Casualties (the Intervention Convention). The Intervention Convention enables a state to intervene beyond its territorial waters if its shores or related interests are threatened by a marine accident. The action taken by the threatened state must be reasonable (arts I, III and V). Indeed, compensation may be required should the action taken be more than is reasonable to counter the potential damage (art. VI) (Ellis, 1995, p. 39).

2.4.3 The Civil Liability and Fund Conventions

Closely allied to the Intervention Convention, in terms of both chronology and issue area connectedness, was the 1969 International Convention on Civil Liability for Oil Pollution Damage (CLC). Having secured a conventional right to intervene in the case of a high seas emergency, the same countries turned their attention to the issue of liability for compensation in the event of damage occurring to their marine environments as a result of said high seas emergency. The CLC established strict but limited liability for compensation for oil pollution damages. Under the CLC, a levy imposed on ship owners generates funds to compensate parties who suffer damages due to oil pollution caused by tanker accidents. The CLC was amended in 1976 and again in 1984 by Protocols that extended the geographical scope of the CLC's application and substantially increased the limits of compensation available upon liability. The 1984 Protocol failed to enter into force because it was unable to secure the required amount of ratifications from the state parties to the CLC.

However, a new Protocol was agreed upon in 1992 which is substantively similar to the 1984 one but procedurally simpler to bring into force because it has less stringent ratification requirements (Ellis, 1995, p. 40).

The 1971 International Convention on the Establishment of an International Fund for Compensation for Oil Pollution Damage (Fund Convention), on the other hand, attempts to balance the burden of pollution costs between the ship owners (already covered by the CLC) and oil companies which often own the cargo that the tankers carry and sometimes own the tankers themselves. The Fund Convention therefore derives its funds from levies imposed on the oil companies as the owners of the means of transport and of the goods transported. The Convention provides that compensation will be paid to the claimant from the International Oil Pollution Compensation Fund (established by the Fund Convention) where the ship owner is not liable under the CLC, or is liable but is unable to meet that liability, or if the pollution damage exceeds the limits of that liability. Like the CLC, it was proposed that the Fund Convention too be amended by protocols in 1976 and 1984. Again, like the CLC, a new Protocol was needed with less stringent ratification requirements, and this was agreed upon in 1992 (Ellis, 1995, p. 40).

The entry into force requirements of both the 1992 CLC and Fund Convention Protocols were fulfilled on 30 May 1995, when Denmark deposited its instruments of ratification with the IMO. The two Protocols have just entered into force 12 months later, on 30 May 1996, only 2½ years after they were adopted. Nine states have ratified both Protocols: Denmark, France, Germany, Japan, Mexico, Norway, Oman, Sweden and the UK. Egypt, meanwhile, has ratified only the CLC Protocol. With the entry into force of these two supplementary 1992 Protocols to the CLC and Fund Convention, the amount of compensation available to the victims of oil pollution from tankers will be more than doubled. The geographical scope of the two Conventions also will be expanded to include damage occurring to property or economic interests within the exclusive economic zone (EEZ), or both, rather than merely the territorial waters of a contracting state party. The provision of higher limits of compensation and wider scope of application thereby ensures the future viability of the compensation system established by these Conventions (*Marine Pollution Bulletin*, News, August 1995, p. 500).

2.4.4 *The MARPOL Convention*

The most comprehensive and important of all the conventions and the first one to provide comprehensive and exhaustive guidelines for ship builders and ship owners to follow in order to actively prevent marine oil pollution, is the 1973 International Convention on the Prevention of Marine Pollution from Ships, as modified by the 1978 Protocol relating to it

(MARPOL 73/78). Annex I of MARPOL is concerned with marine oil pollution and came into force at the same time as the main body of the 1973/78 MARPOL. Under MARPOL, a ship owner is prohibited from discharging oil or oily mixtures into the sea unless certain criteria are met or an exception applies. Its main objective is to eliminate completely intentional pollution of the marine environment by oil and to minimize accidental discharge.

Unlike the above conventions and the voluntary industrial agreements described below, which are primarily concerned with compensation measures to be taken once an accident has occurred, the 1973/78 MARPOL treaty includes preventative provisions relating to the design, construction and maintenance of oil tankers. Once a state has adopted MARPOL, new ships constructed in that state are required to be built according to those guidelines. Unfortunately, MARPOL has not been very successful in either phasing out or appropriately regulating 'ageing ships', i.e. those that were built in the 1970s or even earlier and which are still sailing under minimal safety and environmental standards (Ellis, 1995, p. 41).

By 1992, concern was growing about the condition of some of the world's ships, especially the large number of ageing tankers and bulk carriers. It was agreed that steps had to be taken to ensure that the maintenance of such older ships should be improved and that the 'quality gap' which had arisen because of IMO's success in introducing strict standards for new ships, should be narrowed. Thus, important new measures to improve the safety of existing oil tankers have recently come into operation, on 6 July 1995. These changes were included in amendments adopted in March 1992 to MARPOL 73/78. These included (1) an enhanced programme of inspections that will apply to all oil tankers aged 5 years or more and (2) important new changes to the construction requirements for tankers of 25 years of age and older, including the mandatory fitting of double hulls or an equivalent design.

Other amendments adopted in March 1992 applied to all new tankers ordered after 6 July 1993. Tankers of 5000 dwt (dead-weight tonnage) and above must be fitted with double bottoms and double hulls extending along the full length of a ship's side. The 'mid-deck' design is permitted as an alternative and other designs may be allowed in due course, provided they ensure the same level of protection against pollution (*Marine Pollution Bulletin*, News, September 1995, p. 578).

2.4.5 The OPRC Convention

Another recently agreed international convention, containing preventative and mitigation measures relating to accidental oil spills, is the 1990 International Convention on Oil Pollution Preparedness, Response and

Co-operation (OPRC), which was concluded after the *Exxon Valdez* stranding off the coast of Alaska. This treaty, designed to help governments combat major oil pollution incidents, was adopted in November 1990 by an inter-governmental conference convened by the IMO. The OPRC Convention has now received the required number of ratifications to enter into force and did so on 13 May 1995 (*Marine Pollution Bulletin*, News, July 1995, p. 432). The Convention's objective is to provide for and organize international co-operation in oil spill response planning and implementation (Ellis, 1995, pp. 41–2). The Convention is designed to facilitate international co-operation and mutual assistance between states in preparing for and responding to a major oil pollution incident. It also encourages states to develop and maintain an adequate capability to deal with oil pollution emergencies. It is concerned with preparedness and response issues related to oil pollution emergencies (i.e. involving petroleum in any form including crude oil, fuel oil, sludge, oil refuse and refined products) which pose a threat to the marine environment, or to the coastline or related interests of the states parties. It is also to be expanded in its application to hazardous and noxious substances pending revision of the Convention to cover such substances (*Marine Pollution Bulletin*, News, July 1995, p. 432).

The main features of the Convention include:

- International co-operation and mutual assistance, in which the contracting parties agree to co-operate, and the rendering of assistance to parties that request it to deal with oil pollution incidents, subject to the capability of parties of other states and availability of relevant resources (art. 7).
- Oil pollution reporting, whereby contracting parties agree to ensure that ships, offshore units, aircraft, seaports and oil handling facilities report oil pollution incidents to the nearest coastal state or competent national authority and advise neighbouring states at risk and the IMO, as appropriate (art. 4).
- Oil pollution emergency plans, which are required for oil tankers of 150 gross tonnage and above, and other ships of 400 gross tonnage and above; any fixed or floating offshore installation or structure engaged in gas or oil exploration, exploitation, production activities or loading or unloading oil; and any seaport and oil handling facility that presents a risk of an oil pollution incident (art. 3).

The Convention further imposes an obligation on each of the contracting states' parties to establish a national system for responding promptly and effectively to oil pollution incidents. This includes, as a basic minimum, the creation of a national contingency plan, designated national authorities and operational focal points responsible for oil pollution preparedness and response, and reporting and handling requests for assistance (art. 6). The

Convention also allocates to IMO an important role as operator of an Oil Pollution Co-ordination Centre. Its purpose is to carry out specific functions such as providing information services, education and training, technical assistance and the co-ordination and mobilization of the international response to major oil pollution incidents (art. 12). Furthermore, in respect of responsibility and consequent liability for compensation of damage sustained by contracting states parties as a result of such an incident, the parties took account of the polluter-pays principle as a general principle of international environmental law in the Preamble to the main text of the Convention. The OPRC Convention has so far been accepted by 21 countries: Argentina, Australia, Canada, Egypt, Finland, France, Germany, Greece, Iceland, Mexico, Netherlands, Nigeria, Norway, Pakistan, Senegal, Seychelles, Spain, Sweden, USA, Uruguay and Venezuela (*Marine Pollution Bulletin*, News, July 1995, p. 432).

2.4.6 The voluntary industrial agreements (TOVALOP, CRISTAL and PLATO)

The negative publicity generated by the *Torrey Canyon* incident and its aftermath had a profound effect on the attitudes of both ship owners and oil companies towards the environmental damage caused by their activities. Moves conducted under the auspices of IMO to establish international conventions to deal with any future oil tanker incidents, which eventually yielded the CLC and Fund Conventions examined above, did not appeal to either group for two slightly contradictory reasons. On the one hand, these groups were concerned that the outcome of the then on-going negotiations at IMO would result in the introduction of strict safety precautions and restrictions with which they felt unable to live (Ellis, 1995, p. 44, citing Springall, 1988, p. 25). On the other hand, because of the negative publicity of *Torrey Canyon* and consequently poor general public perception of the oil and shipping industries as a whole, these groups felt constrained to take constructive action among themselves in order to rejuvenate public confidence in their activities (White, 1993, p. 57). Two voluntary industrial regimes were therefore formulated. These were the 1969 Tanker Owners' Voluntary Agreement concerning Liability for Oil Pollution (TOVALOP) and the 1972 Contract Regarding an Interim Supplement to Tanker Liability for Oil Pollution Damage (CRISTAL).

The TOVALOP preceded the CLC but was established for the same purpose: to compensate those who suffered injury as a result of a tanker oil spill. The TOVALOP was created by major oil companies and their ship owner subsidiaries and was funded by a levy on the tanker owners. It came into effect on 6 October 1969 when the owners of 50% of the world's oil tanker tonnage became parties. The scheme is vessel-specific and its objective is to reimburse governments for expenses reasonably incurred to prevent or clean up oil spills that occur due to the negligence of any ship owner in the scheme. Under TOVALOP, the tanker owners and bareboat charterers, who are the parties to the Agreement, assume certain obligations for which they might not otherwise be legally liable. For TOVALOP to apply it is not necessary to demonstrate that the tanker owner or bareboat charterer was at fault and there are only a very limited number of circumstances in which a party will be totally free of any obligations under the Agreement. As a result, compensation can be obtained by claimants without recourse to legal proceedings which may prove lengthy, although the TOVALOP party does not thereby waive any rights of recovery from third parties whose fault may have caused, or at least contributed to, the incident. Thus, TOVALOP facilitates the payment of compensation without in any way shifting the actual responsibility for the spill or prejudging the issue of ultimate liability (White, 1993, p. 58). In this sense, it may be seen as a somewhat limited application of the polluter-pays principle, due to the predetermined limits in the amount of compensation available.

Closely related and in fact complementary to TOVALOP, and preceding the 1971 Fund Convention but established for the same purpose, is the 1972 Contract Regarding an Interim Supplement to Tanker Liability for Oil Pollution Damage (CRISTAL). The CRISTAL supplemented the compensation for pollution damage made available by the tanker owner under the TOVALOP and CLC regimes. The CRISTAL is funded by the cargo owners and was intended to balance the financial liabilities that may arise between ship-owning and cargo interests in such a manner that the cargo interests only contribute when the contribution of the ship-owning interests is insufficient (Ellis, 1995, pp. 44–5, citing Abecassis and Jarashow, 1985), p. 311). Any company engaged in the production, refining, marketing, storing or trading of oil, or which receives oil in bulk for its own consumption or use, can become a party to the CRISTAL contract. Thus major consumers such as power stations are parties to CRISTAL, in addition to oil companies and traders. Also, a high proportion of the total volume of oil transported by sea is subject to the terms of the CRISTAL contract (White, 1993, p. 63).

Finally, the Pollution Liability Agreement among Tanker Owners (PLATO) was created in 1985 and was intended to supplement a new and improved CRISTAL. The objective behind PLATO was to commit tanker owners voluntarily to assume much higher liabilities than those found in the 1984 Protocols to the CLC and Fund Convention. It has never come into effect, however.

It may be seen from the above discussion that most of the international law (in the form of international conventions) that has developed in response to marine pollution problems can be characterized as either preventative or remedial in its thrust. In most cases, the primary purpose has been either to impose reasonable precautions in order to reduce the frequency of incidents and to prevent or minimize damage should it occur, or to provide for the availability of compensatory remedies and other forms of relief to the victims of such damage after it occurs. The remedial branch of marine pollution law appears to have developed mainly on a strict no-fault liability basis, subject to previously determined compensation limits, as evidenced by the 1969 CLC and 1971 Fund Conventions and related 1992 Protocols, and also the voluntary industrial agreements, TOVALOP and CRISTAL. Although this type of strict no-fault liability system may appear to be easier in its application and more equitable in its results, it is still subject to certain limitations. The chief deficiency of such a no-fault liability, compensation fund approach is that it is simply unfeasible and unacceptable in practice unless certain maximum or ceiling compensation limits are prescribed. The existence of such limits, however, means that in many situations some or most of the victims cannot be fully compensated or reimbursed (Rémond-Gouilloud, 1981, pp. 199–200).

The preventative branch of marine pollution law, on the other hand, is more directly concerned with the imposition of regulatory controls, through research or review requirements, higher technical standards, operational conditions, professional criteria, licensing techniques, guarantees of accountability and other modes of management. This is manifested in the MARPOL 73/78 Treaty, for example. As ocean management in general and marine pollution control in particular become more sophisticated, new law has become increasingly regulatory in its orientation (Rémond-Gouilloud, 1981, p. 199). Different kinds of problems tend to arise in the preventative branch of marine pollution law. One familiar difficulty is that of extra territoriality. Here the question is how to enforce national or international measures outside national territorial limits. The traditional response has been to do so by application of flag state jurisdiction but, as we have seen, the dependence on flag states has tended to frustrate rather than facilitate the application of measures beyond territorial limits (Rémond-Gouilloud, 1981, p. 200).

Three innovations have emerged to cope with this problem of flag state intransigence, but each is limited in its scope and effectiveness. First, the 1969 Intervention Convention provides for unilateral intervention by a contracting state party in the case of an emergency in the high seas involving a vessel of another contracting state. It permits drastic precautionary measures to be taken by the intervening state if there is sufficient risk of serious environmental damage to the latter's coastal interests, but the terms of the Convention are narrowly defined and they are binding only on those states that have chosen to ratify it. However, more supportive and less restrictively worded language is contained in the

1982 UN Convention on the Law of the Sea (art. 221) (Rémond-Gouilloud, 1981, p. 200).

Second, the dependence on flag state initiatives has been reduced by the possible extension of coastal state jurisdiction seaward out to the 200 nautical mile limits of the exclusive economic zone, but the entire question of the coastal state's authority over marine pollution control within this extended maritime jurisdictional limits is complicated and clouded by uncertainties due to the nuanced trade-off diplomacy which was engaged, in order to achieve a balance between the extension of coastal state rights and the preservation of residual freedoms of the seas, by the delegations during the UNCLOS III negotiations, resulting in a highly compromised conventional text, which accorded coastal states 'jurisdiction' with regard to, *inter alia*, the protection and preservation of the marine environment [art. 56(1(b(iii)] (Rémond-Gouilloud, 1981, p. 200).

Third, the 1982 UNCLOS further diminished the traditional exclusivity of flag state jurisdiction, as noted above, by introducing the additional and concurrent resort to port state jurisdiction for the purposes of vessel-source pollution control. The effectiveness of this innovation, however, depends on a particular set of fortuitous circumstances, namely, that after any pollution incident, the culprit vessel chooses, or is obliged, to enter the port of a state which is both able and willing to initiate investigations and, furthermore, that the flag state, if it chooses to assert its primacy, acts competently and effectively (Rémond-Gouilloud, 1981, p. 200).

Even where the preventative and remedial elements are combined in a single approach to marine pollution control, as is the case with the 1982 Law of the Sea and 1990 OPRC Conventions, a variety of political and technical difficulties often prevent a successful outcome of the treaty negotiations or result in later disputes over the interpretation of treaty provisions.

2.4.7 State practice

Having discussed the various international regimes in this issue area and noted the various challenges they have presented to the oil and shipping industries, it is important to note also an emerging trend among coastal states to institute unilateral controls over vessel-source oil pollution, even within their exclusive economic zones, beyond their territorial waters. Indeed, unilateral action or the threat of such action on the part of coastal states has proved to be the impetus that propels international oil pollution onto the international agenda. Given industry's and consequently the maritime states' strong aversion to the divergent multiplicity of regulations with respect to international shipping, the threat of unilateral action should readily prompt a willingness by other states to at least discuss international regulation (Mitchell, 1993, pp. 222–3). Coastal state enforcement of more stringent controls, stricter by far than those provided by current conventional regimes, represents an alarming development for oil and shipping industry concerns alike, raising as it does the spectre of different legal standards within the waters of each coastal state traversed during the course of a single voyage! The higher levels of vessel standards established by such states could in effect prevent many vessels from entering their waters or ports.

In order to illustrate this trend, the following two sections will examine two national examples drawn from the developed and developing worlds, namely the USA and Malaysia.

2.4.8 United States of America

In the USA, the enactment of the 1990 Oil Pollution Act (OPA '90) may be seen to represent a milestone in the introduction of comprehensive oil spill legislation, not just within the USA but also throughout the world, bearing in mind the crucial role that the USA plays in the world's maritime oil transport market as one of the largest importers of crude oil. Unilateral action in the form of OPA '90 represented an economic hammer blow to vested interests in the shipping and oil industries. The passing into law of this Act drew a predictable outcry from these industries in relation to the increased transportation costs involved in compliance with its requirements. Now that the dust has settled somewhat on this issue, it is possible to observe more closely the role played by the unilateral imposition of stricter legal controls in the development and utilization of new, more environmentally friendly, technologies within the maritime oil transportation industry in order to ensure the avoidance of unlimited civil liability (in both federal and state jurisdictions), the imposition of punitive fines and even the threat of criminal sanctions under OPA '90. Of particular interest in this context is the emergence (at about the same time) of a new theoretical explanation in favour of tough environmental regulation which asserts that it can be good for a country's long-term economic competitiveness. This so-called 'revisionist' view of environmental regulation of industry is predicated on the idea that higher standards will stimulate industrial innovation, making companies generally fitter and more competitive (Wallace, 1995, citing Porter, 1991, p. 96). The gradual acceptance of the revisionist view of environmental regulation at the highest levels of the US government in the early 1990s result in a number of US Environmental Protection Agency (EPA) initiatives based on the belief that appropriate regulations can improve the competitiveness of US industries (Wallace, 1995, p. 4). In retrospect, therefore, it is possible to discern a change in the US political climate at the time which allowed OPA '90 to gain acceptance from an otherwise resistant maritime oil transport industry.

The more immediate impact leading to the passage of OPA '90 through the US Congress was the catastrophic *Exxon Valdez* tanker oil spill in March 1989, off the Alaskan coastline. Interestingly, prior to the *Exxon Valdez* spill, the enactment of any type of comprehensive oil spill legislation (at least at the federal level in the USA) had proved impossible, despite the obvious need for supplementation, if not overhaul, of existing oil pollution legislation. Even more interesting is the fact that many of the ultimately unsuccessful bills contained language implementing the 1984 Protocols to the 1969 CLC and 1971 Fund Conventions, mentioned above (Wetterstein, 1992, p. 75). The growing environmentalism in the USA led the US government to champion more stringent measures at the international level than had previously been considered, backing their proposals with threats of unilateral action (Mitchell, 1993, p. 186). It is important to note, however, that the USA never became part of the international conventional regime between states (as opposed to voluntary industrial agreements discussed above), for establishing liability and compensation for oil pollution damage. Although the USA played an active part in the negotiation of the 1984 Protocols to the 1969 CLC and 1971 Fund Conventions, it eventually declined to become a party to this compensation arrangement, thereby causing it to fail to come into force until recently, when amended versions of the Protocols to which the USA finally agreed were opened for signature in 1992, and have just come into force on 30 May 1996 (Franckx, 1995, pp. 257–8).

The OPA '90 is comprehensive in its scope and provides, *inter alia*, for increased civil and criminal liability for oil spills (accidental or otherwise), the establishment of an oil spill compensation fund, enhanced regulation of spill clean-ups and stricter prevention procedures (nationally and regionally), manning standards, double hull tankers and oil pollution research and development. The OPA '90 also amends other US Federal Acts to bring them into conformity with it, although it covers only 'oil' pollution, as defined in the Act (Wetterstein, 1992, pp. 76–7). The most noticeable feature of this legislation is that it establishes on tanker owners, operators or demise charterers strict and almost unlimited liability for compensation. The imposed liability is also strict, joint and several (Wetterstein, 1992, p. 80). The high liability limits provided in the Act can rather easily be set aside in certain circumstances. Almost all contemporary oil spills would fall into this category, rendering the nominal limitations on liability almost meaningless (Franckx, 1995, p. 258, citing Garick, 1993, p. 277). Furthermore, there is an explicit provision stating that damage to natural resources is compensatable under OPA '90. Thus, it is apparent that OPA '90 provides for compensation for damage to the ecosystem itself, as opposed to damage only to human activities. The compensation covers not only the costs of removal, i.e. the costs of cleaning up spilled oil (including personnel and material costs, and also the operational costs for the US

Coast Guard) but also 'the cost of restoring, rehabilitating, replacing or acquiring the equivalent of, the damaged natural resources'. In addition, the diminution in value of those natural resources pending restoration and the reasonable cost of assessed natural resource damages are recoverable. However, it is clear that natural resources damages may only be recovered by the US (Federal) Government, a US State, an Indian tribe or a foreign government (Wetterstein, 1992, pp. 145–7). Finally, the Act requires evidence of sufficient financial responsibility in order to cover the maximum liability limits provided for in the Act (Franckx, 1995, p. 258).

Geographically, the liability rules of OPA '90 are applicable when there is a discharge of oil, or a substantial threat of a discharge of oil, from a vessel or a facility into or upon the navigable waters of the USA, including its adjoining shorelines, territorial waters and 200 nautical miles EEZ (Wetterstein, 1992, pp. 77–8). These provisions, therefore, apply to foreign ships exercising their right of innocent passage through both the US territorial and EEZ waters, even if the destination of their cargo is not in fact the United States (Franckx, 1995, p. 258, footnote 30). The fact that the Act explicitly provides that it does not pre-empt individual states law means that its provisions merely constitute a Federal minimum, to which states are allowed to add further requirements (Franckx, 1995, p. 258).

Another salient feature of this Act is certainly that it requires all tanker vessels entering the EEZ of the USA to be double hulled by the year 2015, thus implementing a much stricter time schedule for the introduction of this requirement than do the regulations recently introduced by the International Maritime Organization (IMO), the international body implicitly recognized by the 1982 UNCLOS as the legitimate rule-making and standard-setting organization for international shipping. This requirement, however, is not applicable to foreign vessels in innocent passage through the US territorial sea and EEZ (Franckx, 1995, p. 258, footnote 32). In this respect, it is important to point out the extent of the divergence of this unilateral US legislation from the norm established by the 1982 UNCLOS, which to date the USA has yet to sign. Generally, coastal states may in respect of their EEZs adopt only laws and regulations for the prevention, reduction and control of pollution from vessels that conform and give effect to generally accepted international rules and standards established through the competent international organizations (IMO) (art. 211.5). In particular, art. 211.6(c) provides that even when coastal states legislate for 'special' areas within their EEZs, 'such additional laws and regulations may relate to discharges or navigational practices but shall not require foreign vessels to observe design, construction, manning or equipment standards other than generally accepted rules and standards'. On the face of it, therefore, it appears that the US OPA '90 legislation requiring double-hulled oil tankers far ahead of the time schedule currently being implemented by the designated 'competent international organization', the IMO, is at variance with the provisions of the 1982 UNCLOS, which are now in force among state parties and which the USA has indicated it will abide by and to which it will soon formally accede.

2.4.9 *Malaysia*

Other legal attempts to induce higher environmental and safety standards for oil tanker operations have also been unilaterally introduced by other countries, including even developing countries which have their own emergent merchant shipping industries. Malaysia is a good example of such a country. Although it has a rapidly developing, newly industrialized economy, Malaysia has for too long suffered environmental degradation arising from its geographical proximity to the Straits of Malacca and Singapore, one of the busiest international waterways in the world. Incidences of tanker grounding and collisions during transit through the shallow waters of the Malacca Strait in particular have become a 'hot' topic in the local press. The focus on accidental and now also on deliberate vessel discharges has generated increasing pressure for the enactment and enforcement of legislation laying down stricter rules regarding these discharges. This was provided for in the 1984 Malaysian Exclusive Economic Zone Act, passed pursuant to its 1981 Proclamation of a 200 nautical mile EEZ, (Hamzah, 1988, Appendix, pp. 26–52). In June 1995, for example, the Royal Malaysian Navy, reportedly for the first time in its history, fired a warning shot in order to stop a Greek-registered, Liberian-owned cargo vessel, which was suspected of having dumped sludge, from escaping into international waters The *World Aretus* was allegedly spotted dumping sludge in Malaysian waters by a Royal Malaysian Air Force surveillance aircraft. It was eventually detained through the combined efforts of several Malaysian Royal Navy vessels. Further reports suggested that the ship's owner was liable for prosecution under both Malaysia's 1974 Environmental Quality Act, which governs Malaysian territorial waters and its 1984 EEZ Act, covering Malaysian EEZ waters. This had been the seventh oil dumping incident detected in Malaysian waters in the South China Sea in a month but the first which had resulted in the arrest and detention of a suspected vessel (Azhar *et al.*, 1995; Tan and Anbalagan, 1995; Tan and Atan, 1995). In a separate incident later in the same year, the owners of a Shanghai-registered cargo ship were fined 33 500 Malaysian ringgit (RM; RM1 ≈ US$0.4) for illegal desludging off the coast of the state of Sarawak on Borneo island, in violation of the Malaysian 1974 Environmental Quality and the 1984 Exclusive Economic Zone Acts (Tan, 1995).

Part IV of the 1984 EEZ Act deals with the protection and preservation of the marine environment of the Malaysian EEZ, defined as extending to a distance of 200 nautical miles from the baselines from which the breadth of the (Malaysian) territorial sea is measured, under section 3(1) of Part II of the Act. Under section 10(1) of Part IV of the Act, the discharge or escape (without licence) of any oil, oily mixture or pollutant into the Malaysian exclusive economic zone from any vessel, land-based source, (offshore) installation, device or aircraft, from or through the atmosphere, or by dumping is an offence liable to a fine not exceeding RM 1 million. In the case of a vessel, the owner or master will each be guilty of an offence [section 10(1)(a)]. Moreover, should the act or omission of a person other than the owner or master above cause the discharge or escape, then that other person will also be guilty of an offence and will be liable to a fine not exceeding RM 1 million [section 10(2)]. This provision will not, however, operate to absolve or relieve the owner or master of a vessel from liability for an offence. Furthermore, section 14(3) of the EEZ Act provides that the owner and master of the vessel from which the pollutant was discharged or escaped also will be liable jointly and severally for all costs and expenses incurred in carrying out all or any of the work required to remove, disperse, destroy or mitigate the damage or threat of damage.

This blanket prohibition against the discharge of all types of oil, oily mixtures and any other pollutant throughout the defined area of the EEZ represents a significant extension, in both quantitative and qualitative terms, of the current international discharge standards for these substances under the appropriate conventional regimes. As noted above in relation to US state practice, art. 211.5 of the 1982 UNCLOS provides coastal states with the power to legislate for the prevention, reduction and control of pollution from vessels, where such legislation conforms and gives effect to generally accepted international rules and standards established through the competent international organization or international convention. In the specific case of oil pollution, for example, the 1973/78 MARPOL treaty provides for a similar total ban on discharges, but only up to 50 nautical miles offshore and thereafter allows discharges of certain permitted concentrations of oil to water, in terms of parts per million (ppm).

This unilateral exercise of Malaysian legislative jurisdiction to enact pollution standards that exceed by far the international norm has been coupled with the equally unilateral exercise of its enforcement jurisdiction over vessels whose discharge or escape of any oil or other polluting substance has either damaged or threatened to damage any segment or element of Malaysia's coastal environment or related interests, including fishing, in its EEZ [section 14(1)]. The Director-General (of Environmental Quality), in addition to being able to issue directions or take action to remove, dispense, destroy or mitigate the damage or threat to damage [under section 14(1)], can also take steps to detain any vessel from which the oil, oily mixture or pollutant escaped or was discharged [section 15(1)]. The adoption of enforcement powers over 'any vessel' in the

Malaysian EEZ also represents a unilateral extension of the internationally prescribed jurisdiction under art. 220 of the 1982 UNCLOS, which allow possible detention of a vessel only where there is clear, objective evidence that the vessel has (in the EEZ) committed a violation of applicable international rules and standards for the prevention, reduction and control of pollution from such vessels (art. 220.3), resulting in major damage or threat of major damage to the coastline or related interests of the coastal state concerned (art. 220.6). It may be seen, therefore, that the blanket prohibition against pollutants under section 10 and the expansive detention powers taken on under section 15 of the 1984 Malaysian EEZ Act represents a unilateral extension of Malaysian jurisdiction beyond that envisaged by both the relevant global (1982 UNCLOS) and sectoral (MARPOL 73/78) legal regimes for vessel-source pollution.

2.5 The 1992 Rio Earth Summit (UNCED): implications for the oil industry of the emerging legal regime governing global climate change

The instruments adopted by the United Nations Conference on Environment and Development (UNCED) represent yet another milestone in the development of international environmental law which has obvious implications for the oil industry. As one leading commentator has noted, UNCED focused international attention on the adoption of global policies whose explicit purpose was to reconcile economic development with environmental protection, most obviously under the concept of sustainable development (Boyle, 1994, p. 173). This is especially the case in relation to the emerging legal regime for the mitigation of global climate change induced by the build-up of so-called 'greenhouse' gases, which may result in incremental global warming. Despite continuing doubts raised by some parts of the scientific community over the scale and real or potential effects of greenhouse gas-induced global climate change, it would appear to be the case that the international community of states is now committed to a policy of reduction and mitigation of such gases. [Regarding the above doubts, see, for example, a leading article in *The Economist* (1–7 April 1995, pp. 13 and 109–11), which notes that despite recent advances in knowledge, scientists still understand little about the world's climate system and are uncertain about the extent of global warming. Latest reports, however, suggest that even this now minority-held scientific uncertainty appears to have been overturned since many of the world's top climatologists now agree that global warming is occurring due, at least in part, to artificial pollution of the atmosphere; see Lean (1995a), reporting on documents summarizing the findings of the latest report of the Intergovernmental Panel on (Global) Climate Change, the official scientific body for the Conference of Parties (COP) of the 1992 Framework

Convention on Climate Change, expected to approve the report at its meeting in Rome in December 1995. This report would appear to suggest that the international consensus on action to limit climate change is definitely moving in the direction of introducing curbs on hydrocarbon-based energy industries, in terms of both their production and their consumption sectors. Further efforts by these industries to contravert the results of this latest report would seem to be futile. Indeed, there are signs within the multinational oil companies of an implied acceptance of the need for a stricter policy on greenhouse gas emissions and greater investment in alternative and renewable sources of energy production (see, for example, The future of energy, *The Economist*, 7–13 October 1995, pp. 27–30)]. Owing to the international nature of the causes and effects of the problem of greenhouse gas emissions, it is now also widely accepted that the required controls must be in the form of international action, rather than unilateral national action (Churchill, 1991, p. 151). Among these gases, carbon dioxide is said to have been responsible for over half of the enhanced greenhouse effect in the past and is projected to continue to be so in the future (Doos, 1991, p. 10). The legal and economic implications of the developing consensus that is being achieved on this problem, in terms of its scientific causes, the remedies advised and policies agreed for implementation of the necessary measures to counter it, represent potentially the greatest threat that has occurred to date to the long-term viability of the fossil fuel industry. This is despite the fact that it is still not wholly certain that there is a global warming problem and furthermore, that even if there is an emerging environmental crisis, it is accepted that it is almost certainly not due to commercial energy production activities alone (Read, 1994, p. 3).

However, even if some scientists claim that global warming is still within the limits of natural variability, and will stay so, and even if all the evidence is not yet available to establish otherwise, this need not, indeed should not, prevent steps from being taken to anticipate, minimize or mitigate adverse effects. Not least among the incentives to act preventively are the much lower costs of preventive action in comparison to the cost of efforts to clean up after damage is done (Nilsson and Pitt, 1994, p. 23). This point is rendered even more significant by the knowledge that the consequences of climate change on a planetary scale are most probably irreversible (Pulvenis, 1994, p. 75). Greenhouse gases could stay in the atmosphere for hundreds of years. It could, therefore, be an error to delay action until the scientific evidence is clear and unambiguous. In this context, it has also been argued that precautionary action should be taken until it becomes certain that there is no problem, in the light of the knowledge that, in the meantime, commercial energy activities are almost certainly involved if there is a problem, and that it is only by changing the nature of these commercial energy activities (as opposed to eliminating them) in a deliberate and gradual way over the next few decades that we can most surely and cheaply deal with the problem (Read, 1994, pp. 3–4).

Therefore, at least in part, the basis for action on global warming, despite the then prevailing uncertainties concerning both the nature and extent of the phenomenon, was the rationale of the precautionary principle or approach (Pulvenis, 1994, p. 79). This precautionary principle or approach is utilized as a justification for the construction of regimes for the protection of the environment from threats that are as yet scientifically unproved. It has been defined in terms of Principle 15 of the 1992 Rio Declaration on Environment and Development, as applying in cases where there are threats of serious or irreversible environmental damage. The precautionary principle or approach provides that where there is a lack of scientific proof in such cases it shall not be deemed a reason for postponing the introduction of cost-effective measures to prevent environmental degradation. The crux of the precautionary principle or approach is that action to prevent serious or irreversible damage should not be delayed until the scientific evidence is clear, by which time, of course, it might be too late to act effectively (Freestone, 1993, p. 23). The precautionary approach, then, is innovative in that it changes the role of scientific data. It requires that once environmental damage is threatened, action should be taken at an early stage to control or abate possible adverse environmental impacts, even though there may still be scientific uncertainty as to the precise effects of the activities in issue. It thus represents an important aid to decision making in situations of scientific uncertainty. Hence the policy response in relation to industrial emissions will be to rely on an assessment of the relative risks of various levels of pollutant emissions (Freestone, 1993, p. 25).

As might be imagined, the consequent reversal of the burden of proof that such a precautionary approach to environmental policy and law making could, in the view of strict environmentalists, entail will have enormous implications for the way companies in general and the oil industry in particular conduct the more unsavoury aspects of their business. This may account for the reasons why the application of the precautionary approach or principle has been rendered subject to cost effectiveness in both Principle 15 of the Rio Declaration and more importantly, in art. 3(3) of the Framework Convention on (Global) Climate Change (Pulvenis, 1994, p. 94). Policies and measures should be cost effective so as to ensure global benefits at the lowest possible cost (art. 3.3). Even if the application of the precautionary approach is to be rendered subject to the requirement that it is cost effective, and there are indications that it may be expensive in strictly economic terms, it has nevertheless been argued to be an appropriate response to the problem of global warming (Read, 1994, p. 2). In other words, even if a cost–benefit analysis shows the utilization of the precautionary approach to be less attractive in any particular case than it was initially thought to be, it could still be arguable that precautionary steps should be taken to anticipate the possible climatic effect of rising carbon dioxide concentrations.

At least one reason put foward in support of this view is that a cost–benefit analysis is only properly applicable when a project under consideration is small in relation to the economy as a whole, so that the consequent penalty, should the policy turn out to be wrong, is not very great. This, however, is clearly not the case in a unique decision-making process such as the formulation of a truly global policy towards global climate change due to global warming (Read, 1994, p. 11).

Another reason for advocating the precautionary approach is linked to the theoretical explanation given in section 2.1 concerning the need to internalize environmental degradation within the overall economic equation where it is currently regarded as an externality. In this sense, global warming too represents an 'externality' within the world market system. This externality, and the requirement for a globally effective policy that arises from it, are no less real because of the uncertainties that still surround the phenomenon. Even in this uncertain context, therefore, it has been argued that precautionary action should be taken until we are certain that there is in fact no global warming problem. Thus, a precautionary approach is needed to ensure that an effective regime is in place even if the uncertainty proves to be well founded (Read, 1994, p. 3).

In the face of this veritable onslaught upon it, it is important that the oil industry does not give the impression of being reactionary in its response. In particular, the alliance which appears to have developed between several Western multinational oil companies and some of the world's largest oil-exporting countries against initial proposals for the curbing of greenhouse gas emissions is not a step in the right direction. [For example, as Nicholas Schoon (1995) reports on the gathering scientific certainty over the probable causes of global climatic change, 'Those with vested interests watch this unfold, then put their own spin on things. The USA's gigantic fossil fuel industry, along with oil exporters like Kuwait and Saudi Arabia, play up the uncertainties. Their lobbyists stoop to suggesting that the scientists exaggerate in order to get their research grants]. The chastening experience of the shipping industry which engaged in an increasingly ineffective and unpopular rearguard action against the imposition of technical and design improvements for environmental protection, as briefly described in section 2.4, should act as a timely reminder to oil companies in particular, of the futility of such negative responses. As Michael Grubb, one of the foremost writers on the particular challenge of global climate change for both governments and their affected industries, has noted, 'Energy industries supply goods and services that make much of modern life possible, but often face strong environmental criticism. The energy industries have incurred high costs as a result of past environmental legislation and some energy companies have vigorously opposed and often helped to block new legislation. The nexus between environmental policy, sustainability and the energy industries is fraught with difficulty and contention, but better understanding of the positions, interests, and possibilities is essential if good policy is to emerge' (Steen, 1994, Foreword, p. vii). In more general terms, Read notes that 'the environmental problems of today cannot be resolved by confrontation. Rather must industry, in joint endeavour with non-government activists, work out ways of using the market to resolve these problems' (Read, 1994, Acknowledgements, p. ix).

There should also be an increasing recognition of the fact that the time has come when environmental awareness is seen as a financial imperative rather than merely an ethical option. [See, for example, the report by Roger Cowe (1995) on how companies pushed into environmental action by legal requirements or restrictions have gone further and discovered the economic benefits that they can reap by going down the 'green' route. He notes that even small companies in non-sensitive industries have discovered the financial benefits of introducing environmentally friendly business practices. The falling prices of renewable energy sources, especially in the electricity generating sector of the energy industry, although, significantly, not in the transport sector, is another reason why oil companies need to broaden their horizons and examine the various alternative energy sources that may soon become more widely available, in direct competition with their staple products; see The future of energy, *The Economist*, 7–13 October 1995, p. 28.] As Choucri notes on a theme similar in content to the 'revisionist' view mentioned earlier, to which we shall return in section 2.6, Conclusions and future prospects, 'New environmental factors generate new challenges for corporate activities, shaping new constraints as well as new opportunities' (Choucri, 1995, p. 189). In other words, proposed new or higher environmental standards should act as a spur to firms to innovate in order to capture new or different markets by applying these standards, rather than resisting their introduction.

As is becoming increasingly clear in this discussion, the international policy response to the problem of global climatic change will have serious implications for the oil industry. In a recent study on energy and the environment, for example, the problem of greenhouse gas emissions, mainly arising from the combustion of fossil fuels, was considered to be the most important of the three broad and distinct environmental concerns in relation to the energy industry (Sharma, 1994, p. 2). The other two relate mainly to the threat of pollution from production and transportation activities, which are discussed in sections 2.3 and 2.4, respectively, and the problem of air pollution. Aside from an examination of the conventional regime for the prevention, reduction and control of global warming itself, this section will focus on the impact of this regime upon the oil industry and the prospects for its future in the light of the challenges it imposes on the industry to help reduce or absorb greenhouse gas emissions.

The problem of global environmental change has necessitated the institution of a new international regime in order to deal with a phenomenon the effects of which are technically beyond the limits of national jurisdiction. As one legal commentator noted at the time of the negotiation of this Convention, 'the political significance of the negotiations is such that they are likely to lead to some degree of institutional reform in relation to the treatment of (international) environmental issues in general' (Plant, 1992, p. 122). This was duly effected at the Earth Summit in Rio during the 1992 UN Conference on Environment and Development (UNCED), at which the Framework Convention on (Global) Climate Change was opened for signature. A major component of the accepted response strategy within this new treaty regime, which is directed at the reduction of the greenhouse effect, will be towards controlling the emissions of carbon dioxide from anthropogenic sources (Doos, 1991, p. 14). Interestingly, the negotiating process of the Framework Convention did not address measures for the absorption of carbon dioxide (the main greenhouse gas emitted by human activities) until fairly late in the day, with the result that provisions for effecting and measuring such absorption were not fully addressed in the Convention (Read, 1994, pp. 4–5).

From the outset, it was also clear that the global nature of both the causes and effects of global warming and consequent climate change meant that the Framework Convention had to provide a structure within which not only states but also multinational or transnational corporations (MNC/TNCs), and many other non-governmental organizations (NGOs) could act to achieve its aims (Nilsson and Pitt, 1994, p. 9).

The ultimate objective of the Framework Convention is to commit nations to the stabilization of greenhouse gas concentrations in the atmosphere at a level that would prevent dangerous anthropogenic interference with the climate system (art. 2). It is important to emphasize the goal of stabilization because, as has been mentioned above, such a goal can be achieved through either reduction or absorption of greenhouse gas emissions. In other words, more absorption is just as useful for this purpose as less emission. Significantly for the oil industry, it has been argued that it is very much easier greatly to increase absorption than it is greatly to reduce gross emissions (Read, 1994, p. 3), especially if this is combined with the provision of alternative and sustainable energy sources, based on commercially grown fuel-wood, for example, using biomass technology. Thus, the oil industry may be able to play a continuing role in the energy market should it focus its attention on growing such fuel-wood (which aids absorption of emissions) and converting it through a biomass process into sulphur-free liquid fuel (Read, 1994, p. 7).

The action to be taken under the Framework Convention by its state parties is dual in nature. First, such action aims at improving knowledge of climate change and thus reducing remaining uncertainties regarding the phenomenon itself. Second, it consists of the adoption of policies and measures aimed at preventing or mitigating climate change and its effects (Pulvenis, 1994, p. 95). These actions are provided for in art. 4 (Commitments) of the Convention, which is concerned with the wide-ranging needs for all countries to prepare inventories of emissions, implement abatement measures that promote the sustainable development of sinks and reservoirs, prepare for the impact of climate change, protect vulnerable areas and review and monitor programmes of mitigation as part of an international co-operative effort in the context of the widest possible promotion of communication, education, training and awareness (art. 4.1) (Nilsson and Pitt, 1994, p. 24).

Although the obligations established in this regard must be complied with by all state parties, including both developed and developing countries, the onus is placed on the developed country parties (and other parties included in Annex I) (art. 4.2), in the sense that a specific benchmark for the stabilization of emissions by these parties is provided, namely the 1990 emission levels [art. 4.2(b)]. Developing country parties are therefore not only actors but also beneficiaries of actions to be carried out in this field. Article 4.3 further provides that the developed country parties (and other parties listed in Annex II) should meet the full costs of any obligations incumbent upon the developing country parties under art. 12 (which deals with the communication of information related to the implementation of the substantive obligations of these parties under art. 4), in addition to the agreed full incremental costs of implementing measures undertaken by developing country parties in accordance with their art. 4 commitments.

However, it is still uncertain what this level of assistance will be or even what 'incremental' really means (Nilsson and Pitt, 1994, p. 25). Article 4.3 refers to the 'full incremental costs' of implementation that are agreed between the developing country party and the international entity or entities referred to in art. 11, in accordance with that article. Article 11 provides that the Conference of Parties to the Convention must decide 'in a predictable and identifiable manner' the amount of funding necessary and available for implementation of this Convention. This leaves much room for further negotiation on the amount of 'costs' to be paid for by developed country parties.

The key requirement here, also reflected in Principle 7 of the 1992 Rio Declaration on Environment and Development, which acknowledges the common but differentiated responsibilities of states, is that the North recognizes its responsibility, through the history of its industrialization and the technologies that have fuelled this, for the situation that has arisen. This requirement has been accepted in the Framework Convention, where the greater ability of the North to pay for whatever response is needed is also noted. Such a response must entail the transfer of efficient and sustainable energy technology to the South (which is also required by art. 4.3), where the bulk of new demand is expected (Read, 1994, pp. 7–8). In this context, the term 'joint implementation' becomes important as a means for enabling developed countries to carry out their additional and differentiated responsibilities in relation to developing countries and further, as a possible means by which the oil industry and other energy sector businesses can increase their environmental viability in a post-global warming future.

The third sentence of art. 4.2(a) of the Framework Convention provides that certain policies and measures may be implemented jointly. The policies and measures in question must be national, or at least regional, in the Annex I country in question and they must serve to mitigate climate change. According to the treaty text joint implementation is merely a policy option available to states, which they may enter into if they wish by special arrangement. However, this provision is widely understood as a prescription or recommendation addressed to industrialized countries to increase the efficiency of their actions to reduce the overall concentration of greenhouse gases within the atmosphere by carrying out such action in other, less developed, countries where the return on investment naturally would be higher in terms of greater greenhouse gas emissions reduction per unit of investment, for example. If and when such joint action is taken, as a matter of course it should be taken in co-operation with the government of the other country and it appears to have been part of the original understanding that a portion of the return, i.e. the reduction of greenhouse gas emissions or the increase in greenhouse gas absorption capacity, should be credited to the industrialized country partner in a manner agreed with the other participating country and consistent with any criteria set down by the Conference of the Parties (COP) to the Framework Convention (art. 7) (Kuik *et al.*, 1994, p. 4).

The main concepts laid down in art. 4.2 initially originated from a report entitled 'Protecting the Global Environment: Funding Mechanisms', prepared for the Ministerial Conference on Atmospheric Pollution and Climatic Change held in November 1989 at Noordwijk, The Netherlands. This report presented the case for a two-pronged approach: a first phase primarily based on domestic action, followed by phase two, joint international action covering all greenhouse gases world-wide. In order to minimize the cost to society, a central role was assigned to an international clearinghouse that was proposed to be established. The clearinghouse concept enables action in one country to be substituted by more efficient action in other countries. In the first phase, the key word was to be 'effectiveness': each country should take whatever action is most effective in its own circumstances. It was foreseen that thereafter the costs of further corrective measures would rise and societal resistance would mount. Therefore, during the second phase the emphasis should shift to joint international action and the key word would be 'efficiency'. It is easy to recognize the first phase in the first two sentences of art. 4.2(a) and the second phase in the third sentence of the same provision (Kuik *et al.*, 1994, p. 5).

The concept of joint implementation, as included in art. 4.2(a) of the 1992 Framework Convention, is an innovative and new element in the machinery of international treaties. However, this should not be taken to mean that there are no international legal precedents for such joint action. Indeed, the practice of joint action in the form of the joint development of shared or disputed offshore oil and gas deposits is a well known concept in contemporary oil and gas law literature (Ong, 1995). The utilization of international rivers, exploitation of shared or transboundary fish stocks, protection of regional marine environments and exploitation of the deep sea-bed area are also subject to various forms of joint action under international law. However, the clearest precedent for joint implementation in its narrow sense, including the 'crediting element', can be found in the 1987 Montreal Protocol (on Substances that Depleted the Ozone Layer) to the 1985 Vienna Convention for the Protection of the Ozone Layer (Kuik *et al.*, 1994, p. 8).

As noted above, several categories of actors may be involved in joint implementation, in various different capacities. They include states, as represented by their governments, inter-governmental organizations, non-governmental organizations, as well as public or private enterprises or both. The principal role in matters of initiative and leadership, negotiation, decision making, risk taking and international responsibility belongs to governments, at least for the time being. The other entities mentioned, including public and private enterprises, currently play important but subsidiary roles, in particular in the management of operations and in an advisory capacity. Other entities may also play a role, e.g. banks and academic and research institutions. The role of private enterprise is controversial in this equation. Some governments and commentators see it as the driving force, initiator and financier of joint implementation projects, playing a typically entrepreneurial role. Others are more wary of the possible implications of allowing companies, which are naturally commercially oriented and profit-motivated, both to set up and to run these joint implementation ventures. This is mainly due to the highly ambivalent perception among these governments and the general public of the environmental credentials of many private enterprises, especially those which are multinational or transnational in their ambit. This is particularly the case in relation to the oil industry, some of whose members are among the world's largest companies. Therefore, the general feeling is that even though all or most of the work of a joint implementation project may be undertaken by these corporations, they do not necessarily need to be the principal parties in the joint venture agreement. They may be engaged as contractors by one state party or the other, or by both state parties acting jointly (Kuik *et al.*, 1994, p. 34). The key issue in respect of the participation of private entities within these joint implementation projects under the Framework Convention is the monitoring of their performance, either by the governments involved in the project or some other inter-governmental, or even non-governmental, entity. Monitoring the performance of these enterprises, and also ensuring the required accountability of states parties themselves for the fulfilment of their commitments under art. 4 of the Framework Convention, may prove to be basic activities of the Conference of Parties to the Framework Convention (Read, 1994, p. 232).

Joint implementation therefore represents an eminently useful method by which the oil industry as a whole can make a substantial and potentially crucial contribution to the stabilization and even reduction of greenhouse gas concentrations within the global atmosphere, thereby also reducing the incidence of global warming. Different companies, depending on whether or not they choose to remain true to the use of their traditional technologies and products which are reliant upon fossil fuels, or to invest in new technologies based upon the production of alternative fuels, will be able to take part in the overall regime. The path they choose to follow will ultimately depend on their different perceptions of their future role in the new energy market, and indeed whether, through diversification, they enter other markets (Read, 1994, pp. 233–4).

The way forward in this scenario, however, still lies within the remit of states parties to the Framework Convention. Joint implementation is always a form of co-operation between two or more states. At the end of the first formal meeting of the parties to the Framework Convention in Berlin, no agreement was reached on reduction in the emission of greenhouse gases to combat global warming. [The following information was taken from Crawshaw (1995).] The Conference of the Parties (COP) did, however, give developed countries a mandate to prepare targets, and proposals for their implementation, by 1997. The COP agreed that all the states parties must, before their next meeting in 1997, set quantified limitation and reduction objectives, within a choice of specified time-frames, viz. by the years 2005, 2010 and 2020. The COP also accepted the argument of developing countries that their carbon dioxide emissions should be allowed to increase as their economies expand, while developed countries should be forced to cut back. Thus, the way forward towards a co-ordinated greenhouse gas stabilization regime, incorporating means for both the reduction of greenhouse gas emissions and their absorption, may now be discernible. It remains to be seen whether the oil industry will be at the forefront of these developments.

Other, related issues, which can only be touched on here due to space constraints, are those of the introduction of economic instruments such as carbon taxes and a market for carbon dioxide permits (Rowan-Robinson and Kimber, 1995, pp. 10–2). Taxation on the consumers of energy from hydrocarbon sources has been heralded as one of the few market-efficient tools for reducing carbon dioxide emissions in order to meet emission reduction targets (Boulton, 1995). In the opinion of one writer, there is a clear case, perhaps on ecological grounds and certainly on economic grounds, for switching taxes away from forms that create unemployment (such as employers' National Insurance contributions and income tax) to forms that inhibit the production and use of polluting energy (Macrae, 1995). This has, however, also been attacked as yet another costly imposition on the oil and other industries that rely on fossil fuel consumption. In this context, it is interesting to note that the hitherto fairly united front presented by the European Union (EU) at the Rio 'Earth Summit' Conference in 1992, on the issue of greenhouse gas emission reductions in order to stave off global warming and hence global climate change, has been undermined by an internal dispute over the proposed introduction of an EU region-wide carbon tax designed to induce consumers to reduce their usage of energy from carbon-based sources. This proposal and an alternative European Commission compromise proposal to impose a tax based on a fuel's carbon content and its energy value, in order both to discourage the use of polluting carbon energy sources and to stimulate wider efficiency in the use of energy from other, non-carbon based sources, have both failed. Instead, the EU member states have adopted the easier option of allowing member states to follow individual strategies on energy taxation, rather than that of the imposition of a common tax (Simonian, 1994). This development is a disincentive to oil companies working in the EU area to examine other fuel options, as it encourages them to carry on with their present wait-and-see attitude, rather than driving them towards exploration of and investment in alternative, non-carbon-based fuel technologies that would result in a lower carbon tax burden (were it to be imposed), in addition to helping to reduce greenhouse gas emissions.

A pilot scheme to sell and buy carbon dioxide permits has also been recommended to the UN's Commission on Sustainable Development in advance of an ambitious scheme to halt global warming by launching an international market in carbon dioxide permits. The project, which would involve a number of European countries and the USA, has been said to offer the possibility of reaching the desired environmental goals at a lower cost than would be possible if each country were limited to the options for reduction available only within its own borders. [The following discussion is heavily derived from an article in the Business and the Environment pages of the *Financial Times* (Lapper and Morse, 1995).]

The scheme's proponents are loosely modelling their proposals on an existing US (national) programme, which aims to cut acid rain by reducing emissions of sulphur dioxide by one-third over a 20 year period. Under the US programme – currently in its fourth year – the US Environmental Protection Agency allocates certificates to coal-burning electric power plants, allowing emissions within legal limits. If the power-generating company chooses to reduce its sulphur dioxide output by, for example, switching to cleaner fuels or installing a costly smokestack scrubber, it is then free to sell its excess allowances to another company that wants to increase its own emissions above the legal limits. Consultants who have been observing the evolution of this domestic sulphur dioxide allowance programme are enthusiastic about its application to reducing greenhouse gas emissions.

A new UN Global Environmental Protection Agency would be the cornerstone of the proposed market and responsible for the initial allocation of carbon dioxide permits to national governments that elect to participate in the pilot scheme. The amount allocated to each country would be based on existing and future acceptable levels of national and global carbon dioxide emissions. It would then be up to each government to channel permits to installations, such as coal-powered power plants, which – according to the UN – are responsible for between 70 and 80% of carbon dioxide emissions. The permits would eventually be traded on three of the world's derivatives exchanges, with one exchange operating in three time zones. If the scheme works, it is hoped that other countries would be tempted to join in, gradually transforming it into a more complete system.

Before this point is reached, however, a number of technical problems must be confronted. In particular, the task of monitoring the effectiveness of the programme could be an onerous one. For the US sulphur dioxide programme, sensors were installed in some 2300 smokestacks. Data from these sensors are collected and analysed at a central location every hour. It has been estimated that up to 10 gigabytes of data will be reviewed every quarter merely for the first phase of smokestack sensors.

Yet there is growing optimism. Economic theory and history, as well as the early experience of the sulphur dioxide entitlement programme, suggest that the proposed tradeable carbon dioxide entitlement programme will provide the lowest cost solution to the global warming problem. Indeed, it has been argued already that the idea of creating a market in tradeable permits should be expanded to encompass not only greenhouse gas (specifically, carbon dioxide) emission entitlements, but also greenhouse gas absorption requirements or obligations, under the concept of a Tradeable Absorption Obligation (TAO). The TAO is part of the market-oriented arrangement discussed above, which is designed to induce the absorption of fuel-wood and biomass technology with minimal uncertainty of effect. The TAO is aimed at forcing the uptake of alternative fuel technologies (of which the fuel-wood and biomass technology is just one option) at a rate prescribed by international policy targets set by the Conference of the Parties to the Framework Convention, with least cost secured through the tradeability of obligations (Read, 1994, p. 168).

2.6 Conclusions and future prospects for environmental regulation of the oil industry

Taking a wider perspective and with a view to some, albeit well grounded, 'crystal ball gazing' as to future trends in environmentalism and their attendant legal implications for the continued progress of the oil and gas industry, this final part of the chapter will return to a theme expressed earlier concerning the need for the industry to be fully adaptable to new environmental constraints in the near future [according to Choucri, there are at least three definitions of environmentalism: (a) a political belief shaping political action; (b) a new 'theory' influencing economic policy; and (c) a significant factor in investment decisions and business practices; see Choucri (1995), citing Choucri, N., Environmentalism, in Krieger (1993)]. As Choucri has noted generally in respect of most corporate activities around the world, 'Today, firms operating across sovereign borders are confronted with a new challenge: how to evaluate and respond to emerging environmentalism' (Choucri, 1995, p. 189). This is especially the case with respect to the bigger, multinational oil and gas companies whose international activities nowadays span the globe and for whom the rising challenge of international environmentalism is especially pertinent to corporate strategy and culture. In the aftermath of the *Brent Spar* debacle, for example, there was much media attention devoted to the effect of this incident on the loss of environmental credibility in the generally sound corporate culture that is inculcated at Shell, which has hitherto reflected with pride on its environmental credentials.

Choucri also warns that, 'If a firm is to compete effectively in an increasingly competitive global market, it cannot misread the signals of growing environmentalism and proceed to conduct business as usual. But, while governments, public interest groups and international organizations are searching for institutional innovation and adaptation in that area, global corporations – with few exceptions – have generally failed to develop a strategy for dealing with the changing business environment due to the emergent environmentalism' (Choucri, 1995, p. 190). One recent survey notes, for example, that despite exhortations from environmental groups and government ministers, companies appear sceptical that, in the absence of legislative threats, it is worth their while to become 'greener'. In other words, businesses do not appear convinced that, without the threats of fines or penalties under environmental regulation, 'greener' can mean richer. This scepticism may also derive from the perception among companies that after a period of strict 'green' legislation, politicians are now more alert to the economic implications of over-penalizing polluters. As the costs of compliance with higher environmental standards have become clearer, governments have become ambivalent about continuing to legislate for such standards. This legislative wavering may be one reason why some companies are less willing than in the past to adopt environmental improvements. Unless companies are convinced that going 'green' for its own sake pays off, their interest in environmental management will continue to be dictated by the volume and rate of legislation and the vigour with which it is enforced. Given governments' current lukewarm attitudes towards more environmental rule-making, the short-term outlook for environmental management may be bleak (Maddox, 1995).

Interestingly, one area of business and industry that has begun to take account of the impact of generally deteriorating environmental conditions on human endeavours is the insurance services sector. Here, it is acknowledged that there are logical reasons for supposing that the carbon that is currently being emitted into the atmosphere from used fossil fuel energy sources, coupled with the diminishment of rainforests, could slowly lead to global warming. Hence, insurance companies (which do have to look at the bottom line) have started to worry about floods and other disasters attributable to climatic changes (though not all are due to global warming) that their actuaries had not expected (Macrae, 1995). Growing evidence of man-made climate change, causing an unprecedented series of natural catastrophes, is posing a threat to the viability of the insurance industry. Indeed, insurance underwriting analysts have even begun to assume that since fossil fuel industries are lobbying against steps to tackle global warming and are likely to emasculate any measures that might be taken, it is probable that the insurance industry will have to take the initiative, either by itself or along with the banking industry, to switch some of its more than US $100 billion (£63 billion) worth of the investments that it currently holds in fossil fuel industries into alternative energy-generating companies (Lean, 1995b).

Higher environmental standards in themselves, especially when taking into consideration the extra costs involved for those who comply, and the possibility of poor enforcement undermining the effectiveness of these stricter rules, do not seem to be the answer. Therefore, the search has been on for some time now to introduce alternative tools and methods for ensuring compliance with pollution controls and, more important, the creation of an atmosphere in which an economically viable environmental ethic can be inculcated within the management of business and industry generally, and the oil industry in particular.

One such alternative is the introduction of voluntary agreements between industry and government for environmental protection. This may be seen as part of a wider policy approach designed to induce industrial self-regulation, by which is meant 'rules which govern behaviour in the market (that) are developed, administered and enforced by the people whose behaviour is to be governed' (Rowan-Robinson and Kimber, 1995, pp. 8–10, citing National Consumer Council, 1986). For example, oil companies operating in the UK sector of the North Sea voluntarily entered into an agreement by which all UK Continental Shelf operators will meet claims for pollution damage arising out of oil spillage or escape and to pay for the cost of remedial measures. The agreement headed off any formal statutory requirement and has since been extended beyond the UK sector (Rowan-Robinson and Kimber, 1995, p. 10).

In the run-up to the recent Berlin Conference of Parties to the Framework Convention on Climate Change, German industry and the Bonn government struck a deal: three-quarters of German industry, including the cement, steel and electricity sectors, agreed voluntarily to cut their carbon dioxide emissions by up to 20% by 2005 to help Germany meet its CO_2 reduction targets. In return, as long as industry kept its side of the bargain, the government agreed to delay plans for a new law to reduce industrial energy consumption. Such voluntary, or negotiated, agreements are being increasingly proposed by European industry as an alternative to more traditional regulatory or fiscal measures for purposes of tackling pollution. Industry commonly complains that environmental legislation is not subject to rigorous cost–benefit analysis and can be difficult to implement. Voluntary agreements, negotiated between industry and governments, allow companies the flexibility to decide for themselves the most cost-effective way of meeting environmental targets. For governments, on the other hand, such agreements can prove quicker to implement than legislation and more likely to attract industry support.

This policy approach is proving popular at the national level. A recent survey by the Union of Industrial and Employers' Confederations of Europe (Unice) shows that more than 130 such voluntary agreements, lasting between 5 and 10 years, have been concluded among their member states. These voluntary agreements are also a main plank of Dutch environmental policy, more than 70 agreements having been signed between industry sectors and the government, tackling limitation of atmospheric emissions and promoting energy conservation. European industry, via Unice, is now proposing that voluntary agreements should increasingly play a practical role in the development of environmental policy at the European level, and there are signs that this strategy is beginning to work. Voluntary agreements also accord with the current preoccupation of European policymakers with the need to safeguard industrial competitiveness and promote growth as Europe struggles to emerge from recession (Plaskett, 1995).

A theoretical explanation for the above approach can be found in the now widely held notion among legal and economic policy analysts that the unilateral imposition of norms is no longer the most desirable model for state intervention in societal processes generally (Koppen, 1994, p. 185). This is especially the case when we consider the lack of success of 'command and control' type regulation in the formulation of environmental policy affecting the business and industry sectors of society (Rowan-Robinson and Kimber, 1995, p. 8). On the other hand, the lack of progress in business attitudes towards achieving higher environmental standards set down by legislation (as noted earlier), coupled with the continuing doubts expressed by environmentalist groups over the outcome of the development of such 'cosy' relationships between government and industry, leave many observers to conclude that industry still needs the spur of binding legislation as a continuing 'incentive' for action to improve environmental standards. Thus, as has been noted recently, there is little agreement over the effectiveness of the different means for securing environmental policy ends (Rowan-Robinson and Kimber, 1995, p. 15).

One way or another, either through increasing corporate liability for environmental harm (as in the case of the 1990 US Oil Pollution Act, for example) in the traditional 'command and control' manner, or through voluntary agreements entered into between government and industry setting down environmental targets for achievement within set time-scales (as in the case of the reduction of greenhouse gas emissions, for example), it can be seen that a new framework of rules governing the various activities of the different sectors of the oil industry is evolving. This new regime is, at least ostensibly, based on the notion of national, regional and global environmental protection from the effects of these activities. However, this system is so fragmented in terms of its application to different sectors of the oil industry that it can by no means be stated that the international community is intent on following any grand design when laying down new rules for the protection of the environment.

In so far as they may be pertinent to the development of different aspects of this new regime for the oil industry, the following new environmental principles have been identified as being imperative to guide the conduct of business towards more environmentally directed policies in the international scene today. They include:

- best practice: using the best available technologies and processes, designed for minimum environmental impact;
- pollution prevention: preventing the generation of pollution, rather than cleaning up afterwards;
- the polluter-pays principle (PPP), whereby polluters pay the full cost of environmental damage caused by the production of goods and services;
- the precautionary principle under which, in situations of scientific uncertainty, actions taken should be based on a cautious evaluation of the risks;

- full-cost pricing of goods and services: taking into account or 'internalizing' not just the full production costs but also a quantification of the attendant environmental costs involved in such production;
- eco-efficiency, whereby practices to increase efficiency based on economic criteria alone are not sufficient and environmentally efficient practices using environmental criteria are also developed and applied;
- eco-labelling: a practical application of and guideline to the above principle of eco-efficiency;
- monitoring compliance and reporting: businesses should measure their performance in terms of their overall impact on the environment; and
- common but differentiated responsibility: involving a recognition of a 'global partnership' for environmental protection among different countries of the world at different levels of their socio-economic development, and therefore assigning more responsibility to the developed or industrialized countries (and by extension, companies based in these countries) as the main contributors to environmental degradation both in the past and at present, but also because they alone often have the resources and technology to resolve environmental problems (Choucri, 1995, p. 196).

As a final point, it may also be noted that the basic thrust of the development of the new environmental principles that are proposed to guide all industrial activity, including the oil industry, is towards subjecting the entire 'life cycle' of the industry to continuous review of its overall environmental impact, beginning from the basic stage of collection of raw material (i.e. drilling for crude oil), to the final stage of finished product (in terms of refined petroleum derivatives for use in various other manufacturing and transport industries and services). This type of policy approach, based on analysing the oil industry's complete 'life cycle' in terms of its overall impact on the environment, could then be utilized to improve progressively the oil industry's environmental performance, through the introduction of an 'innovation-based' industrial strategy designed to respond to the new challenge presented by the need to protect the environment. Multinational companies such as 3M and DuPont, for example, have found that by anticipating sources of pollution and eliminating them at the design stage, they can save on raw materials, waste remediation equipment and pollution charges. This has the effect of reducing their capital requirements, increasing their efficiency and saving money (Wallace, 1995, pp. 265–6).

An essential element of this innovation-based strategy depends on the ability, willingness and capacity of the traditional regulators of the oil industry, i.e. national governments and, sometimes, international organizations and institutions, to create an open, honest and high-quality dialogue with the industry. Generally, firms are more comfortable innovating when risks are reduced. Risks are lower when environmental policy is stable and credible over the long term, and when regulatory processes are based on open, informed dialogue and executed by competent, knowledgeable regulators (Wallace, 1995, Summary and Conclusions, pp. xvii–xx). Stable long-term policies and effective dialogue can therefore create a culture which rewards proactive, responsible firms and reduces the risks of investing in innovative solutions (Wallace, 1995, p. 266).

In concluding this chapter it seems apt to quote Wallace's conclusions in his *Environmental Policy and Industrial Innovation*, as this approach appears to be the best way forward for the oil industry today, faced as it is with the further challenge of protecing the environment: 'The long-term challenge of sustainable development is an opportunity for governments to make environmental policy more stable and less reactive. New industry–government working relationships such as flexible 'voluntary' agreements and contracts are devolving greater responsibility to firms, while increasing dialogue. This leads to more flexibility to innovate, lower compliance costs and less opposition to environmental policies. Sustainable development will require such politically sustainable environmental policies and continuous pressure, and opportunity, for industry to innovate' (Wallace, 1995, Summary and Conclusions, p. xx).

References

Abecassis, D.W. and Jarashow, R.L. (1985) *Oil Pollution from Ships*, 2nd edn, Stevens, London.

Azhar, S., Lau, L. and Sam, A. (1995) Cargo ship held: warning shot fired to stop vessel suspected of dumping. *The Star*, 1 June, 4.

Birnie, P. (1990) Protection of the marine environment in joint development, in *Joint Development of Offshore Oil and Gas*, Vol. II, (ed. H. Fox), British Institute of International and Comparative Law, London, pp. 202–22.

Birnie, P. (1993) Protection of the marine environment – the Public International Law approach, in *Liability for Damage to the Marine Environment* (ed. C.M. De La Rue), Lloyd's of London Press, London, pp. 1–22.

Birnie, P. and Boyle, A. (1992) *International Law and the Environment*, Clarendon Press, Oxford.

Boyle, A. (1994) Economic growth and protection of the environment, in *Environmental Regulation and Economic Growth* (ed. A.E. Boyle), Oxford University Press, Oxford, pp. 173–88.

Boulton, L. (1995) Higher carbon tax heats up debate. *Financial Times*, 14 June, 20.

Broadus, J.M. and Vartanov, R.V. (eds) (1994) *The Oceans and Environmental Security: Shared U.S. and Russian Perspectives*, Island Press, Washington, DC.

Brown, P. (1995) Survey boosts seabirds. *The Guardian*, 17 October, 9.

Brownlie, I. (1974) A survey of international customary rules of environmental protection, in *International Environmental Law* (eds L.A. Teclaff and A.E. Utton).

Choucri, N. (1995) Corporate strategies toward sustainability, in *Sustainable Development and International Law* (ed. W. Lang) Graham and Trotman, London, and Martinus Nijhoff, Dordrecht, pp. 189–201.

Churchill, R. (1991) Controlling emissions of greenhouse gases, in *International Law and*

Global Climate Change, (eds R. Churchill and D. Freestone), Graham and Trotman, London, and Martinus Nijhoff, Dordrecht, pp. 147–63.
Collins, D.M. (1987) The tanker's right of harmless discharge and protection of the marine environment, *Journal of Maritime Law and Commerce*, **18**, 275.
Corzine, R. (1995) Oil groups try to avoid deep water: the industry is determined there will be no repeat of last June's *Brent Spar* fiasco. *Financial Times*, 15 August, 8.
Cowe, R. (1995) Saving more than the planet: green code lesson means money in the bank. *The Guardian*, 17 October, 19.
Crawshaw, S. (1995) Way cleared towards greenhouse gas targets. *The Independent*, 8 April, 8.
Demsetz, H. (1967) Toward a theory of property rights. *American Economic Association Papers and Proceedings*, **57**, No. 2; cited by Pearson (1975, p. 18).
Doos, B.R. (1991) Environmental issues requiring international action, in *Environmental Protection and International Law* (eds W. Lang, H. Neuhold and K. Zemanek), Graham and Trotman, London, and Martinus Nijhoff, Dordrecht, pp. 1–54.
Dzidzornu, D. and Tsamenyi, B.M. (1991) Enhancing international control of vessel-source oil pollution under the Law of the Sea Convention, 1982: a reassessment, *University of Tasmania Law Review*, **10**, 269.
Ellis, E.J. (1995) International law and oily waters: a critical analysis, *Colorado Journal of International Environmental Law and Policy*, **6**(1), 31–60.
Franckx, E. (1995) Coastal state jurisdiction with respect to marine pollution – some recent developments and future challenges, *International Journal of Marine and Coastal Law*, **10**(2), 253–80.
Freestone, D. (1993) *The Road from Rio: International Environmental Law after the Earth Summit*, University of Hull Press, Hull.
Freudenburg, W.R. and Gramling, R. (1994) *Oil in Troubled Waters: Perceptions, Politics and the Battle over Offshore Drilling*, State University of New York Press, Albany, NY.
Gaines, S.E. (1991) The polluter-pays principle: from economic equity to environmental ethos, *Texas International Law Journal*, **26**, 463.
Garick, J. (1993) Crisis in the oil industry: Certificates of Financial Responsibility and the Oil Pollution Act of 1990, *Marine Policy*, **17**, 272.
Hamzah, B.A. (1988) *Malaysia's Exclusive Economic Zone*, Pelanduk Publications, Petaling Jaya.
Jones, G. and Saunders, M. (1994) Abandonment of offshore petroleum production installations, in *European Community Energy Law: Selected Topics* (eds D.S. MacDougall and T.W. Walde), Graham and Trotman, London, and Martinus Nijhoff, Dordrecht, pp. 239–59.
Kirgis, F.L., Jr (1995) Shipping, in *United Nations Legal Order*, Vol. 2, (eds O. Schachter and C.C. Joyner), Cambridge University Press, Cambridge, pp. 715–51.
Koppen, I. (1994) Ecological covenants: regulatory informality in Dutch waste reduction policy, in *Environmental Law and Ecological Responsibility: the Concept and Practice of Ecological Self-Organization* (eds G. Teubner, L. Farmer and D. Murphy), Wiley, Chichester, pp. 185–205.
Krieger, J. (ed.) (1993) Oxford *Companion to Politics of the World*, Oxford University Press, Oxford.
Kuik, O., Peters, P. and Schrijver, N. (eds) (1994) *Joint Implementation to Curb Climate Change: Legal and Economic Aspects*, Kluwer Academic Publishers, Dordrecht.
Lapper, R. and Morse, L. (1995) Market makers in CO_2 permits. *Financial Times*, 1 March, 18.
Lascelles, D. (1995) Swamped by a sea of public anger: *Brent Spar* means that businesses must include public opinion in environmental plans. *Financial Times*, 22 June, 21.
Lean, G. (1995a) Global warming is leading to climatic upheaval, say scientists. *The Independent on Sunday*, 15 October, 5.
Lean, G. (1995b) Insurers urged to go greener. *The Independent on Sunday*, 7 May, Business Section, 1.
Low, K.S. (1990) Tukau oil and Duyong gas platforms: Malaysia, in *Energy Systems and the Environment: Approaches to Impact Assessment in Asian Developing Countries* (eds P. Hills and K.V. Ramani), Asian and Pacific Development Centre, Kuala Lumpur, pp. 213–38.

Macrae, N. (1995) Save the world: vote for an econut. *The Sunday Times*, 2 April. section 3, 4.

Maddox, B. (1995) Green light turns amber. *Financial Times*, 21 June, I.

Mitchell, R. (1993) Intentional oil pollution of the oceans, in *Institutions for the Earth: Sources of Effective International Environmental Protection* (eds P.M. Haas, R.O. Keohane and M.A. Levy), MIT Press, Cambridge, MA, pp. 183–247.

Nanda, V.P. (1995a) *International Environmental Law and Policy*, Transnational Publishers, Irvington-on-Hudson, NY.

Nanda, V.P. (1995b) Environment, in *United Nations Legal Order*, Vol. 2, (eds O. Schachter and C.J. Joyner), Cambridge University Press, Cambridge, pp. 631–69.

National Consumer Council (1986) *Self-Regulation*, National Consumer Council, London.

Nilsson, S. and Pitt, D. (1994) *Protecting the Atmosphere: the Climate Change Convention and its Context*, Earthscan Publications, London.

Ong, D. (1995) Southeast Asian state practice on the joint development of offshore oil and gas deposits, in *The Peaceful Management of Transboundary Resources*, (eds G.H. Blake, W.J. Hildesley, M.A. Pratt *et al.*), Graham and Trotman, London, and Martinus Nijhoff, Dordrecht, pp. 77–96.

Pearson, C.S. (1975) *International Marine Environment Policy: the Economic Dimension*, Johns Hopkins University Press, Baltimore, MD.

Plant, G. (1992) Institutional and legal responses to global environmental change, in *Global Environmental Change and International Relations* (eds I.H. Rowlands and M. Greene), Macmillan, Basingstoke, pp. 122–44.

Plaskett, L. (1995) Agreements via the voluntary route. *Financial Times*, 3 May, 13.

Porter, M.E. (1991) America's green strategy, *Scientific American*, April, 96.

Pulvenis, J.-F. (1994) The Framework Convention on Climate Change, in *The Environment After Rio: International Law and Economics* (eds L. Camiglio, L. Pineschi, D. Siniscalco and T. Treves), Graham and Trotman, London, and Martinus Nijhoff, Dordrecht, pp. 71–110.

Read, P. (1994) *Responding to Global Warming: the Technology, Economics and Politics of Sustainable Energy*, Zed Books, London.

Reguly, E. (1995) Taxpayer may have to foot bill for next 30 years. *The Times*, 22 June, 9.

Rémond-Gouilloud, M. (1981) Prevention and control of marine pollution, in *The Environmental Law of the Sea* (ed. D.M. Johnston), IUCN Environmental Policy and Law Paper No. 18, International Union for Conservation of Nature and Natural Resources (IUCN), Gland, Switzerland, pp. 193–202 (Introduction to Chapter 3).

Rowan-Robinson, J. and Kimber, C. (1995) The oil industry and mechanisms for securing environmental protection, *Contemporary Issues in Law*, **1**(3), 1–16.

Schneider, J. (1981) Prevention and control of marine pollution, in *The Environmental Law of the Sea* (ed. D.M. Johnston), IUCN Environmental Policy and Law Paper No. 18, International Union for Conservation of Nature and Natural Resources (IUCN), Gland, Switzerland, pp. 203–17 (II. Pollution from Vessels, Chapter 3).

Schoon, N. (1995) The right climate for tax on fuel. *The Independent*, 16 October, 21.

Sharma, S. (ed.) (1993) *Energy, the Environment and the Oil Market: an Asia–Pacific Perspective*, Institute of Southeast Asian Studies (ISEAS), Singapore.

Simonian, H. (1994) EU rift takes steam out of energy taxes. *Financial Times*, 22 December, 2.

Springall, R.C. (1988) P & I insurance and oil pollution, *Journal of Energy and Natural Resources Law*, **6**, 25.

Steen, N. (ed.) (1994) *Sustainable Development and the Energy Industries: Implementation and Impacts of Environmental Legislation*, Royal Institute of International Affairs, London.

Stephenson, M.A. (1992) Vessel-source pollution under the Law of the Sea Convention – an analysis of the prescriptive standards, *University of Queensland Law Journal*, **17**, 117.

Tan, E. and Anbalagan, V. (1995) Polluting vessel detained by RMN after 20 hour chase. *The New Straits Times*, 1 June, 4.

Tan, E. and Atan, H. (1995) Ship owner can be fined RM 1 million. *The New Straits Times*, 3 June, 1.

Tan, R. (1995) Foreign ship operators fined for desludging. *The Sun*, 31 October, 8.

Taverne, B. (1994) *An Introduction to the Regulation of the Petroleum Industry: Laws,*

Contracts and Conventions, Graham and Trotman, London, and Martinus Nijhoff, Dordrecht.
Wallace, D. (1995) *Environmental Policy and Industrial Innovation: Strategies in Europe, the USA and Japan*, Earthscan, London.
Wetterstein, P. (1992) *Environmental Impairment Liability in Admiralty: a Note on Compensable Damage under U.S. Law*, Abo Akademi University Press, Abo.
White, I.C. (1993) The Voluntary Oil Spill Compensation Agreements – TOVALOP and CRISTAL, in *Liability for Damage to the Marine Environment* (ed. C.M. De La Rue), Lloyd's of London Press, London, pp. 57–69.

Further reading

Brubaker, D. (1993) *Marine Pollution and International Law: Principles and Practice*, Belhaven Press, London.
Choucri, N. (ed.) (1993) *Global Accord: Environmental Challenges and International Responses*, MIT Press, Cambridge, MA.
Cutajar, M.Z. (1995) Comment on the paper by Gunther Handl: the FCCC experience, in *Sustainable Development and International Law*, (ed. W. Lang), Graham and Trotman, London, and Martinus Nijhoff, Dordrecht, pp. 45–8.
De La Rue, C.M. (ed.) (1993) *Liability for Damage to the Marine Environment*, Lloyd's of London Press, London.
Fankhauser, S. (1995) *Valuing Climate Change: The Economics of the Greenhouse*, Earthscan Publications, London.
Gavouneli, M. (1995) *Pollution from Offshore Installations*, Graham and Trotman, London, and Martinus Nijhoff, Dordrecht.
Grubb, M. (1995) Seeking fair weather: ethics and international debate on climate change, in *International Affairs*, Special Issue on Ethics, the Environment and the Changing International Order, **71**(3), 463–96.
Grubb, M., Koch, M., Munson, A. *et al.* (1993) *The 'Earth Summit' Agreements: a Guide and Assessment: an Analysis of the Rio '92 UN Conference on Environment and Development*, Earthscan Publications, London.
Haas, P.M., Keohane, R.O. and Levy, M. (eds) (1993) *Institutions for the Earth: Sources of Effective International Environmental Protectios*, MIT Press, Cambridge, MA.
Hayes, P. and Smith, K. (eds) (1993) *The Global Greenhouse Regime: Who Pays?*, Earthscan Publications, London, and United Nations University Press, Tokyo.
Liberatore, A. (1994) Facing global warming: the interactions between science and policy-making in the European Community, in *Social Theory and the Global Environment* (eds M. Redclift and T. Benton), Routledge, London, pp. 190–204.
MacDougall, D.S. and Walde, T.W. (eds) (1994) *European Community Energy Law: Selected Topics*, Graham and Trotman, London, and Martinus Nijhoff, Dordrecht.
Morgan, H. (1993) The potential for seabed resource development, in *Maritime Change: Issues for Asia* (eds R. Babbage and S. Bateman), Allen and Unwin, St Leonards, Australia, pp. 89–95.
Nollkaemper, A. and Hey, E. (1995) Implementation of the Law of the Sea Convention at regional level: European Community competence in regulating safety and environmental aspects of shipping, *International Journal of Marine and Coastal Law*, **10**(2), 281–300.
Oxman, B.H. (1995) Law of the sea, in *United Nations Legal Order*, Vol. 2 (eds O. Schachter and C.C. Joyner), Cambridge University Press, Cambridge, pp. 671–713.
Patterson, M. (1994) The politics of climate change after UNCED, in *Rio: Unravelling the Consequences* (ed. C. Thomas), Frank Cass, Ilford.
Pearce, D. (1992) Economics and the global environmental challenge, in *Global Environmental Change and International Relations* (eds I.H. Rowlands and M. Greene), Macmillan, Basingstoke, pp. 60–87.
Ramani, K.V., Hills, P. and George, G. (eds) (1992) *Burning Questions: Environmental Limits to Energy Growth in Asia–Pacific Countries during the 1990s*, Asian and Pacific Development Centre, Kuala Lumpur.

Rowlands, I.H. (1995) *The Politics of Global Atmospheric Change*, Manchester University Press, Manchester.

Rowlands, I.H. and Greene, M. (eds) (1992) *Global Environmental Change and International Relations*, Macmillan, Basingstoke.

Sebenius, J.K. (1993) The Law of the Sea Conference: lessons for negotiations to control global warming, in *International Environmental Negotiation* (ed. G. Sjostedt), Sage Publications, Newbury Park, CA, pp.189–218.

Teubner, G., Farmer, L. and Murphy, D. (eds) (1994) *Environmental Law and Ecological Responsibility: the Concept and Practice of Ecological Self-Organization*, Wiley, Chichester.

Weale, A. (1992) *The Politics of Pollution*, Manchester University Press, Manchester.

3 Environmental control technology in petroleum drilling and production

A.K. WOJTANOWICZ

3.1 Introduction

Environmental control technology (ECT) is a process-integrated pollution prevention technology. Within the broader scope of environmental technology that includes assessment of environmental impact, remediation and prevention, ECT relates mostly to prevention and risk assessment. Historically, developments in preventive techniques came after analytical and remediation measures, which have been found to be inadequately reactive and progressively expensive.

Reactive techniques focus on impacts and risk. With reactive pollution control, the positive action is entirely linked to the environmental objective. History provides ample evidence that reactive strategies do little more than transfer waste and pollution from one medium to another. *Preventive* action seeks root causes of pollution generation. It often requires modification of technology that has no apparent linkage to an environmental objective and is intrinsically more comprehensive than reactive strategies [1].

In principle, ECT is a process-engineering approach to the prevention of environmental damage resulting from industrial (oilfield) operations. The approach draws on the modern theory of 'clean production', a term coined by the United Nations Environmental Program's Industry and Environmental Office (UNEP/IEO) in 1989 [2].

The clean production theory, in its broadest sense, delineates an approach to industrial development that is no longer in conflict with the health and stability of the environment, a kind of development that is sustainable. In the narrowest sense of the theory, clean production signifies a preventive approach to design and management of 'environmentally controlled' industrial processes. The approach seeks to reduce 'downstream' or end-of-pipe solutions to environmental problems by looking 'upstream' for reformulation and redesign of the processes or products. It also involves a broader, integrated, systematic approach to waste management.

Within the parameters of clean production, then, oilfield environmental control technology allows an examination of drilling, well completion and production as environmentally constrained processes containing inherent mechanisms of environmental impact. These mechanisms include the generation of waste, induction of toxicity or creation of pathways for pollutant migration. Identification and practical evaluation of these mechanisms constitute two parts of the ECT scope. A third part involves the development (at minimum cost) of new methods and techniques to meet environmental compliance requirements without hindering productivity.

Naturally, ECT tackles a large spectrum of oilfield technologies, such as closed-loop drilling systems, subsurface injection, borehole integrity, toxicity control in petroleum fluids, downhole reduction of produced water and use of land for on-site storage and disposal of oilfield waste. In this chapter, basic concepts of the ECT approach are presented first. Then, the ECT approach is used to analyze oilfield processes of drilling and production and to describe developments of environmental control components in these technologies.

3.2 Environmentally controlled oilfield processes

For 100 years, oilfield science and technology have been continually improving. The oil industry has evolved from one that was interested mainly in inventing tools and equipment to one that is not only economically, but also environmentally, conscious. In the 1980s, low oil prices forced oilfield technology to focus on economic efficiency and productivity. Simultaneously, environmental regulatory pressure added a new factor to petroleum engineering economics: the cost of working within the constraints of an environmental issue. The industry has been absorbing this cost while also revising oilfield technology to reduce the environmental impact.

Conceptually, the perception of environmental problems and solutions is an evolutionary process of shifting paradigms of waste management as depicted in Fig. 3.1. Over time, concepts regarding what is the best strategy for waste management have changed from 'disposing at will' (followed by remediation), to dilution/dispersion of waste below the assimilative capacity of the environment, to controlling the rate or concentration of pollutants at the waste discharge ('end-of-pipe' treatment), to developing truly preventive technologies.

In the petroleum industry this shift of paradigms was recently described as a transition from a PCD (produce–consume–dispose) approach to a WMT (waste management technology) approach and, finally, to a preventive ECT approach [3]. The large quantities of waste fluids and slurries (drilling muds and produced waters), and their associated wastes that are created during everyday oilfield activities have been conventionally

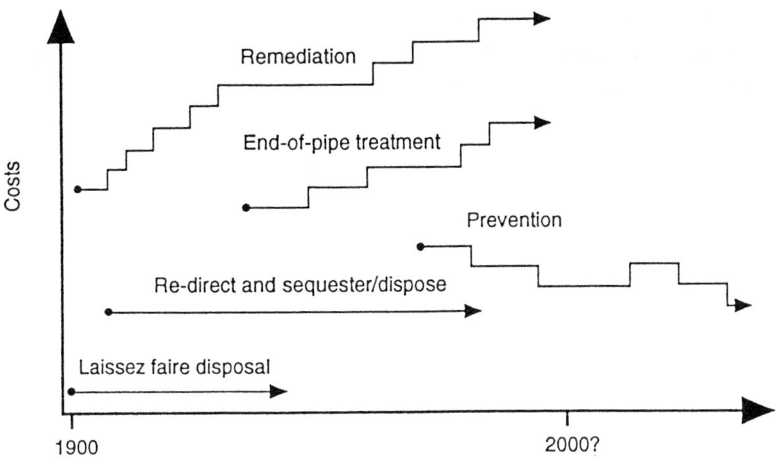

Figure 3.1 Waste management strategy paradigm shift [1].

perceived as unavoidable. This perception is typical of the PCD approach. Not only does this approach assume a proportional relationship between the production stream rate (oil/gas) and the volume of waste, but it also assumes that the flow of materials is open so that the waste must be discharged from the process into the environment. Such an attitude has prevailed for most of the modern history of petroleum engineering.

In the early 1980s, evidence of health and environmental hazards in the oilfield was accumulated and made public, which triggered serious public concerns and resulted in regulatory pressures [4–8]. Public opinion has been documented in several surveys. Growing public pressures (and private lawsuits) prompted regulatory activities. Since the mid-1980s in the USA, for example, oilfield waste has been identified, its volume and toxicity evaluated and its disposal methods scrutinized [9–11]. This scrutiny, together with the industry's PCD-dominated environmental paradigm, resulted in the rapid development of waste management programs (the WMT approach). Indeed, at the time, clean-ups were prioritized over preventive measures in an effort to employ the existing waste disposal industry rather than to rethink the whole oilfield process again and identify environmental control techniques.

This seemingly logical paradigm was founded on three fundamental arguments: (1) waste must be managed because there is no other way to protect the environment; (2) waste has no value so its management is the most efficient solution; and (3) waste is external to the oilfield process. In fact, all these arguments lack substance: (1) the environment can be efficiently protected by reducing waste volume and/or its toxicity (source reduction and source separation); (2) oilfield waste does sometimes have value; for example, in California, production sludge is processed to recover crude, and in Alaska the drilled cuttings gravel is used for road construction [12]; and (3) waste becomes external only if it is released from the process; for instance, the annular injection of spent drilling mud leaves no drilling waste.

Within the petroleum industry, a change in the environmental paradigm from the PCD syndrome to the preventive approach of environmental control has recently emerged as a result of high disposal costs. The cost of waste management has grown steadily in response to increasing volumes of oilfield waste. Interestingly, the amount of regulated waste has grown much faster than oil and gas production because regulated waste volume has been driven mainly by regulations rather than by production rates.

In principle, the environmental control paradigm in petroleum engineering involves three concepts: (1) the fundamental purpose of petroleum engineering is not to protect the environment but to maximize production while preventing environmental impact; (2) compliance problems can be eliminated when environmental constraints are introduced into the production procedures; and (3) any stream of material is off-limits to regulatory scrutiny and can be controlled by oilfield personnel as long as it remains within the oilfield process. In practice, this attitude requires an understanding of environmental impact mechanisms and the willingness to redesign the process.

The environmental control paradigm presented above is a philosophical concept which needs a practical methodology. Such a methodology would give a designer some guidelines regarding how to analyze an industrial process and where to put efforts to make the process 'cleaner' (or 'greener', as some put it).

3.2.1 Scope and characteristics of oilfield ECT

This overview of ECT methodology includes a definition, objectives and characteristic features, general ECT methods and a description of basic steps needed to develop a specific technology. ECT is defined as a technical component of an industrial process that is functionally related to the interaction between the process and environment. Such interaction involves pollution and other adverse effects (impacts) on environmental quality. The objective ECT is to prevent this interaction by controlling the impact mechanisms. The three important features of ECT are integration with the process, specific design and association with productivity.

These three features make ECT different from the technologies of waste management. The difference requires further discussion in relation to oilfield applications. First, however, we must recognize the difference between waste and the process material stream. This difference draws on two facts: (1) where the material is with respect to the process; and (2) what the material's market value is. This concept assumes that no waste exists inside the process – just material streams. On leaving the process (i.e. crossing the process boundary) a stream of material becomes either a product (including by-products) or waste. The difference stems from the market value of the material. Having a positive market value, the material becomes a product. Material with zero value becomes waste. When the value is negative, the material becomes regulated waste (regulated waste requires expenditures for proper disposal).

In view of the above, WMT becomes extraneous to the process because it operates outside the process boundaries and within the environment. WMT involves processing and disposing of the waste as it is discharged from a well site or production plant. Expertise in waste management technologies lies mostly outside the petroleum engineering field. Over the last 10 years, the oil industry has been offered several waste management technologies, providing considerable understanding of the available services. Examples of alternative WMT for production operations are land farming, incineration, road spreading, commercial waste injection facilities and brine demineralization plants. The WMT for drilling operations, other than those for production, include offshore hauling of drilling fluids and cuttings for onshore disposal. These techniques abate pollution without interfering with oilfield procedures; therefore, they provide no incentive for process improvement. Also, the implementation of WMT requires no expertise in petroleum engineering and does nothing to prevent waste generation.

In contrast to WMT, ECT is an integral part of petroleum engineering. It addresses all of the mechanism and control techniques that relate to adverse environmental effects, such as generation of the waste volume and its toxicity, subsurface migration of toxicants and damage to the land surface. The objective of ECT is to minimize, through process improvements, interactions between oilfield processes and the environment. Therefore, the ECT concepts draw exclusively from petroleum engineering expertise. However, development of specific techniques may require expertise outside of petroleum engineering, such as solid–liquid and liquid–liquid separation, environmental science and environmental law, risk analysis and economics.

The use of outside expertise to develop ECT for petroleum engineering includes, of course, some waste management techniques. Indeed, both technologies are bound to draw from the same pool of science. This may sometimes create an impression that ECT is merely a part of WMT. There is, however, a distinct difference between the two. For example, dewatering of abandoned oilfield waste pit slurries, highly diluted with rainfall/run-off water, is a WMT and does not require any oilfield expertise. However, the inclusion of the dewatering component within the closed-loop mud system is an ECT. In this application, dewatering becomes intrinsic to the drilling process; it requires an in-depth knowledge of mud engineering. It also poses a research challenge since drilling fluids, unlike waste water, contain high concentrations of surface active solids.

ECT overlaps with WMT in the area of subsurface injection, which has long been perceived as a waste disposal option in various industries. In this case, however, the petroleum engineering expertise in borehole technology has merely been extended to other applications. Further, when subsurface injection is used in the oilfield for recycling produced water or annular injection of drilling fluids, the method is (1) intrinsic to the oilfield process and (2) requires oilfield expertise to perform, thus making it an ECT.

There is a strong affiliation between ECT and process-control measures. Similar to process control projects, ECT requires a considerable knowledge of oilfield processes in order to identify the chain reactions that lead to the environmental impact. As an example, let us consider the cause-and-effect relationship between the seemingly unrelated phenomena of drilling mud inhibition and the environmental discharge of drilling waste from the well-site. In fact, there is a strong functional relationship between the degree of drilled cuttings dispersion in mud and the waste mud volume. There is also a close analogy between ECT and process-control methods when solving design problems. In process-control design one must prioritize objective function and consider constraints imposed on the design. Similarly, any practical design of ECT must consider the environmental regulations as constraints, while also prioritizing productivity measures (such as daily production or cost per foot).

In this chapter, the term 'environmental control' is preferred over 'pollution prevention' because it implies broader objectives and suggests the process control-related means to accomplish these objectives. Oilfield operations create the potential for ecological damage that can hardly be viewed as 'pollution', though this damage may set the scene for pollution. Examples of such ecological impact include land subsidence or damage to subsurface zonal isolation resulting from a poor annular seal or from fracturing a confining zone. Characteristically, the destruction of interzonal isolation will not result in pollution if there is no sufficient pressure differential across confining zones.

In summary, any WMT may become ECT if it becomes integrated with the oilfield process. Such an integration requires (1) containing the process within clearly defined environmental boundaries and (2) placing the WMT within these boundaries.

3.2.2 Methodology of ECT design

A conceptual schematic diagram of an environmentally controlled industrial process is shown in Fig. 3.2. Any process including oilfield

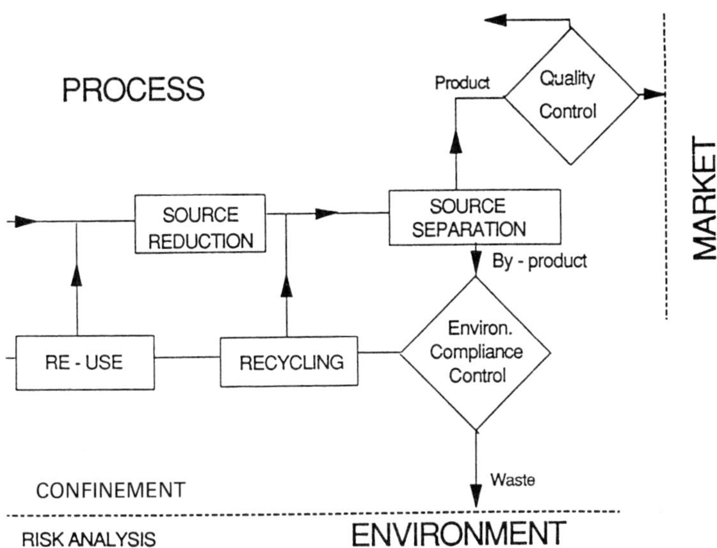

Figure 3.2 Conceptual flowpath of environmentally controlled process.

operations can be visualized as such an entity having both market and environmental boundaries. Of course, manufacturing processes are best fitted to this schematic because their boundaries are visible and clearly defined. Nevertheless, petroleum drilling and production can also be visualized using the material flowpath in Fig. 3.2. In contrast to manufacturing, oilfield processes do not have readily perceived environmental boundaries, particularly in the subsurface environment. However, they may generate subsurface pollution, which implies a flow of pollutants across a subsurface environmental boundary. The presence of such a boundary is implicit in the issues of borehole integrity and migration across confining (sealing) zones into underground sources of drinking water. Oilfield technologies related to these issues are discussed later.

Although ECT must be specifically designed for each industrial process, its methodology includes general techniques such as source reduction, source separation, recycling, confinement, beneficial use (reuse), environment risk analysis and life-cycle assessment. Figure 3.2 depicts the concepts that underlie these methods.

Source reduction involves restricting the influx of pollutants into the process or inhibiting reactions that produce toxicants within the process (examples: slim-hole drilling; subsurface water 'shut-off'; low toxicity substitution).

Source separation means the removal of pollutants from the process material stream before the stream leaves the process across the environmental boundary and becomes a waste (examples: surface or downhole separators of petroleum and water; segregated production of oil and water; reserve-pit dewatering).

Internal recycling involves closing the loop of a material stream within the process (examples: drill solids control systems; annular injection of cuttings; downhole separation and disposal of produced brines).

Internal reuse involves employing potential waste within the process (examples: mud-to-cement technology; reservoir pressure maintenance through produced water reinjection; water flooding with produced brines).

Containment means prevention of an uncontrolled transfer across the environmental boundary caused by leaking, leaching, breaching or cratering (examples: mechanical integrity tests; shallow well shut-in procedures; anti-gas migration cements; annular pressure monitoring during subsurface injection).

Environmental risk analysis (ERA) consists of analytical methods for predicting localized environmental impact (end point) for a given variant of process design (emission point). Generally, these are mathematical models (and software) of flow, transport, mixing and disperson. ERA for oilfield operations involves simulation models of flow across leaking confining zones, channeling outside unsealed boreholes and disposal fracture propagation.

Life-cycle assessment (LCA) is another analysis method for economic production strategies that considers concurrently the productivity and pollution aspects of the production process. In petroleum production the LCA approach qualifies for macro-analysis of petroleum development projects in environmentally sensitive areas, economic impact analysis of environmental regulations or, on a smaller scale, for designing environmental management of a single drilling well or production site [13].

Conceptually, process modification through additions of the environmental-control components requires a systematic approach that can be summarized in the following steps:

- define environmental boundary of the process;
- identify inherent mechanisms of environmental impact;
- consider ECT methods and create options for process modification;
- evaluate technical performance (upstream and downstream) of each ECT option;
- calculate net ECT cost;
- decide on process modification.

The difficulty in defining subsurface environmental boundaries for oilfield drilling and production has been discussed above. The surface boundary is somewhat easier to define, but the decision is still based upon subjective judgement rather than scientific definition. In drilling operations, for example, reserve pits were initially included in the drilling fluid circulation systems (hence the name 'reserve') and considered part of the drilling process. Later, the pits were often used as a waste dump that belonged to the environment. After well completion, reserve pits were either abandoned [4] or opened and spread on the surrounding land. Today, on modern rigsites, reserve pits during drilling are carefully isolated from the surrounding environment and are closed promptly after well completion using various environmental techniques described in section 3.9. In this modern approach, reserve pits are considered part of the drilling process rather than as part of the environment; they reside within the environmental boundary that surrounds the whole rigsite and underlays the bottoms of the pits.

Being an integral part of the process, each ECT component not only improves environmental compliance (downstream performance), but also affects the process productivity (upstream performance). Thus, evaluation of ECT performance should include both the upstream and downstream effects. The most typical example here is the screening of various oilfield chemicals in search of those chemicals that give a combination of the highest performances both upstream and downstream. In one such study [14], five different biocides used to prevent microbically induced corrosion, souring (generation of hydrogen sulphide) or fouling (plugging) of petroleum production installations were evaluated. The evaluation method involved assessment of upstream performance, i.e. the effectiveness of these chemicals in reducing production of H_2S or soluble sulfides (by-product of bacterial growth). Downstream performance was evaluated by modelling transport and the fate of these chemicals for five scenarios of their possible emissions from the production process to the environment.

The net cost of an ECT component is the sum of the ECT cost, value of lost (or gained) production due to ECT and savings in compliance costs due to ECT. Typically, the use of ECT would result in some productivity losses. In drilling, for example, the use of water-based, low-toxicity mud substitute for an oil-based mud would result in a slower rate of drilling. However, some ECT components show potential for improvement of both productivity and environmental compliance. One example here is the new production technique of *in situ* water drainage, described later. Potentially, this method may increase petroleum production while reducing both the amount and contamination level of produced water.

3.3 Environmental components in drilling and production processes

A fundamental notion in the ECT approach is that petroleum production, being a process of extraction of minerals from the environment, comprises inherent mechanisms of environmental impact that result from disruption of the ecological balance. The objective of this chapter is to identify these mechanisms and discuss the present level of understanding.

The disruption of the ecological balance (environmental impact) through drilling operations (excluding the wellsite preparation work) occurs in two ways: (1) surface discharge of pollutants from an active mud system and (2) subsurface rupture of confining zones (that hydrodynamically isolate other permeable strata) to provide a potential conduit for vertical transport of pollutants.

The regulatory definition of pollutant (in contrast to the popular perception based on health hazards) includes seemingly non-toxic elements such as total suspended solids (TSS), biological oxygen demand (BOD), pH and oil and grease (O&G) (the list of conventional pollutants in the USA includes TSS, BOD, pH, fecal coliform and O&G).

3.3.1 Mechanisms of drilling waste discharge

Volume and toxicity are two environmental risk criteria for evaluating drilling waste discharge. The flowpath of the drilling process and its environmental discharge mechanisms is shown in Fig. 3.3. The process material stream comprises two recycling loops, the solids-control (drilling mud) loop and the volume-control (water) loop. Conventional drilling operations employ only the solids-control loop. Theoretically, the solids-control loop could be 'closed' so that all drill cuttings may be removed in their native state, and the mud may be recycled in the system. In reality, however, some cuttings are retained in the mud system and some drilling fluid is lost across the separators so that the loop is always open, thus contributing to surface discharge. The excessive build-up of drilling mud from loop 1 passes over to the second stage process depicted as the water loop 2 in Fig. 3.3 [15]. The objective of the water loop process is to reduce the volume and recover the water phase of drilling mud. The process has been developed from the principles of industrial sludge dewatering and it employs two mechanisms of mud dewaterability: soil destabilization and cake expression. Dewatering is discussed in more detail later.

The largest volume of drilling-related wastes is spent drilling fluids or muds. The composition of modern drilling fluids or muds can be complex and vary widely, not only from one geographical area to another, but also from one depth to another in a particular well as it is drilled. Muds fall into two general categories: water-based muds, which can be made with fresh or saline water and are used for most types of drilling, and oil-based muds, which can be used when water-sensitive formations are drilled, when high temperatures are encountered, when pipe sticking occurs or when it is necessary to protect against severe drill string corrosion. Recently, there has been a rapid development of a third category of drilling fluids,

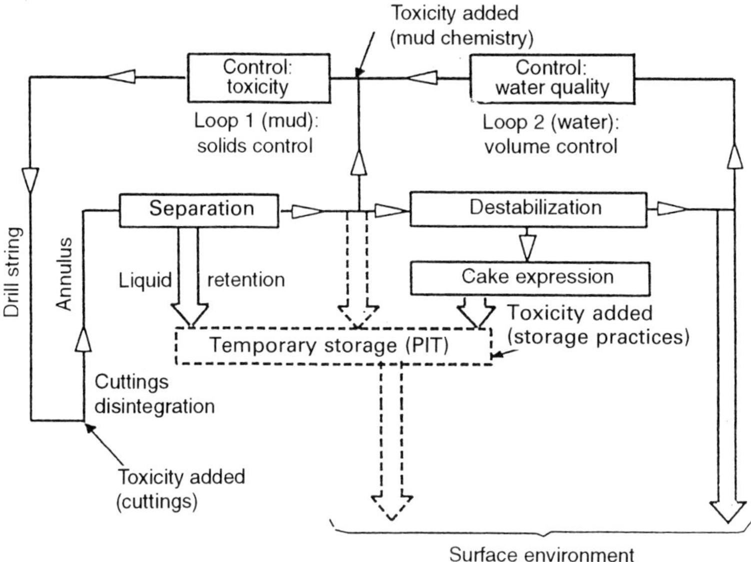

Figure 3.3 Flowpath of drilling process in relation to environmental discharge.

synthetic muds. These muds are formulated with synthetic organic compounds instead of mineral or diesel oil and are less toxic than oil-based muds.

Drilling muds contain four essential parts: (1) liquids, either water or oil or both; (2) active solids, the viscosity/filtration building part of the system, typically bentonite clays; (3) inert solids, the density-building part of the system, such as barite; and (4) additives to control the chemical, physical and biological properties of the mud.

Drill cuttings consist of inert rock fragments and other solids materials produced from geological formations encountered during the drilling process and must be managed as part of the content of the waste drilling mud. Other materials, such as sodium chloride, are soluble in fresh water and must be taken into account during disposal of drilling muds and cuttings.

The most general classification of drilling waste includes primary waste and an associated waste. The classification considers the origin and volume of generated waste. Drilling wastes with low toxicity constitute primary waste. The category of primary drilling waste comprises drilling muds and drill cuttings. Associated drilling waste may include rigwash, service company wastes such as empty drums, drum rancid, spilled chemicals, workover, swabbing, unloading, completion fluids and spent acids.

Large volumes of primary drilling waste are generated during the drilling process as a result of volumetric increase in the mud system. The volumetric

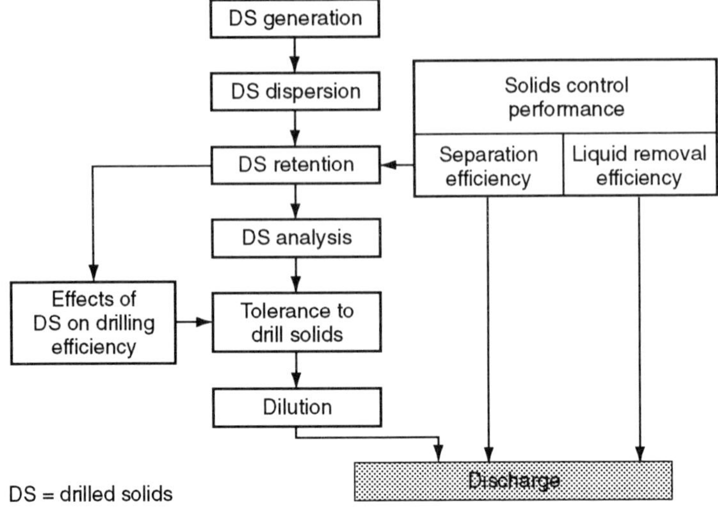

Figure 3.4 Chain of causality in generation of primary drilling waste [15].

increase of the active drilling fluid (loop 1 in Fig. 3.3) is inherent in the drilling process. The volume build-up mechanism is a chain reaction shown in Fig. 3.4 [15]. The chain reaction begins with the dispersion of reactive cuttings into the drilling fluid environment. The dispersion results in the decrease of cuttings size from their initial size to the few-microns size range. Most currently used separators do not work efficiently with small solids, i.e. they remove only a small fraction (or none) of these solids. The resulting build-up of fine solids affects the ability of the drilling fluid to perform its functions, which, in turn, hinders drilling process performance (low drilling rate, hole problems).

The minimum acceptable drilling performance relates to a certain maximum concentration of solids or solids tolerance. Solids tolerance varies for different mud systems and densities. Low-solids/polymer systems display the lowest level of solids tolerance (4%), whereas the dispersed systems display the highest (15%). Also, the increase in mud density reduces its tolerance to solids. (Specific values of solids tolerance for various muds have been compiled in various empirical nomograms.) Dilution with fresh mud (or water) is used to keep the solids concentration below the solids tolerance level. The dilution results in a steady build-up in the mud system volume and a subsequent overflow of loop 1 in Fig. 3.3. In conventional drilling operations, the overflow of loop 1 becomes a waste discharge stream. Its volume may exceed by several-fold the actual borehole volume. Table 3.1 shows the estimated discharge volumes of waste mud per barrel of the drilled hole [16]. It is evident that the volume

Table 3.1 Mud used per hole drilled[a]

Mud type	Mud/hole (v/v)
Lignosulfonate	6–12
Polymer	4–8
Potassium (KOH)/lime	3–6
Oil-base	2–4

[a]After Ref. 16.

Table 3.2 Effect of roller cone insert bit type on initial size of cuttings[a]

Bit type	Chip volume) (mm^3)	Height[b] (mm)	Diameter[c] (mm)	R/R_2[d]	T/T_2[e]
Very soft	825	5	26	2.5	2
Soft[f]	504	4	22	1	1
Soft[g]	495	3	26	2	2

[a]After Ref. 17.
[b]Minimum measured.
[c]Calculated for cylindrical chip.
[d]Relative drilling rate, related to bit No. 2.
[e]Relative bit life, related to bit No. 2.
[f]Slim, wedge-shaped inserts.
[g]Thick, short, scoop-shaped chisels.

build-up mechanism is most active for dispersed lignosulfonate systems. Characteristically, these systems are the most tolerant to solids.

Disintegration of drilled solids takes place during annular transport from the drilling bit to the flowline. As a result, cuttings become smaller. This size reduction of cuttings is the first factor contributing to cuttings retention in the mud system. The size of cuttings depends upon (1) the initial size resulting from the bit action, (2) bottomhole cleaning efficiency and (3) the mechanical strength of cuttings in the mud environment. Besides a qualitative understanding of the effects of bit type and pressure differential across the rock face, very little is known about the initial size of cuttings. An example of the actual initial size of cuttings generated by various types of cone bits is shown in Table 3.2 [17]. Data support the common knowledge that the harder is the bit type, the smaller are the cuttings. However, there is no predictive model based on drilling mechanics that would relate initial cuttings size to bit geometry and rock strength. A preliminary study in this area determined the relationship between the specific energy of rock destruction, total mechanical energy of a bit and cuttings size [18].

The effect of bottomhole cleaning on the initial size of cuttings can be inferred from the experimentally verified response of the drilling rate to the bottomhole hydraulic energy generated by bit nozzles. It is generally assumed that in soft rock drilling, the bit flounder point represents an offset of poor cuttings removal from under the bit [19]. The remaining cuttings undergo additional grinding, which results in size reduction. The flounder point can be determined experimentally using the drill-off test. Further cuttings destruction can be prevented by adjustment of the mechanical energy to the hydraulic energy at the bottom of the hole.

Size reduction of cuttings is caused by loss of cohesion due to hydration of their rock matrix. Cuttings originating from non-swelling rocks (sand, limestone) are unlikely to lose their initial cohesion on their way up the borehole annulus. It has been proved, however, that even these inert solids undergo disintegration under conditions of shear, as shown in Table 3.3 [20].

The major mechanism controlling cuttings disintegration stems from the hydration energy of their source rock, usually shale. The disintegration has been correlated with several variables measured in various tests of cuttings hydration rate, such as (1) the swelling test (measured: linear expansion); (2) capillary suction time test, CST (measured: time of water sorption); (3) cation-exchange capacity test, CEC (measured: dye adsorption); (4) activity test (measured: electrical resistance of water vapor); and (5) rolling test (measured: weight loss of drill cuttings of a certain size) [21–24]. The drawback of these tests is that they do not provide a direct measurement of drill cuttings properties (strength, size). However, they do determine other variables that correlate with these properties.

The proposed single property of shale cuttings representing their strength is the storage modulus of viscoelasticity [25]. The storage modulus is a measure of the energy stored and recovered under conditions of oscillating stresses. It can be measured using an oscillatory viscometer and a compacted 'drill cutting' platelet after various exposure times of a cutting to drilling mud. Figure 3.5 shows the strength of a shale cutting after 18 h of exposure to various concentrations of salts (KCl) and polymer in the drilling fluid.

The initial strength of cuttings and their tendency to become hydrated

Table 3.3 Shear disintegration of inert solids in mud[a]

Shear treatment	Particles smaller than 2 μm (volume fraction)						
	Barite A (green)	Barite B (orange)	Barite C (orange)	Barite D (buff)	Barite E (orange)	Itabarite	Ilmenite
None	6.6	8.0	5.3	8.8	12.6	4.3	0.3
Ultrasound (1 min)	13.3	13.2	12.1	16.9	12.8	15.7	0.6

[a]After Ref. 20.

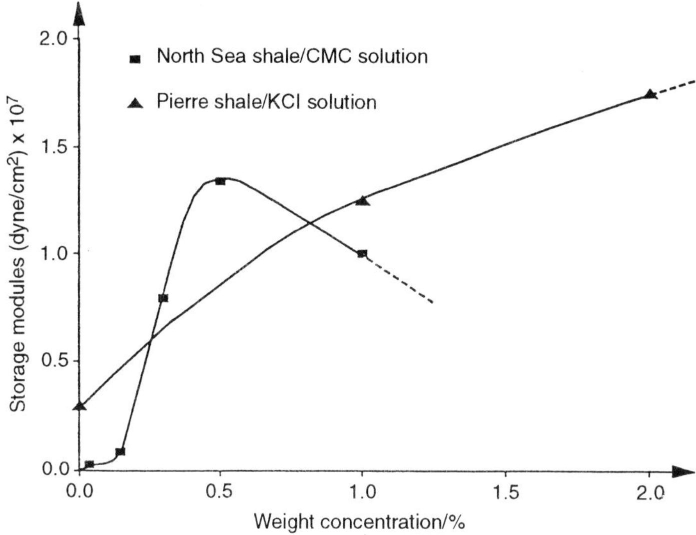

Figure 3.5 Strength of shale cutting in various mud environments (1 dyne = 10^{-5} N) [25].

can be inferred from the mineralogy of shales with respect to depth. The disintegration rate of shale cuttings results from the mineralogical composition of the shale and can be directly related to geological structures in the drilling area. For example, Fig. 3.6 shows the drilled-depth correlations of the illite concentration (low-reactivity clay) and shale water content for the offshore Louisiana Gulf Coast [26].

The depth-related reactivity of shales can also be observed in the size of cuttings coming from the well. An analysis of the size distribution of solids at the flowline versus drilling depth shows different rates of cuttings disintegration during their annular transport, as evidenced by Fig. 3.7 [27]. Also shown in Fig. 3.7 is a correlation between size of mud solids at the flowline and at the pump suction (i.e. upstream and downstream of solids control system). Such correlations are more useful than measurements of the rock hydration rate because they not only identify well sections with water-sensitive rocks but also provide data that can be used to evaluate solids control systems.

The separation efficiency of a solids-control system is limited by the size of the solids in the drilling mud entering the separators. This limitation is the next factor contributing to solids retention in the mud system. The plots in Fig. 3.7 show a comparison of solids size in drilling mud samples taken from the flowline and the suction tank. In the three sections of the well (2300–2800, 5000–5600 and 6150–7215 ft; 1 ft = 0.3048 m), the efficiency of cuttings removal was evidently almost zero. The most likely

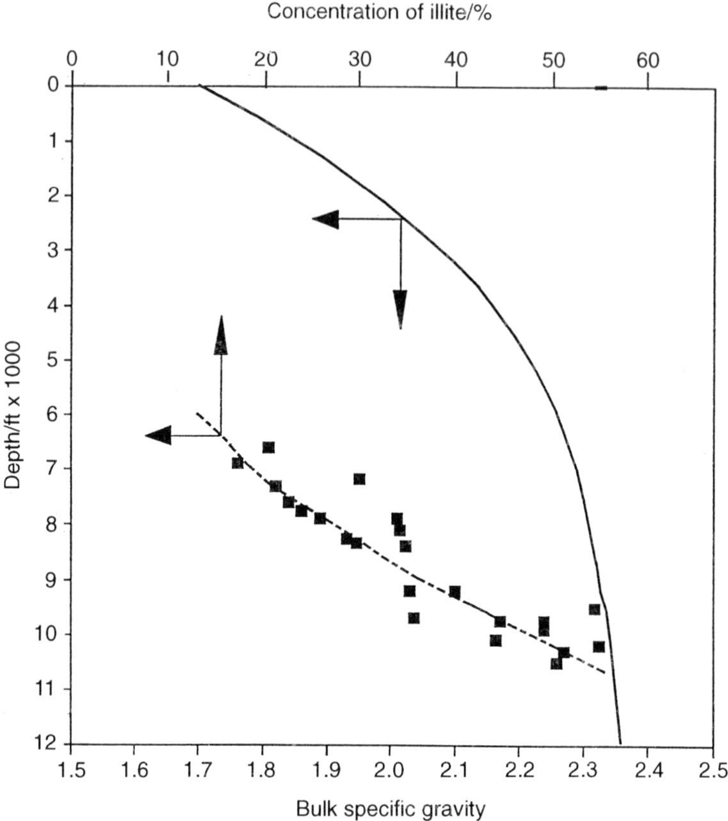

Figure 3.6 Shale reactivity indicators versus depth for Louisiana Gulf Coast [21].

reason is that the size of the solids was below the removal range of the surface separators. Thus the drilling fluid loop in these sections was 'wide open' because the only way to control mud solids was to dilute the mud system and generate an excessive volume.

There is an important misconception about the performance of solids control separators. The widely recognized concept of the subsequent size exclusion of solids holds that the shale shaker removes cuttings > 120 μm, desander 50 μm, desilter 15 μm and a centrifuge 3 μm. However, the actual performance is not only lower than the theoretical one, but it is also affected by the feed mud rheology and operational parameters of a separator. As an example, Fig. 3.8 shows the theoretical and actual grade separation curves for a 4 in (10 cm) hydrocyclone [20, 27, 28]. Both the

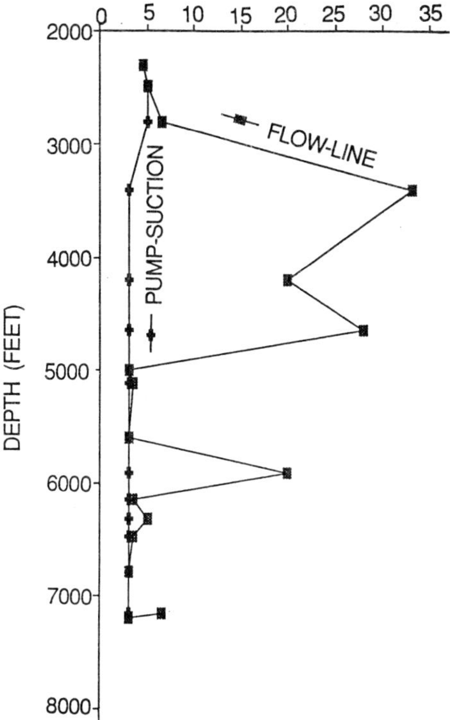

Figure 3.7 Depth related size of cuttings upstream (flowline) and downstream (pump suction) from solids-control separators [20].

laboratory and the field data indicated poor performance of hydrocyclones with weighted mud systems; this raised some questions regarding the applicability of mud cleaners. Reportedly, the 50% cut made by the 100-mesh screen was smaller than the cut for the 4 in hydrocyclone [28]. Note however, that when comparing separators, the grade efficiency should be considered together with the load capacity. The liquid conductance of vibrating screens has been proved to decrease rapidly with increasing mesh size and mud viscosity [29]. In contrast, the operator can increase the volume processed by the hydrocyclones simply by adding more cones.

The separation efficiency of centrifuges is highly dependent upon the type of separated solids. The theoretical values of 50% cut, 3–4 μm, claimed by manufacturers are relevant only for the barite-recovery

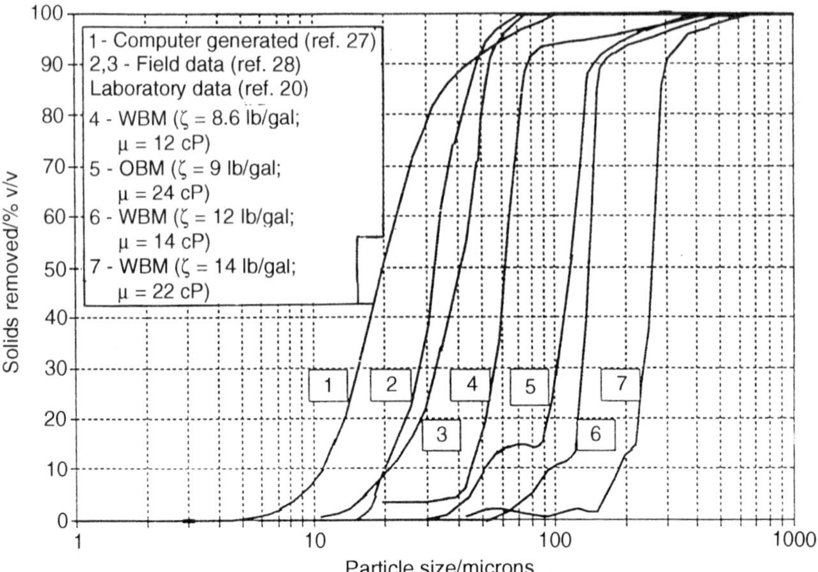

Figure 3.8 Theoretical and actual performances of 4 in hydrocyclones: effects of mud and type and rheology (1 lb = 0.454 kg; 1 gal = 3.785 dm^3; 1 cP = 10^{-3} N s/m^2)

application of centrifuges. Much poorer separation is obtained for low-gravity (reactive) solids, as shown in Fig. 3.9 [30]. The inability of the decanting centrifuge to control fine solids in the mud system during the double-stage centrifuging was observed in both field [28] and full-scale laboratory tests [30].

3.3.2 Sources of drilling waste toxicity

There are three contributing factors of toxicity in drilling waste: the chemistry of the mud formulation, inefficient separation of toxic and non-toxic components and the drilled rock. Typically, the first mechanism is known best because it includes products deliberately added to the system to build and maintain the rheology and stability of drilling fluids. The technology of mud mixing and treatment is recognized as a source of pollutants such as barium (from barite), mercury and cadmium (from barite impurities), lead (from pipe dope), chromium (from viscosity reducers and corrosion inhibitors), diesel [from lubricants, spotting fluids, and oil-based mud (OBM) cuttings] and arsenic and formaldehyde (from biocides).

Inefficient separation of toxic components from the drilling waste

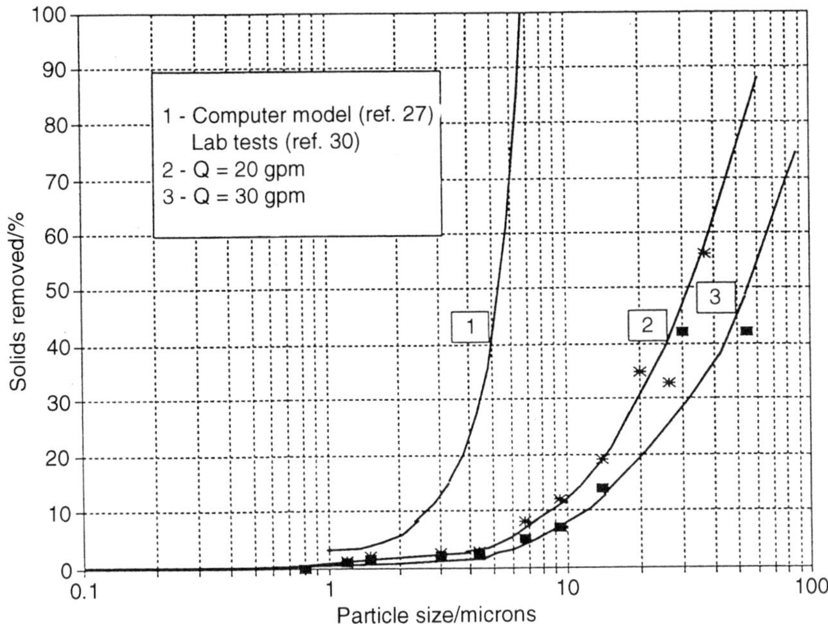

Q = centrifuge throughout, gallons per minute

Figure 3.9 Theoretical (inert solids) and actual (active solids) performance of decanting centrifuge.

discharge stream becomes another source of toxicity through retention of the liquid phase on OBM cuttings, use of spotting pills or indiscriminate practices of on-site storage. Removal of the liquid phase from cuttings separated by the solids control equipment becomes particularly important while using diesel-based drilling fluids (DOBM). Field data show that the total oil-based mud discharge rate jointly for the mud cleaner and centrifuge is 10 bbl/h [28]. Also, the OBM removal performance is different for various separators as shown in Table 3.4 (the highest for mud cleaners, and lowest for centrifuges) [28, 31, 32].

Research revealed that the OBM retention on cuttings is smaller for the mineral oil-based than for diesel-based OBMs, as evidenced by field data in Table 3.5 [33, 36]. The hypothetical mechanisms of oil retention on solids have been attributed to adhesive forces, capillary forces and oil adsorption and were identified as the amount of oil removed from OBM cuttings using centrifugal filtration, *n*-pentane extraction and thermal vaporization, respectively. The conclusion has been forwarded that 50% of

Table 3.4 Liquid discharge and oil retention on cuttings from oil-based muds (OBM) for various separators

Reported data	Oil content (% w/w)/OBM discharge rate (gal/min)[a]		
	Shale shaker	Mud cleaner	Centrifuge
Ref. 32	12.3/NR	14.1/NR	8.4/NR
Ref. 28	NR/NR	NR/4.2	NR/0.7
Ref. 31	11.1–16.5/NR	NR/NR	3–10.2/NR

[a]NR = not reported.

Table 3.5 Oil retention on OBM cuttings[a] vs type of oil[b]

Drilling fluid	Well			
	1	2	3	4
Diesel OBM	20.0	13–16	9.8	10.8
Mineral OBM	7.9	10.3	NR	NR

[a]Per cent by dry weight of discharge from shale shaker.
[b]Compiled from Refs 33–36.

Table 3.6 Toxicity difference between active and waste drilling fluids[a]

Toxicant	Active mud	Detection rate (%)	Reserve pit	Detection rate (%)
Benzene	No	–	Yes	39
Lead	No	–	Yes	100
Barium	Yes	100	Yes	100
Arsenic	No	–	Yes	52
Fluoride	No	–	Yes	100

[a]Based on Ref. 9.

the oil–solids bond could be attributed to adhesive/capillary forces, 29% to weak adsorption and 20% to strong adsorption, i.e. 20% of oil on cuttings could not have been removed with *n*-pentane extraction. The adhesive mechanism was also explained using the wettability preference of drilled rock. The preference was evaluated by measuring the adhesion tension of thin-cut plates of quartz and shales immersed in OBM. The results showed that the rocks immersed in diesel OBM became strongly oil-wet, whereas for the mineral OBM, the initially oil-wet surfaces tended to reverse their wettability and became water-wet.

Indiscriminate storage/disposal practices using drilling mud reserve pits can contribute toxicity to the spent drilling fluid, as shown in Table 3.6. The data in Table 3.6 are from the US EPA survey of the most important toxicants in spent drilling fluids. In the survey, sample taken from active drilling mud in the circulating system were compared with samples of spent drilling mud in the reserve pit [9]. The data show that the storage/disposal practices were a source of the benzene, lead, arsenic and fluoride toxicities in the reserve pits because these components had not been detected in the active mud systems.

The third source of toxicity in the drilling process discharges is the type of drilled rocks. A recent study of 36 core samples collected from three areas (Gulf of Mexico, California and Oklahoma) at drilling depths ranging from 3000 to 18 000 ft revealed that the total concentration of cadmium in drilled rocks was more than five times greater than the cadmium concentration in commercial barites [37]. With a theoretical well discharge volume in a 10 000 ft well model, 74.9% of all cadmium in drilling waste was estimated to be contributed by cuttings, whereas only 25.1% originate from the barite and the pipe dope.

3.3.3 Waste generation mechanisms in petroleum production

Petroleum production involves the extraction of hazardous substances, crude oil and natural gas, from the subsurface environment. Therefore, by its very nature, production technology involves pumping and processing pollutants. Any material used in conjunction with the production process and exposed to petroleum becomes contaminated. In essence, there are two mechanisms of pollution in the production process: generation of contaminated waste and leakage of material streams from the process to the environment. All non-petroleum materials entering the production process are either naturally occurring subsurface substances, such as formation waters and produced sand, or deliberately added chemicals facilitating production operations. Inside the process, these materials are mixed into the stream of petroleum, then separated into three final streams at the process output: marketable oil or gas products, produced water and associated waste. This simplified analysis is depicted in Fig. 3.10 and discussed below.

The mechanisms of waste generation are related to production operations. *Downhole production operations* include primary, secondary and tertiary recovery methods, well workovers and well stimulations. Primary recovery refers to the initial production of oil or gas from a reservoir using only natural pressure to bring the product out of the formation and to the surface. Most reservoirs are capable of producing oil and gas by primary recovery methods alone, but this ability declines over the life of the well.

Eventually, virtually all wells must employ some form of secondary recovery. This phase of recovery is at least partially dependent on artificial lift methods, such as surface and subsurface pumps and gas lift, but typically also involves injection of gas or liquid into the reservoir to

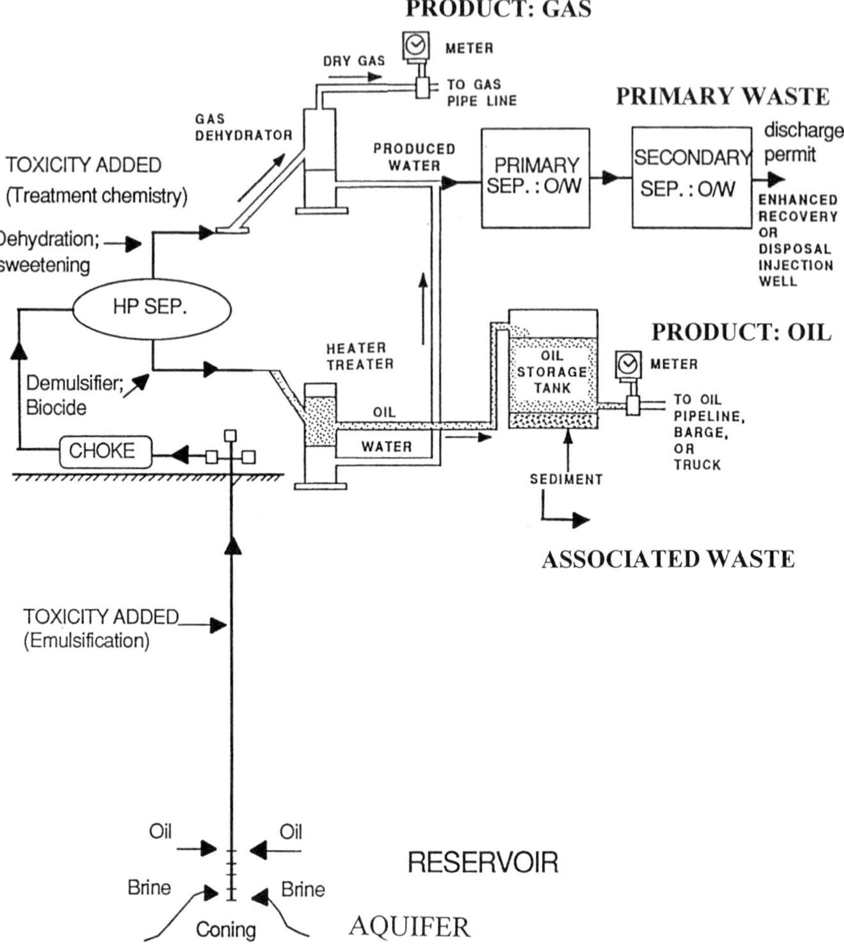

Figure 3.10 Waste generation mechanisms in petroleum production process.

maintain pressure within the producing formation. Water flooding is the most frequently employed secondary recovery method. It involves injecting treated fresh water, sea water or produced water into the formation through a separate well or wells.

Tertiary recovery refers to the recovery of the last portion of the oil that can be economically produced. Chemical, physical and thermal methods are available and may be used in combination. Chemical methods involve injection of fluids containing substances such as surfactants and polymers. Miscible oil recovery involves injection of gases, such as carbon dioxide and natural gas, which combine with the oil.

When oil eventually reaches a production well, injected fluids from secondary and tertiary recovery operations may be dissolved in formation oil or water or simply mixed with them. The removal of these fluids is discussed below in conjunction with surface production operations.

Workovers and stimulations are another aspect of downhole production operations. Workovers are designed to restore or increase production from wells whose flows are inhibited by downhole mechanical failures or blockages, such as those caused by sand or paraffin deposits. Fluids circulated into the well for this purpose must be compatible with the formation and not adversely affect permeability. Stimulations are designed to enhance the wells productivity through fracturing or acidizing. Fluids injected during these operations may be very toxic (hydrochloric acid, for example) and may be produced partially back to the surface after petroleum production is resumed. Other chemicals may be periodically or continuously pumped down a production well to inhibit corrosion, reduce friction or simply keep the well flowing. For example, methanol may be pumped down a gas well to keep it from becoming plugged with ice.

Surface production operations generally include gathering the produced fluids (oil, gas, gas liquids and water) from a well or group of wells and separating and treating the fluids.

During production operations, pressure differentials tend to cause water from adjoining formations to flow into the producing formation (water breakthrough or water coning). The result is that, in time, production water/oil ratios may increase steeply. New wells may produce little, if any, water; mature wells may produce more than 100 barrels of water for every barrel of oil. Virtually all of this water must be removed before the product can be transferred to a pipeline (the maximum water content permitted is generally less than 1%). The oil may also contain completion or workover fluids, stimulation fluids or other chemicals (biocides, fungicides) used as an adjunct to production. These, too, must be removed. Some oil–water mixtures may be easy to separate, but others may exist as fine emulsions that do not separate by gravity settling. Conventionally, gravity settling has been performed in a series of large or small tanks (free water knock-outs, gun barrels, skim tanks), the large tanks affording longer residence time to increase separation efficiency (API separators). When emulsions are difficult to break, heat is usually applied in so-called 'heater treaters'. Whichever method is used, crude oil flows from the final separator to stock tanks. The solids and liquids that settle out of the oil at the tank bottoms ('produced' sand) must be collected and discarded along with the separated water.

Natural gas requires different techniques to separate out crude oil, gas liquids, entrained solids and other impurities. These separation processes can occur in the field, in a gas processing plant, or both. Crude oil, gas liquids, some free water and entrained solids can be removed in simple separation vessels. Low-temperature separators remove additional gas liquids. More water may be removed by any of several dehydration processes, frequently through the use of glycol, a liquid desiccant or various solid desiccants. Although these separation media can generally be regenerated and used again, they eventually lose their effectiveness and must be discarded.

Both crude oil and natural gas can contain the highly toxic gas hydrogen sulfide (200 ppm in air is lethal to humans). At plants where hydrogen sulfide is removed from natural gas, sulfur dioxide (SO_2) release may result. Sulfur is often recovered from the SO_2 as a commercial by-product. Hydrogen sulfide (H_2S) dissolved in crude oil does not pose any danger, but, when it is produced at the wellhead in gaseous form, it poses serious occupational risks through possible leaks or blowouts. These risks are also present later in the production process when the H_2S is separated out in various 'sweetening' processes. The amine, iron sponge and selexol processes are three examples of commercial processes for removing acid gases from natural gas. Each H_2S removal process results in spent, or waste separation, media which must be disposed of.

Production waste is broadly classified as either primary or associated waste. Most of the materials used and discarded from production operations fall into the associated waste category. A listing of associated waste is shown in Table 3.7. This waste is characterized as having low volume and high toxicity. Produced water is a primary production waste having a very large volume and relatively low toxicity compared with associated waste. In 1989, the daily average discharge of produced water from all North Sea production operations was 355 000 m^3/day, with oil and gas production rates of 535 000 m^3/day and 267×10^6 m^3/day, respectively [38]. During 1990, Gulf of Mexico oilfield operations produced 866.5 million barrels of water [39], while the total US production of water from oil and gas operations was 14 billion barrels [40]. Because of these large volumes, produced water is the major production waste stream with potential for environmental impact.

Table 3.7 Associated production waste

Oily wastes: tank bottoms, separator sludges, pig trap solids
Used lubrication or hydraulic oils
Oily debris, filter media and contaminated soils
Untreatable emulsions
Produced sand
Spent iron sponge
Dehydration and sweetening wastes (including glycol amine wastes)
Workover, swabbing, unloading, completion fluids and spent acids
Used solvents and cleaners, including caustics
Filter backwash and water softener regeneration brines

The system analysis of the production process in Fig. 3.10 clearly shows that formation water enters the process downhole through the petroleum producing perforations, where it begins to mix with hydrocarbons. The water may flow into the hydrocarbon formation through processes of coning or fingering. The process kinetics of mixing oil and water under conditions of variable temperature and pressure during the two-phase flow in the well have not yet been investigated. In this process, formation water becomes contaminated by dispersed oil and soluble organics. The time required to reach an equilibrium concentration of fatty acids and other polar, water-soluble components of crude oil in produced brine is expected to be significantly shorter than the time of the two-phase flow [41]. Thus, a maximum level of contamination is reached before the brine is separated from oil. In addition to hydrocarbons, all treating chemicals used in surface operations are mixed into the water, thus adding to the final toxicity of produced water discharge. Characteristically, most of the recent research regarding composition and toxicity of produced water has focused solely on the end-point product of the above mixing mechanism while disregarding subsequent stages of water contamination on its way from the aquifer to the environmental discharge point.

3.3.4 Sources of toxicity in produced water

As discussed above and depicted in Fig. 3.10, toxicity of produced water results from two factors: properties of formation water in its natural state and toxicity contributed by the very process of production. Sources of produced water toxicity that has been added to the water during the production process include hydrocarbons and treating chemicals. Water toxicity has been shown to increase along its flowpath across the production process [9]. Table 3.8 compares toxic components in a typical oilfield production waste stream at the mid-point and at the end-point of the production process. As can be seen, the hazard of benzene and pH toxicity increases along the process flowpath. Also, three additional

Table 3.8 Toxicity increase of produced water across production process[a]

Pollutant	Mid-point	Detection rate (%)	End-point	Detection rate (%)
pH	6.4, 6.6, 8.0	–	2.7, 7.6, 8.1	–
Benzene	Yes[b]	60	Yes[b]	76
Phenanthrene	No	–	Yes[b]	24
Barium	No	–	Yes	87
Arsenic	No	–	Yes	37

[a]Based on Ref. 9.
[b]Detected concentration was 1000 times greater than that hazardous to humans.

toxicants, phenanthrene, barium and arsenic, are detectable at the end-point but are absent in the mid-point samples.

Prior to production, formation waters may display some level of toxicity which is usually unknown. Unlike toxicity of produced water, the *in situ* toxicity of oilfield brines has not been investigated. The most likely sources of toxicity in formation water prior to production are salt and radionuclides.

The lack of hydrocarbon contamination of the formation water column underlying the oil column was recently evidenced in a pilot study in which water was produced separately from, and concurrently with, oil using a dually completed well [42, 43]. No polyaromatic hydrocarbons (PAHs) or oil and grease were detected in that water. Therefore, conventional concurrent production of petroleum and water was concluded to be the sole source of hydrocarbon contamination of produced water, at least in water-drive reservoirs where the oil column is separated from the water column. The contamination may take two forms: dispersed oil and soluble oil (mostly non-hydrocarbon organic material).

Dispersed oil consists of small droplets of oil suspended in the water. As a droplet moves through chokes, valves, pumps or other constrictions in the flowpath, the droplet can be torn into smaller droplets by the pressure differential across the devices. This is especially true of flow viscosity oils and condensates. Precipitation of oil from solution results in a water fraction with smaller droplets. These small droplets can be stabilized in the water by low interfacial tension between the oil and the produced water. Small droplets can also be formed by the improper use of production chemicals. Thus, the addition of excess production chemicals (such as surfactants) can further reduce the interfacial tension so that coalescence and separation of small droplets becomes extremely difficult.

Oilfield deoiling technology, discussed later in this chapter, is designed to remove dispersed oil. Failure to remove small oil droplets results in the presence of dispersed oil in produced water discharges. (The total maximum concentration of oil and grease, O&G, in these discharges varies in different areas. In the USA, for example, the daily maximum O&G concentration is 42 mg/l, while under the Paris Convention the maximum dispersed oil concentration is 40 mg/l.)

Soluble oil includes organic materials such as aliphatic hydrocarbons, phenols, carboxylic acids and low molecular weight aromatic compounds. The concentration of dissolved oil in produced water depends upon the type of oil. However, it is also related to technological factors, such as the type of artificial lift techniques (mixing energy of petroleum in water) and stage of production (encroachment of formation water into petroleum-saturated zone).

The concentration of dissolved organics may in some cases reach the maximum regulatory limit for offshore discharge (O&G 29 mg/l monthly

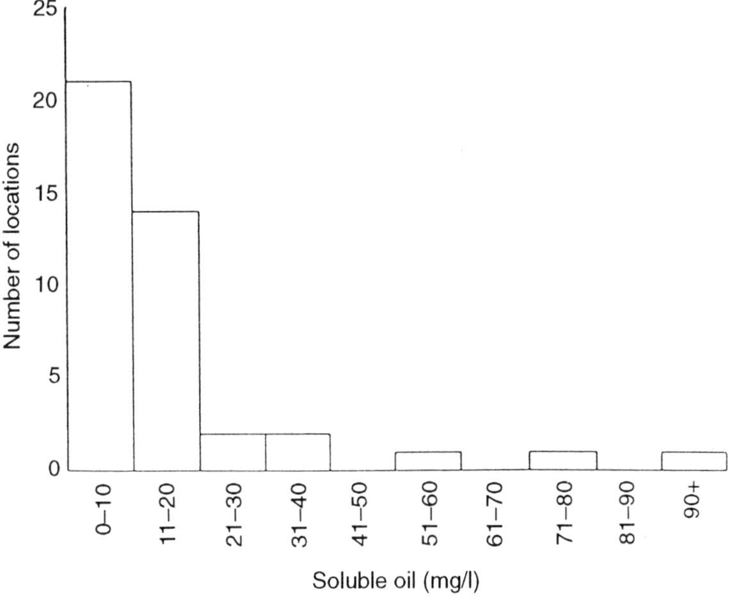

Figure 3.11 Concentration of soluble oil in produced water [44].

Table 3.9 Phenols and volatile aromatics in produced water[a]

Production	Concentration (μg/l)	Toxicant			
		Phenols	Benzene	Toluene	C_2-Benz.
Gas	Average	4 743	5 771	5 190	700
	Std dev.	5 986	4 694	4 850	1 133
	Maximum	21 522	12 150	19 800	3 700
	Minimum	150	683	1 010	51
Oil	Average	1 049	1 318	1 065	221
	Std dev.	889	1 468	896	754
	Maximum	3 660	8 722	4 902	6 010
	Minimum	0	2	60	6

[a]From Ref. 45.

average), as shown in Fig. 3.11 [44]. Most of the contribution to these concentrations comes from phenols and volatile aromatics, as shown in Table 3.9 [45].

At least one study has shown that the toxicity of soluble oil is not significant. The soluble oil fractions of two different produced waters were tested for toxicity and found to have acute toxicities of 15.8 and 4.8% [45,

46]. One of the reported characteristics of these components is that they are easily biodegraded. Therefore, low levels of dissolved organic materials are easily assimilated by the receiving ambient water. In addition to locally increasing BOD, the components of soluble oil each have a different fate in the environment [46].

Heavy metals in produced waters may be either present in formation water or added through the production process. Metals that may contribute to toxicity include barium, cadmium, chromium, copper, lead, mercury, nickel, silver and zinc. Typically, their concentrations in produced water may be in the range of thousands of μg/l while their concentration in sea water varies from trace to tens of μg/l. Heavy metals have been reported to pose little harm in the marine environment [46, 47]. They may settle out in marine sediments, thus increasing the sediment metal concentrations. However, they are tightly adsorbed to other solids and have much lower bioavailability to marine animals than do the metal ions in solution.

Radionuclides found in produced waters are often referred to as naturally occurring radioactive material (NORM). The source of the radioactivity in scale deposits from produced water comes from the radioactive ions, primarily radium, that coprecipitate from produced water along with other types of scale. The most common scale for this coprecipitation is barium sulfate, although radium has also been found in calcium sulfate and calcium carbonate scales.

Studies of soluble radionuclides in produced water have been summarized recently [45]. Early studies of wells in Oklahoma, the Texas panhandle and the Gulf of Mexico coastal area showed ^{226}Ra levels ranging from 0.1 to 1620 pCi/l (1 Ci = 3.7×10^{10} Bq) and ^{228}Ra levels ranging from 8.3 to 1507 pCi/l. Recent studies conducted by the State of Louisiana, Offshore Operators Committee and the US Environmental Protection Agency showed ^{226}Ra level ranges of 0–930, 4–584 and 4–218 pCi/l, respectively, and ^{228}Ra level ranges of 0–928, 18–586 and 0–68 pCi/l, respectively. These levels are considerably lower than those from early findings. Also, reported research provides no evidence of the impact of radionuclides on fish or human cancers exceeding that resulting from a background concentration of radium.

Treating chemicals used in production operations can be classified according to types of production operations and the purpose of the treatment, as production liquid treating chemicals, gas processing chemicals and stimulation or workover chemicals. The production liquid treating chemicals are those routinely added to the produced oil and water (including waters used for water flooding). Chemically, these compounds are complex mixtures manufactured from impure raw materials. However, when looked upon as a source of toxicity in produced water these chemicals can be broadly analyzed according to their function, initial

Table 3.10 Classification of toxicity grades[a]

Classification	LC_{50} value (ppm)
Practically non-toxic	>10 000
Slightly toxic	1000–10 000
Moderately toxic	100–1000
Toxic	1–10
Very toxic	<1

[a]From Ref. 47.

toxicity, solubility in water and treatment concentration. Obviously, all the above factors will control individual contribution of these chemicals to the final toxicity of produced water discharge. For the purpose of reference. Table 3.10 shows the general grading of toxicity using lethal concentration values representing the 50% mortality rate (LC_{50}) [47]. The following analysis summarizes findings regarding production chemical use and toxicity [48].

Biocides control bacterial growth, particularly sulfate-reducing bacteria that cause corrosion or fouling. Aldehydes, quaternary ammonium salts and amine acetate salts are the most commonly used biocides. All the biocides are highly water soluble. Intermittent slug treatments at 50–200 ppm of formulation are used to obtain good control with a minimum total biocide usage. The LC_{50} values for biocides may vary from less than 1 to above 1000 ppm.

Scale inhibitors control deposition of common oilfield scales of calcium carbonate, calcium sulfate, strontium sulfate and barium sulfate. Three generic chemical types – phosphonates, phosphate esters and acrylic-type polymers – comprise 95% or more of the chemical being used. All formulations are highly water soluble. A minimum concentration, typically 3–10 ppm, must be present at all times to prevent scale deposition. After squeeze treatments (relatively uncommon) the concentration of compound in the produced water may be as high as 5000 ppm for a few days. The LC_{50} values for scale inhibitors fall within the range 1000–11 000 ppm.

Corrosion inhibitors include compounds of the amide/imidazoline, amine or amine salt, quaternary amine and heterocyclic amine types. Oil-soluble inhibitors generally are preferred for oil production because of their great effectiveness. Continuous treatment with 10–20 ppm may be used in oil wells or pipelines. The initial LC_{50} values for corrosion inhibitors may be below 1 ppm. Most typical values, however, are from 1.2 to less than 10 ppm.

Emulsion breakers improve the separation of oil from water. The most common compounds are oxyalkylated alkylphenol–formaldehyde resins, polyglycol esters and alkylaryl sulfonates. Almost all formulations contain more than one of these generic types, as well as a surfactant. Virtually all components of these formulations are very insoluble in water and distribute into the oil phase. Typical use concentrations are about 25–100 ppm based on oil, with perhaps only 0.4–4 ppm distributing into the produced water. Initial LC_{50} values for emulsion breakers range from 3.8 to 80 ppm.

Reverse breakers are used to help remove droplets of oil from the produced water before discharge into the ocean. The two most common generic types are low molecular weight (2000–5000) polyamines and polyamine quaternary ammonium compounds. Both types are highly water soluble. Some formulations also include moderately high concentrations of aluminium, iron or zinc chlorides. Dosages of 5–25 ppm may be required, with perhaps half distributing into the discharged water. Minimum initial values of LC_{50} for reverse breakers can be below 1 ppm. Coagulants and flocculants are used to enhance the oil-water separation process. They are polymers similar to reverse breakers, but have a wider range of molecular weights, from 0.5 to 20 million. They are water soluble and used in concentrations from 5 to 10 ppm. Their LC_{50} values in the salt water environment are from 2 to 14 800 ppm. They are, however, more toxic to freshwater organisms.

Surfactants are used for cleaning equipment, tanks and decks. The two most common types are the alkylaryl sulfonates and the ethoxylated alkylphenols, both of which are widely used in other industrial and household applications. Oil-soluble versions are available for maintenance of tank and vessel internals. The LC_{50} values for surfactants may be as low as 0.5 ppm.

Paraffin inhibitors prevent solid hydrocarbons from forming or sticking to the walls of the system, thereby controlling accumulations of solid hydrocarbons in the system. Vinyl polymers, sulfonate salts and mixtures of alkyl polyethers and aryl polyethers are the most common compounds. Paraffin solvents are used to remove accumulations of deposits. The solvents are usually refinery cuts and may be primarily aliphatic or aromatic, depending on the nature of the deposits. Inhibitors are usually added in the 50–300 ppm range, while the solvents may range from a few percent in a stream to near 100% in cleaning out a vessel. All these materials are far more soluble in the oil than in the produced water. The LC_{50} values range from 1.5 to 42 ppm.

Gas treating chemicals include hydrate inhibitors and dehydration agents. A typical hydrate inhibitor is methanol, which has LC_{50} values from 8000 to 28 000 ppm. Also, glycol dehydration is a closed-loop process that may produce leaks. However, glycol toxicity is low, with LC_{50} values from 5000 to 50 000 ppm.

Stimulation and workover chemicals include hydrochloric acid (HCl) and workover brines. If properly used, these fluids should not contaminate produced water. Acids should be caught separately and neutralized, while

Table 3.11 Toxicity of treatment chemicals and their potential concentration in produced water[a]

Function type	Use concentration (ppm)	Discharge concentration (ppm)	LC_{50} concentration (ppm)
Scale inhibitor	3–10 normal	3–10	1200–>12 000,
	5000 squeeze[b]	50–500	90% >3000
Biocides	10–50 normal	10–50	0.2–>1000,
	100–200 slug	100–200	90% >5
Reverse breakers	1–25 normal	0.5–12	0.2–15 000, 90% >5
Surfactant cleaners	Not measured	Not measured	0.5–429, 90% >5
Corrosion inhibitor	10–20 water[b]	5–15	0.2–5, 90% >1
	10–20 oil[b]	2–5	2–1000,
	5000 squeeze[b]	25–100	90% >5
Emulsion breakers	50 oil	0.4–4	4–40, 90% >5
Paraffin inhibitor	50–300	0.5–3	1.5–44, 90% >3

[a]After Ref. 48.
[b]Water indicates solution of a water-soluble inhibitor; oil means that the inhibitor is mostly oil soluble; squeeze is the maximum concentration of inhibitor in returns from the well after squeeze or batch treatment.

toxic brines (e.g. zinc bromide) should be collected and reconditioned for reuse.

The potential effect of treating chemicals on produced water toxicity is summarized in Table 3.11 [48]. The 'discharge concentration' is an estimated concentration range in the discharge pipe. The top four chemicals are all water soluble and expected to be primarily in the water phase. The biocides are the only type in which the discharge concentration is likely to be above the LC_{50} values, and then only for periodic, short durations. The corrosion inhibitors are the most complex type, as compounds and formulations are made to be water soluble, oil soluble or mixed soluble/dispersible. The water-soluble compounds are most likely to resemble biocides chemically but are most commonly added to injection water or gas pipelines and are not discharged to the ocean continuously. The oil-soluble corrosion inhibitors are at or below the LC_{50} value, except possibly for short periods after squeeze or batch treatments.

The *salinity* of produced water can vary from very low to saturation, depending on geology and the production process. It is believed that the impact of discharging fresh or brackish produced water into the ocean would be the same as for rain [45]. This view is supported by observations from platforms that discharge produced water with very high salt contents show that there is a lively aquatic life community present. Also, dilution of a 200 000 mg/l salt water solution, such as produced water, in a 35 000 mg/l ocean occurs very quickly. Therefore, the concentration of salt in produced water discharged offshore has little potential to cause a harmful impact on aquatic life.

3.4 Control of drilling fluids volume and toxicity

This section presents environmental control technologies for drilling fluids. The technologies have been developed to control some of the pollution mechanisms that have been discussed above.

A steady build-up of the mud system volume, as shown in the previous section, is inherent in the drilling process and results from both the disintegration of cuttings during their transport to the surface and the limited efficiency of cuttings removal by the solids control separators. For water-based muds, this mechanism can be controlled by adding a second (dewatering) loop to the mud processing system so that the mud–water phase can be recycled and the volume of drilling waste minimized. Ultimately, disposal of this waste depends upon the toxicity of mud systems used to drill the well. Therefore, the properties of mud systems that are directly related to pollution are dispersibility, dewaterability and toxicity. In a 'clean' drilling process these properties must be controlled. Also such a process requires improvements in mud solids removal efficiency.

3.4.1 Control of mud dispersibility

In mud engineering, several conventional methods can be used to inhibit swelling of shales. These methods have been developed primarily to combat the borehole instability problems. In addition, these methods usually prevent disintegration of cuttings, thus providing a basis for development of dispersibility control systems. Most of known inhibitive muds, however, are too toxic to be environmentally acceptable. Table 3.12 lists the inhibitive drilling fluids together with values of their toxicities, as reported by various sources. The data indicate a general trend, suggesting that the stronger the inhibitive properties are, the more toxic the mud becomes.

Potassium/polymer muds have traditionally been the best water-based system with the lowest dispersibility. Unfortunately, in the USA, the toxicity limitation of a minimum LC_{50} value of 30 000 ppm essentially eliminated potassium from use in the Gulf of Mexico and other offshore areas of the outer continental shelf [50, 52]. High-salt (NaCl) polymer muds, instead of the more effective potassium systems, are now being used in the Gulf of Mexico. However, potassium muds are being used in the

Table 3.12 Drilling mud dispersibility vs toxicity[a]

Mud type[b]	Mysid shrimp LC_{50} (ppm)
PHPA (9.6 lb/gal)	>1 000 000
PHPA (14.3 lb/gal)	>1 000 000
PHPA/salt water (20% NaCl, 14.5 lb/gal)	140 000
PHPA/sea water (13.5 lb/gal)	>1 000 000
Sea water lignosulfonate (generic No. 2)[c]	621 000
Freshwater lignosulfonate (generic No. 8)[c]	300 000
Lime base (generic No. 3)[c]	203 000
KCl/polymer (generic No. 1)[c]	33 000
Cationic mud system	>1 000 000
Freshwater CLS (2% diesel)	5 970
Freshwater CLS (2% mineral oil, 15% aromatics)	4 740
Freshwater CLS (2% mineral oil, 0% aromatics)	22 500
Mineral oil-based mud[d]	180 000

[a]After Ref. 49.
[b]CLS = chromium lignosulfonate.
[c]Generic muds [47, 50].
[d]After Ref. 51.

North Sea and elsewhere where regulations are not biased against addition of potassium to sea water. To reduce the dispersibility characteristics of potassium muds in the North Sea, a variety of additives based on glycol and glycerol chemistry have been developed and are being used successfully [53–55].

One feature of polymer mud systems is that they typically operate at low pH levels relative to lignosulfonate muds that are highly dispersive. Lignosulfonate requires an alkaline additive for activation, such as sodium hydroxide (caustic soda), and the pH ranges from 9 to 11.5. The lower pH of polymer muds appears to be an important feature that helps reduce cuttings disintegration when cuttings are circulated to the surface. However, a number of high-pH lime muds are being used to take advantage of low dispersibility arising from the presence of insoluble lime [$Ca(OH)_2$] [56–58].

The recently developed 'cationic' mud systems appear to have low dispersibility and toxicity. These systems are usually formulated using non-reactive sepiolite or attapulgite clay cationic polymeric extender and cationic inhibitors so that the solids in suspension are positively charged. Negatively charged reactive cuttings are encapsulated by adsorption of the cationic inhibitor on their surfaces, thus preventing their disintegration. Another formulation of the cationic mud system employs a solids-free combination of pregelatinized starch and hydroxyethylcellulose (HEC) for viscosity and fluid loss with cationic polymer and 10% KCl for dispersibility control. Because the system is solids free, it has been developed exclusively for slim-hole drilling with high rotating speeds and annular transport velocities [59–61].

No toxicity data have been reported on the recently developed inhibitive mud system known as the mixed metal-layered hydroxide compound MMLHC (or MMH) fluid system [62–64]. However, the system formulation clearly implies low toxicity, employing a low concentration of bentonite (10 lb/bbl) and an inorganic MMLHC (<1 lb/bbl). Microscopically, the MMLHC compound contains discrete layers built of metal ions surrounded by hydroxide ions. The layers are positively charged and are smaller than clay platelets. The clay inhibition is based on an ion-exchange mechanism (similar to that of KCl systems) with the MMLHC exchange capacity being more than three times greater than that of sodium bentonite. However, not only are the particles of bentonite inhibited from swelling through the exchange of sodium ions for the metal ion hydroxide platelets, but they are also aggregated around MMLHC particles owing to their excess of positive charge. The practical result of this interparticle association is the development of gel structure and excellent solid suspension ability. Field applications confirmed the non-dispersive behavior of MMH drilling fluids through the following observations: (1) no washouts; (2) no viscosity increase; (3) clean borehole; (4) small volume of clean shaker cuttings; and (5) low MBT values. Also, the retention of simulated cuttings on a 6-mesh screen was over 80% by weight.

One of the most interesting and promising new water-based muds that has been successfully field tested recently is based on highly concentrated solutions of methyl glucoside (30–70% by weight). Laboratory studies indicate that this fluid may indeed possess the low dispersability property achievable by oil muds [65].

3.4.2 *Improved solids control – closed loop systems*

The overall efficiency of cuttings removal by the solids control system, E_s, can be expressed as

$$E_s = E_1 f_1 + (1 - E_1) E_2 f_2 + (1 - E_2) E_3 f_3 + (1 - E_3) E_4 f_4 \quad (3.1)$$

where E_1–E_4 are solids removal efficiencies (by volume) of the shale shaker, desander, desilter and centrifuge, respectively, and f_1–f_4 are volume fractions of mud processed by these separators. The equation has little practical use because the efficiencies E_1–E_4 are dependent upon separators' inputs, which in turn depend on the variable content of the flowline mud. However, equation (3.1) is useful for the design of a new system configuration, and also for the evaluation of solids control separators at work. In the latter case, the efficiency of each separator should be determined using API procedures [66]; then the overall efficiency should be calculated from equation (3.1). The API procedures are difficult to use; the physical meaning of symbols used for the separator's efficiency must be understood since some of them are given as volume fractions while others are expressed as weight fractions. A revised version of the API procedures is currently being prepared.

There are a few direct methods available at the wellsite to determine the overall efficiency of cuttings removal. The methods are based either on density measurements or water dilution records. Calculation of the overall separation efficiency using mud density measurements at the suction pit usually takes a long time (a day) and requires several cycles of mud circulation. The other method, measurement of the density difference between flowline mud and suction pit mud, does not give enough accuracy with the use of a mud balance. Alternatively, determination of reactive cuttings in the mud using the retort and the Methylene Blue tests does not have the precision required to detect the increase of clay concentration before it affects the mud rheology. An interesting method has been presented to determine a solids control index (SCI) from the monitored water dilutions required to control drilled solids [67]. (SCI can be converted to the separation efficiency through the equation $E_s = 1 - \text{SCI}$.) Although very practical, the method requires monitored water usage for dilutions and cannot be used for weighted mud systems.

Several attempts have been made to develop a mathematical computerized model of cuttings removal [68–72]. All of these attempts use the steady-state material balance approach with known and constant values of separation efficiencies of system components. They do not consider the relationship between the separation efficiency and particle size distribution, solids throughput and liquid-phase properties of the processed mud stream. Also, practical verification of the models is limited because no solids control instrumentation is available on drilling rigs. More successful efforts have been made to develop experimental models of single separators: hydrocyclones [73, 74], shale shakers [75, 76] and centrifuges [77], together with the analytical and field-deployable techniques for evaluation of the separators' performances [78–80].

Emphasizing the efficiency of solids removal may lead to the generation of excessive volumes of drilling waste. For any separator, whether shale shaker, hydrocyclone or centrifuge, a strong correlation exists between solids separation efficiency and volume removal of the associated mud liquid phase. Hydrocyclones, for example, when operated at 0.6 solids separation efficiency, may remove up to nine times more liquids than solids, as shown in Fig. 3.12 [74, 81]. Generally, any increase in E_s would result in increasing values for liquid removal, represented by the liquid removal ratio, R (the ratio of the volume of removed liquid to the volume of removed solids). The correlation between E and R is unique for solids control equipment and drilling mud used in the well. Theoretical

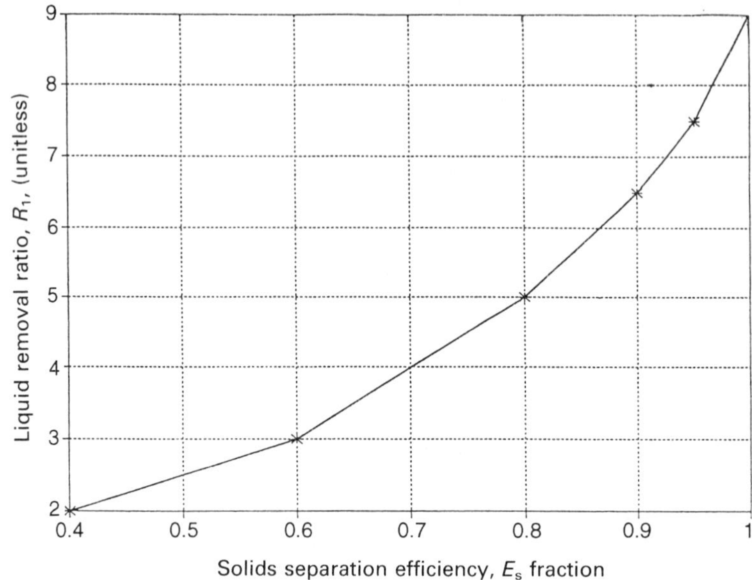

Figure 3.12 Typical relation between efficiency of solids removal and liquid removal ratio for hydrocyclones. (Modified from [74].)

calculations indicate that maximizing the efficiency of solids separation may result in up to a 50% increase of drilling waste volume [81]. Hence there is an optimum value of E_s that gives a minimum volume of waste.

Over the last decade, considerable improvements have been made in solids-control separators [82]. One significant improvement is in shale shakers and screens. Drilling rigs are now equipped with two or more linear-motion shale shakers. Some rigs may have as many as ten shakers, several of which are used as scalping shakers upstream of the fine-screen linear-motion shakers. The linear-motion shakers are often fitted with screens having an equivalent mesh size of 150 or more, which results in the removal of fine particles. The dramatic reduction in the size of the particles that can be screened from the drilling fluid has led to improved drilling-fluid performance and to a reduction in the volume of fluid required for drilling a well and discharged at the end of drilling the well. In addition to shake shakers and screens, the importance of the entire mechanical solids-removal system in reducing waste volumes from drilling operations has become better understood, which has resulted in the development of closed-loop drilling systems [83, 84].

The closed-loop system approach requires that the drilling waste should be disposed of at the drilling site and not taken out of the loop for off-site disposal [84]. From the standpoint of ECT methodology, closed-loop system technology integrates on-site disposal techniques with the drilling

Table 3.13 Development and performance of closed-loop drilling systems[a]

Performance measure	1983	1984–85	1986	Closed-loop condition
Surface hole removal efficiency (%)	15	46	68	81
Production hole removal efficiency (%)	20	67	80	89
Surface hole mud and disposal costs ($)	10 200	7 800	6 300	4 500
Production hole mud and disposal costs ($)	25 600	14 300	8 300	4 800
Total costs ($)	35 800	22 100	14 600	9 300

[a]After Ref. 84.

process (the environmental boundary is drawn around the drillsite, reserve pits and land treatment area). The drilling mud loop is partially closed through improved efficiency of the solids-control separators. The loop is finally closed through ultimate disposal on-site within the process boundaries. Table 3.13 shows the improvement in cuttings separation (hole removal) efficiency and economics resulting from the closed-loop system approach [84]. Closed-loop technology employs high-quality solids control separators in various configurations. Sometimes these systems are provided as skid-mounted tandems known as unitized solids control systems. Two types of unitized systems are available: one built by the solids control equipment vendors and the other custom designed and built by operators.

3.4.3 Dewatering of drilling fluids: 'dry' drilling location

An ECT alternative to closed-loop systems is a zero-discharge, or 'dry', drilling location at which no disposal on-site is permitted. A dry drilling location requires advanced technology for mud processing to minimize the volume and cost of on-site storage and off-site disposal [81]. One such technology is mud dewatering [85]. The dewatering component incorporates technology for separating water from water-based muds for reuse in the mud system. It also significantly reduces the volume of liquid waste that is destined for ultimate disposal.

A schematic diagram of the mud processing system with the dewatering component is shown in Fig. 3.13. After flowing out of the well, drilling mud is initially processed by solids-control separators (classification) and recycled back to the well. Since cuttings removal is not complete, a continuous increase of mud contamination by solids occurs. The contamination is controlled through additions of freshly mixed mud so that the mud system is steadily replaced with the new one. The rate of mud replacement is directly proportional to the rate of contamination of the system with fine cuttings. As a result, the rheological and filtration properties of drilling fluids are constant. Also unchanged is the mud system

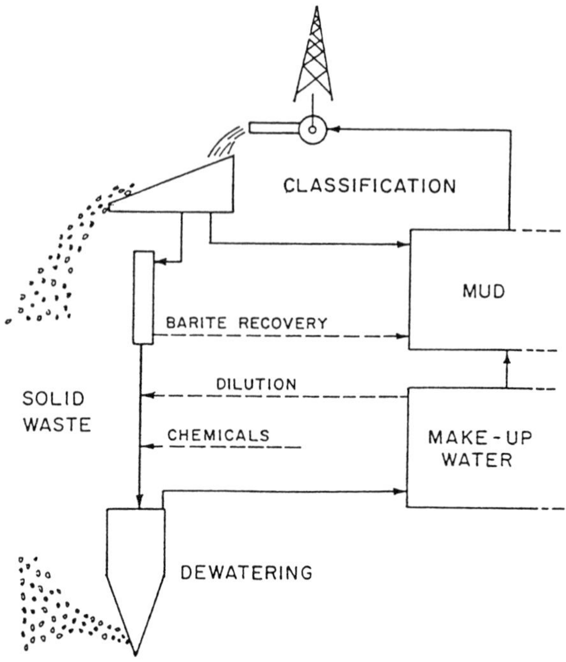

Figure 3.13 Principles of drilling mud dewatering.

chemistry, which is closely maintained to its original formulation. In order to maintain a constant volume of the surface mud, the rate of mud replacement must be balanced with the mud discharge rate. Therefore, part of the mud stream, after being processed by the solids control system, is diverted and treated by the dewatering component. First, the weighting material (barite) is removed and recycled back to the mud system. Second, the diverted mud is diluted with water to improve the chemical treatment which follows. Third, the diluted mud is treated with chemicals. The treatment transforms the mud from a stable suspension into a mixture of water-soaked flocculates and free water. The flocculates readily release water under a squeeze. The last stage of dewatering involves centrifugation of flocculates, resulting in a dense, solid cake (underflow) and solids-free water (overflow). The volume of underflow is significantly smaller than the feed mud volume. Also, returning the overflow water to the mud dilutions reduces water consumption and saves on chemicals dissolved in the mud–water phase.

Dewaterability involves the ability of drilling fluid suspensions to destabilize and release their water phase. The treatment consists of two stages: (1) chemical destabilization, in which a uniform liquid suspension is converted to two phases, free water and wet structure of solids (flocculates);

Table 3.14 Dewaterability of drilling fluids[a]

Mud system	Density (lb/gal)	Water removal (% v/v)	Cake solids (% v/v)	Volume reduction (% v/v)
Spud	9.2	65	43	72
Salt/polymer	13.5	65	66	28
Lime	9.6	63	47	62
CLS/unweighted	9.1	59	49	79
KCl/polymer	11.6	48	53	30

[a]After Ref. 88.

and (2) mechanical expression, in which additional water is released by squeezing the solid structure. Like other properties of muds (e.g. water loss, viscosity and gel strength), dewaterability combines complex physical mechanisms. However, it can be determined simply by measuring relative volume reduction after the squeeze [86, 87].

Dewaterabilities of various drilling fluids are presented in Table 3.14 [88]. The data indicate that, theoretically, the volume of waste drilling mud can be reduced by 1.4–4.8-fold. On the other hand, the data show that the presence of inert solids (barite) may distort the dewatering performance. For example, a high solids content in the dewatered salt/polymer mud may create the illusion of high performance and 'dry cake'. However, the actual performance is low, a mere 1.34-fold volume reduction. Therefore, in field applications, barite should be separated from drilling fluid prior to dewatering.

The inverse effect of reactive solids on dewaterability was observed in laboratory tests [89] and documented in field tests as shown in Fig. 3.14 [90]. Evidence shows that mud solids with a high cation-exchange capacity (CEC) produce moist cakes. However, the data do not show the simultaneous effects of mud inhibition and cuttings CEC on the cake's moisture. Moreover, the high moisture level in dewatering cakes has been often misinterpreted for low dewatering efficiency. In fact, the volume reduction ratio R_{vr} is a function of both the cake moisture M and the fraction of water phase in the dewatering mud, f_w, as

$$R_{vr} = \frac{1 - M}{1 + M(SG_m - 1) - f_w} \tag{3.2}$$

where SG_m is the specific gravity of the mud. Equation 3.2 indicates that a significant volume reduction can be obtained even with wet cakes (large M) for the low-solids mud systems (large f_w). For example, if the dewatering of a mud with 4% cuttings (and 10 lb/bbl commercial solids) produces a

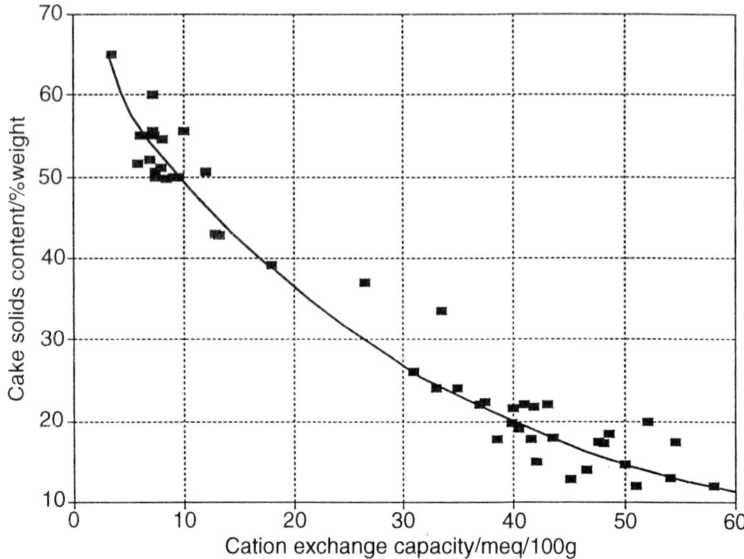

Figure 3.14 Effect of drilling mud solids reactivity on dewatering cake dryness [90].

solid cake having only 30% solids by weight (14% solids by volume), the volume reduction is still a significant 2.5-fold.

Selection of the best chemical treatment for a drilling mud has been repeatedly reported as a difficult design problem. Typically, the only selection method is the tedious trial-and-error approach. A solution to this problem has been developed using the theory of multiple factorial experiments [86]. In this method, the number of experiments required to find the best treatment (dilution, coagulant, flocculent, error) is reduced to nine points (nine-point test). In principle, the nine-point test is a simultaneous optimization of three variables of chemical treatment.

The second stage of the dewatering process, cake expression, is critical for reducing the volume of waste mud because it generates almost all of the water recovered in the process. Characteristically, for drilling fluids, the content of water in the flocculated structure of solids is greater than that in untreated drilling mud.

The cost of mud dewatering has been considered a key factor of the process design and control in all reported applications [85, 90–92]. The decision regarding whether or not to use the mud dewatering process should be based upon calculations of economics because (1) the dewatering process may be non-economical for a well, when traditional solids control system is efficient enough, or when savings due to volume reduction with the dewatering process cannot offset its cost; and (2) at certain stages of well drilling, the dewatering component should be disconnected because its cost breaks even with the off-site disposal cost.

3.4.4 Control of drilling fluids toxicity

Remarkable progress was made during the 1980s and early 1990s in the development of technical measures to control the toxicity of environmental discharges from drilling operations. The methodology of toxicity control includes testing methods, low-toxicity substitutes and source separation techniques.

Toxicity testing of drilling fluids is currently required in the USA, North Sea and other offshore drilling areas. Various tests have been adapted from conventional bioassays, measurements involving living organisms, for marine, freshwater or sediment toxicities. Organisms used in marine toxicity testing are oysters, shrimp (white, brown, grass or Mysid), crabs, fish and clams. Freshwater assays involve fish such as sheepshead minnows, bluegill, rainbow trout and daphnia. Typically, bioassays are conducted in licensed laboratories under controlled environmental conditions (light–dark cycles, temperature, salinity, pH, etc.), over time periods from a day to a week, and use organism populations carefully grown to meet sensitivity standards. Because of these reasons, the laboratory tests, rather than field-based toxicity tests, have been incorporated into environmental discharge regulations.

For example, the 96 h Mysid shrimp bioassay for drilling fluids was adapted from the US Army Corps of Engineers procedure for measuring the toxicity of dredged materials in compliance with ocean dumping criteria [47]. The test has been included in general permits for offshore dumping of drilling waste to the waters of the US outer Continental Shelf (OCS) since the early 1980s. The Mysid shrimp LC_{50} value of 30 000 ppm has been set as the limiting toxicity to maintain the general permit for drilling mud discharge offshore, together with ‘no sheen’ and ‘no free oil’ requirements, and concentration limits for mercury (1.0 mg/l) and cadmium (3.0 mg/l) in barite. Companies that discharge mud with LC_{50} value smaller than 30 000 ppm are subject to penalty because acute toxicity increases as the LC_{50} decreases.

The Mysid shrimp bioassay has been criticized for its imprecision and inconvenience in practical applications [49, 52, 93, 94]. The test's turnaround time may be as long as 2–3 weeks, which is comparable with the well's drilling time. Major problems for operators in using the 96 h LC_{50} test is just how to comply because results are not known for days or weeks following a mud or cuttings discharge. Operators currently comply with the regulations by setting an internal margin of safety based on LC_{50} tests run previously on the mud type they are using. This safety level may be set 60 000 ppm higher than the regulatory limit of 30 000 ppm or even higher, reflecting the fact that LC_{50} test results are highly variable and that some cushion is needed for unexpected events [95].

A considerable effort has been made to develop a new field-deployable test of toxicity, a rapid bioassay [96–100]. The three basic requirements for such test are a short (few hours) completion time, feasibility for use at well sites and correlation with the Mysid shrimp bioassay. The Microtox toxicity test is a promising alternative for rapid bioassay. One concept was to use the test as a statistical tool to predict on the offshore drilling platform whether the mud's Mysid shrimp toxicity would exceed (or not) its limiting value of 30 000 ppm (with a probability level of 98–100%) [98]. The mud passing such a test can be discharged overboard. The method could probably be further refined by introducing an element of calculated environmental (and economic) risks.

Another rapid toxicity test, cumulative bioluminescence, showed promise for further developments [100]. The test measures the total cumulative flux of light generated by a stirred suspension of algae plants in a controlled solution of drilling mud. The preliminary research results showed sensitivity of the test to progressive changes of mud toxicity. Also observed was a drastic improvement in the correlation with the Mysid test for higher mud toxicities (below the Mysid LC_{50} value of 300 000 ppm).

A computer program is also available for estimating the LC_{50} based on mud composition [101]. In conclusion, several useful methods are currently used for quickly checking a mud for compliance before discharge, but a fully reliable field test is not available. Quick checks and computer estimates cannot be substituted for a full 96 h LC_{50} test [53].

Low-toxicity substitutes include either completely new mud systems, or replacement of individual mud treatment chemicals with low-toxicity alternatives. The idea of replacing diesel OBM with mineral oil-based mud (MOBM) was initially derived from toxicity measurements made in the UK. These measurements showed that the toxicity of mineral oil is five times lower than that of diesel oil [51]. Other comparisons of mineral and diesel oil toxicities in sea-water emulsions showed mineral oil to be at least 14 times lower in toxicity [102]. The difference has been attributed to reduced content and different types of aromatic hydrocarbons in mineral oils. Aromatics are particularly toxic because of their rapid bioaccumulation rates. Toxic effects of monocyclic and polynuclear aromatics are dependent upon their water solubilities [103]. Mononuclear and dinuclear aromatics are the most toxic. Other polynuclear aromatics (with higher molecular weight) contribute little to toxicity because their solubility in water is low. Because mineral oils do not contain volatile monocyclic aromatics, their main toxic component is dinuclear aromatics.

Currently available mineral oils with no aromatics may be almost non-toxic, with the Mysid shrimp LC_{50} value over 10^6 ppm. However, some presence of aromatics is necessary for stability and suspending properties of invert emulsions. Therefore, a toxicity trade-off is needed for the MOBM formulations. The reported toxicities of MOBM are different, as shown in Table 3.12. The LC_{50} value of 180 000 ppm does not compare well with the values of 22 500 and 4740 ppm reported for two freshwater muds having 2% mineral oil with 0% and 15% aromatics, respectively. One explanation might be a different concentration of aromatics in the base mineral oils. Also, higher toxicities of MOBMs than their base mineral oils may result from the toxic nature of primary and secondary emulsifiers used in these muds.

The low-toxicity substitutions have also been used to solve the metal toxicity problem in drilling muds. Chromium lignosulfonate contains 2–4% by weight of trivalent chromium. Because it is considered a heavy metal, chromium presents an environmental problem. Even though toxicity tests have usually not indicated an adverse effect caused by the presence of chromium in lignosulfonate, considerable effort has been made to reduce the chromium content or replace the chromium with another cation. Chromium lignosulfonates have been replaced with modified sulfonates of the less toxic metals, such as iron, manganese, calcium, potassium, titanium and zirconium. Most of these substitutes have shown certain deficiencies in performance when compared with chromium-based thinners, particularly in the thermal gelation after hot oven rolling. One of these new products, based on titanium lignosulfonate, has been reported as not showing any increase in gel strength, yield point and plastic viscosity when the weighted freshwater muds are heat-aged [104]. Also, the reported field applications indicated that the viscosity control performance with this new thinner (measured by the treatment dosage, lb/bbl, required to maintain a low value of yield point) was equivalent to the conventional chromium lignosulfonate performance.

A whole new class of non-toxic drilling fluids has been developed within the past few years. These muds are formulated with a variety of synthetic organic base fluids. The resulting so-called synthetic-based muds possess most of the performance properties of oil-based muds but avoid most of the environmental problems of diesel and mineral oil muds [105–107]. The chemistry of the synthetic-based fluids that are currently commercially available includes an ester derived from palm kernel oil, a diether, a food-grade paraffin and a poly-α-olefin [108]. The chemistry of the components of the synthetic-based muds, other than the base fluid, is usually different from those in mineral-oil muds.

Synthetic-based muds are not toxic. They pass the toxicity test required for offshore discharges. (Problems have been reported in passing the US-based sheen test for these muds.) The disadvantage of the synthetic-based muds is their high cost, typically several hundred dollars per barrel. However, this high cost is offset by cost reductions arising from the use of a high-performance, high-penetration-rate fluid and the ability to handle cuttings disposal on-site without special equipment. The main technical uncertainty associated with these fluids is the threat of lost circulation. Losses can be extremely expensive because lost fluid cannot be returned to the service company at the end of the well for credit, reconditioning and reuse [52].

Spotting fluids used for freeing stuck drillstrings have been traditionally based on diesel or mineral oil and are notorious for adding toxicity to the mud systems. Starting in the late 1980s, suppliers and chemical companies began to develop spotting fluids formulated without diesel or mineral oil [109]. Effective low-toxicity, water-based spotting fluids are now available that, after freeing a pipe, can be incorporated into the water-based mud system without causing a significant change in the toxicity so that overboard discharges of mud and cuttings can be continued [110].

Other low-toxicity substitutes for miscellaneous drilling chemicals, such as biocides, lubricants, defoamers and corrosion inhibitors, have also been developed recently.

The *source separation* approach has been used to reduce oil-related toxicities of offshore drilling discharges. Typical applications include removal of oil from OBM cuttings and separation of diesel spots from water-based muds. Table 3.15 gives a summary of the maximum oil retention values for OBM cuttings using various separation techniques. Considerable controversy exists regarding the performance of centrifuges, with the lowest and highest values of oil retention being 3% and 10.25%, respectively (the typical reported values fall within the range 5–8%). The best-performing separation technique, vacuum distillation, has been commercially applied in the oilfield. Three vacuum distillation plants for OBM cuttings have been reported as working efficiently in the North Sea [114].

Table 3.15 Separation techniques for oil removal from OBM cuttings[a]

Separation method	Oil retention (%)
Shale shaker[b]	11.1–16.5
Mechanical cuttings washer[c]	9.4
Centrifuge[d]	3.0–10.25
Incinerator	0.0005–3
Solvent extraction	0.2
Vacuum distillation	0.01–0.05
Ultrasonic cleaning[e]:	
diesel washed	3–5
unwashed mineral oil	8–15

[a]After Ref. 111.
[b]After Ref. 103.
[c]After Ref. 112.
[d]After Ref. 32.
[e]After Ref. 113.

Characteristically, most of the research and development work regarding OBM cuttings cleaning methods has been done in Europe for North Sea applications [112, 113, 115, 116]. European regulations specify the maximum oil content on OBM cuttings with different values for different types of oils: 3% and 10% in Norway and 5% and 15% in the UK for diesel oil and mineral oil, respectively. In the USA, however, the general permit regulations place a ban on the overboard discharge of OBM cuttings, regardless of whether they come from diesel OBM or MOBMs. Therefore, the techniques listed in Table 3.15 are unlikely to be implemented in most of the US outer Continental Shelf unless the controversy about the toxicity of MOBM is resolved.

The principles of source separation can also be used for handling toxic spotting fluids. After a stuck pipe has been freed, the spotting fluid is circulated out of the hole and is separated from the water-based drilling fluid, and drilling operations continue. A separation technique for diesel-based spotting fluids was pilot-tested in the USA under the 1 year diesel pill monitoring program (DPMP) [117]. The program allowed participating operators to use a diesel pill that had been separated from the remaining mud by 50 bbl buffers on each side. After the diesel spot had been used in the well, the pill and the buffers were separated from the mud and sent ashore for toxicity testing, while the remaining mud was allowed to be discharged overboard regardless of diesel content. The purpose of DPMP was to create a database to determine toxicity limitations for diesel oil.

The results of DPMP showed that only about 70% of the spot was actually separated; the rest was incorporated into the drilling fluid. The remaining 30% has been proven to increase the toxicity of the water-based mud to the extent that it cannot be discharged even if a mineral-oil spot has been used. DPMP generated data that disqualified this separation technique and resulted in the ban on dumping mud after using diesel-based spotting fluids. Although this separation technique may still work for mineral oil-based spots, operators frequently haul all of the mud and cuttings to the shore instead of taking the risk of non-compliance following use of mineral-based spotting fluids [50].

3.5 Produced water cleaning – source separation technology

This section presents a brief overview of the source separation technology for removing pollutants from oilfield produced waters to comply with environmental discharge limitations. The technology is categorized according to the type of pollutant as deoiling, organic treatment and demineralization. Deoiling involves separation of free oil suspended in the continuous water phase. The objective of organic treatment is to remove dissolved oil. The demineralization process is designed for removing salinity from produced water.

Limitations regarding the discharge of produced water to surface waters vary considerably in different countries. For land production operations the most restrictive limitation is prohibition of discharge. In this case, the only two alternatives for final disposal are either subsurface injection or evaporation to dryness followed by disposal of the solid material in permitted landfills. However, in arid areas having little surface water, discharge of produced water may be allowed under limitations on salinity (within a few thousand ppm of chlorides) and O&G (below 30 mg/l). In this case, the discharged water is used for beneficial purposes, such as crop irrigation or livestock watering.

In offshore production, a simple approach to regulating overboard discharge may address only maximum O&G concentrations in the discharge with little consideration given to other pollutants. In fact, such an approach has been typical for early regulatory initiatives in many countries. In this approach, the objective was to lower the O&G concentration in produced water and was subject to the discretion of regional authorities. For example, the O&G discharge limits would vary for geographical areas within the following values: 48 mg/l for the Gulf of Mexico, 40 mg/kg for the UK/North Sea, 30 mg/l for Australia and 15 ppm for the Red Sea and the Mediterrranean Sea [118–120].

Produced water discharge limitations have undergone, and are continuing to undergo, steady evolution. A conventional regulatory approach to the produced water effluent guidelines has been changed from one based solely upon the total O&G concentration to one which discriminates between the limiting constituents and specifies maximum concentrations for each constituent separately. For example, Table 3.16 shows effluent limitations for discharging produced water to the saline inland, coastal and offshore state waters of Louisiana, effective at 1 January 1995 [121].

If this regulatory trend continues, more sophisticated (and expensive) technology for water cleaning will be needed. Some believe that the costs associated with such development may result in the technology shift from the source-separation approach to subsurface injection (recycling–containment) or subsurface reduction (source reduction) of produced water. These methods are discussed later in this chapter.

3.5.1 Deoiling

In the early 1980s, the conventional systems of produced water treatment were exclusively designed for oil removal and employed a two-stage configuration. In these systems, the primary stage would incorporate either a gravity settler (skim tank, gun barrel) or a coalescer (parallel/corrugated plates, serpentine path), and the second stage would employ a flotation unit.

All gravity settlers are settling tanks designed to provide sufficiently

Table 3.16 Produced water discharge limitations to saline waters of Louisiana[a]

Pollutant	Discharge limitation
Benzene	0.0125 mg/l (daily maximum)
Ethylbenzene	4.380 mg/l (daily maximum)
Toluene	0.475 mg/l (daily maximum)
Oil and grease	15 mg/l (daily maximum)
Total organic carbon	50 mg/l (daily maximum)
pH	6–9 standard units
Total suspended solids	45 mg/l (daily maximum)
Chlorides	Dilution required at a ratio of 10:1 (ambient water : produced water). All other prescribed parameters must be within acceptable limits prior to dilution
Dissolved oxygen	4.0 mg/l (daily minimum)
Toxicity (acute and chronic)	1 toxicity unit[b]
Soluble radium	60 pCi/l (2.2 Bq/l)
Visible sheen	No presence

[a]After Ref. 121.
[b]Toxicity unit is defined as the ratio of discharged effluent concentration to concentrations producing either lethality (acute toxicity) or no observable effects (chronic toxicity).

quiescent flow conditions so that free oil rises to the water surface and coalesces into a separate oil layer to be mechanically removed. In addition, particulates coated with heavy oil may settle to the bottom and are removed as a sludge or underflow. Chemicals such as de-emulsifiers and/or coagulents may be added to improve separation.

Serpentine-path coalescers convert small oil droplets to larger ones. The process of oil coalescence can be realized by forcing the oil–water mixture to flow through a permeable pack of a granular or fibrous material. The idea is attractive, but there are a number of practical difficulties (one of which is the occurrence of both droplet coalescence and droplet fragmentation in the permeable pack). In practice, this technique is not often used for reduction of the oil concentration in produced water.

A plate coalescer consists of an assembly of parallel plates, through which the oil-in-water emulsions flow. The presence of the plates leads to a reduction in the settling distance of the oil droplets and to coalescence on the plates' surfaces. To enhance the removal of the collected oil, the plates are inclined and corrugated. The main advantages of plate coalescers are their simplicity, low maintenance and lack of moving parts. Their limitation is that oil droplets below a minimum size, reportedly around 8 μm, cannot be separated. However, also reported was a practically achievable minimum size of oil droplets in the range 20–30 μm [122].

The induced gas flotation process disperses fine gas bubbles into a reaction chamber to suspend particles that ultimately rise to the surface and form a froth layer. Oil droplets and oil-coated solids, which are suspended in the water, attach to these bubbles as they rise to the surface, are trapped in the resulting foam and are removed when the foam is skimmed from the surface. Flotation cells for deoiling produced water utilize two different methods to induce gas into the produced water. The most common method is mechanical and uses a rotating impeller positioned inside a stator at the base of a draft tube. The rotation of the impeller creates a vacuum which draws gas down the draft tube. The gas is then ejected from the impeller through the stator, which disperses the gas in the form of fine bubbles. The second type of gas induction uses hydraulic ejectors to aspirate gas into the produced water. This requires recirculation of a portion of the treated water for use as the motive force to aspirate the gas.

The oil removal performance of conventional water treatment systems has been evaluated in field [123–125] and laboratory studies [126–127]. The results provided a general assessment of this technology: (1) there was no removal of dissolved organic fractions; (2) the minimum oil concentration at the output of gravity settlers was 113 mg/l; (3) the mean oil concentrations in effluents from over 50% of the flotation units tested were above the regulatory limit of 48 mg/l; and (4) the design of a system should incorporate an actual brine and crude produced from a reservoir.

The field survey data [124] were further analyzed [3]. The objective was to determine a relationship among the system variables, such as water flow rate, the oil content in the feed water and the oil content in the effluents from primary and secondary separators. A multiple regression analysis was used to model the simultaneous changes of the recorded variables. The results indicated a lack of any statistically meaningful correlation between the variables. The oil-separation performance, measured as the effluent oil concentration, appeared insensitive to varying input rates and oil contents. Several factors explain this insensitivity. First, the system was operated at a fraction of its nominal throughput (insensitivity to the flow rate). Second, the separation efficiency was a possible maximum (insensitivity to the influent oil content). Additionally, the mean value of the effluent oil content was below the compliance level of 48 mg/l (monthly average) for only five out of ten systems, and the daily values fluctuated closely to the compliance limit of 72 mg/l (daily maximum). Further reduction of oil content at the process end-point was concluded to be accomplished only by adding an efficient separator downstream from the flotation unit.

Also, the statistical analysis provided an interesting insight into the performance of the primary separation devices. The study revealed that, during most of the test, the primary separation was redundant. As shown in Fig. 3.15, the flotation units were capable of reducing the oil content in produced water to levels of 10–60 mg/l for influents containing less than 800 mg/l oil. This performance was not significantly dependent either on the feed oil content or the flow rate. The plot in Fig. 3.15 also indicates that, for the same range of the input oil, the primary-stage separator

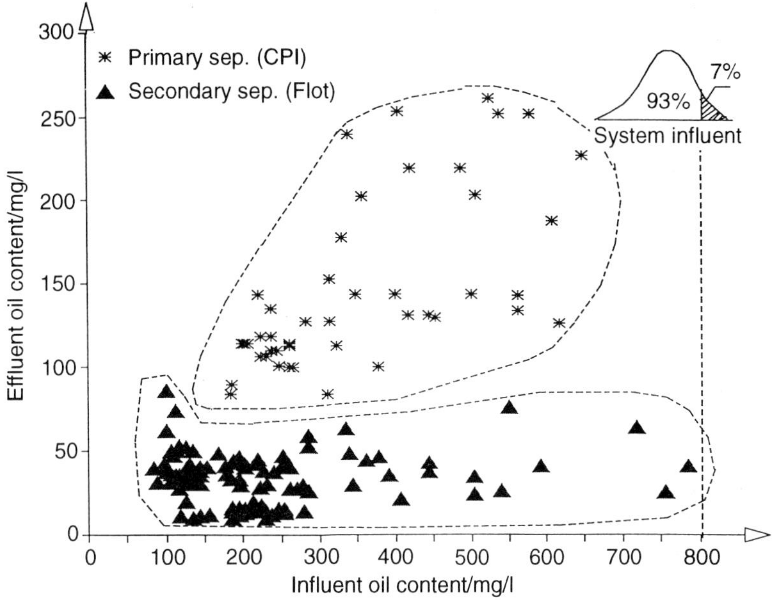

Figure 3.15 Redundance (93%) of primary treatment of produced water [81].

effluents had oil content levels well above those for flotation units. Moreover, the field data used in this analysis show that system input oil contents smaller than 800 mg/l were very common (93% of all input samples contained less than 800 mg/l oil). Therefore, the actual use of gravity settlers and coalescers was minimal.

The logical steps in the future development of deoiling systems for the oilfield production process appear to be: (1) the improved control of effluents from heater treaters using API separators to stabilize oil concentration below 500 mg/l; (2) design of the first-stage separation (e.g. flotation unit) to reduce the oil content to a range of 10–50 mg/l; and (3) addition of a new, high-quality separator to the second stage of the process.

Several new technologies show promise for the oilfield surface process application. A list of these technologies, together with their tested efficiencies of oil removal, is presented in Table 3.17 [119, 120, 128–134]. This table has been compiled using information from various sources, ranging from rigorous scientific laboratory projects [120] to commercial publications [129]. Therefore, the data in Table 3.17 should be viewed as the best estimates of the performances for each method. In addition, the oilfield applicability of the methods either has not been fully analyzed or is controversial. For example, the use of hydrocyclones requires a stable

Table 3.17 Environmental performance of modern techniques for deoiling produced waters

Technology	Influent oil (mg/l)	Effluent oil (mg/l)
Vortoil hydrocyclone [119][a]:		
35 mm	43	11
60 mm	408	16
Colman–Thew hydrocyclone [120][b]	100	12
	1000	100
Rotary hydrocyclone [128][c]	100	15
	1000	35
Disk-stack centrifuge [129][d]	<1000	5
Crossflow microfiltration [130, 131][e]	28–583	5
High-gradient magnetic separation [132][f]	190–240	23
Electrolytic treatment:		
[133][g]	1000–2000	3–11
[134][h]	500–5000	TR[i]

[a]Field tests offshore; flow rate up to 11 gpm/cone.
[b]Laboratory tests; constant size of oil droplet in influent, d_{50} = 35 μm; flow-rate range 21.5–37.4 gpm/cone.
[c]Prototype test offshore (mean value of results from two platforms); flow rate 26–36 gpm/cone; rotary speed 1900 rpm.
[d]Commercial data for oily water only; flow rate 29 gpm; rotary speed 5000 rpm.
[e]Offshore field test; permeate flux 850 gpd/ft^2; flow rate 3 gpm per two units in series.
[f]API separators effluent tests.
[g]Bench- and pilot-scale experiments; wastewater from manufacturing plant.
[h]Bench-scale experiments; Nigerian light crude + sea-water emulsion.
[i]TR = no residual turbidity; 100% removal claimed.

input pressure and a constant feed rate, both of which cannot be easily achieved at the output of free water knock-outs (FWKO) [135]. There is an ongoing discussion among oilfield service companies on the superiority of various modern deoiling technologies; hydrocyclones, centrifuges, membrane filters, diffusion-barrier filters, etc. [136].

Cost performance of the deoiling technology is shown in Fig. 3.16 [137]. Unit cost curves are presented for five options of deoiling technology versus water production rate: deep bed filter; gas flotation; hydrocyclone; and API separator, with and without chemical conditioning. The unit costs presented in Fig. 3.16 have been calculated using the following assumptions regarding removal efficiency: 5 mg/l O&G concentration in effluents from induced gas flotation or API separator, 98% removal efficiency for deep bed filtration and 80% removal efficiency for hydrocyclones. These assumptions are not universal but represent average performances of these technologies.

From Fig. 3.16, the least-cost deoiling treatment is apparently the API separator, followed by the hydrocyclone, deep bed filter, induced gas flotation and the API separator with chemical polymer addition. The higher cost for the API separator with chemical conditioning results from

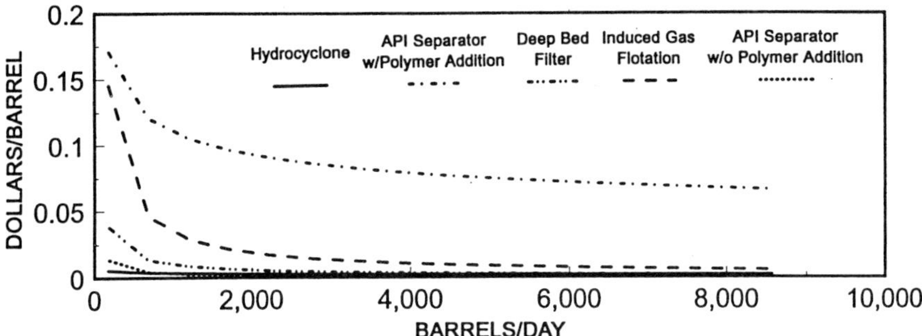

Figure 3.16 Cost of deoiling technologies for produced water (1994 dollars; 20 year project life; 5% discount rate) [137].

the use of the chemicals. However, the selection of a deoiling technology should be based on technical performance as well as cost. Technical performance determines the lower limit of O&G concentration that each treatment technology can attain and is dependent on the removal efficiency and influent O&G concentration. Moreover, these two factors, removal efficiency and influent quality, are inter-related. Therefore, selection of the specific deoiling treatment process would require consideration of the upstream quality of the process influents and the downstram quality of the effluents. The effluent O&G concentration may be either subject to discharge permits or determined by downstream pretreatment requirements.

3.5.2 Removal of dissolved organics

Two technologies, bio-oxidation and granular carbon adsorption, have been recently selected as the most promising options for removal of organic material dissolved in produced waters. These technologies have been included in the computer-aided engineering model for calculation of the cost of different produced water treatments for the natural gas industry [138].

The bio-oxidation process for produced water has been adapted from the biological fluidized bed reactor (FBR) process for treatment of municipal wastewater. FBR for produced water is an aerobic reactor employing aerobic bacteria to biodegrade dissolved organics. The process consists of passing the produced water to be treated upwards through a bed of fine-grained media, such as sand, granular activated carbon or ion-exchange resins, at a velocity sufficient to impart motion to, or 'fluidize', the media. This occurs when the drag forces caused by the liquid moving past the individual media particles are equal to the net downward force exerted by gravity (buoyant weight of the media). This is referred to as the point of incipient fluidization (defined either as the point at which fluidization occurs or the maximum bed porosity achievable prior to fluidization occurring). Greater fluid upflow velocities (flux rates) cause the bed of media to expand beyond the point of incipient fluidization.

Fluidization of fine-grained media allows the entire surface of each individual particle to be colonized by bacteria in the form of a biofilm. Surface areas of the order of 300 m^2/m^3 of bed are common in FBR systems. This results in accumulation of biomass concentrations of 5–50 000 mg of volatile suspended solids (VSS) per liter of fluidized bed, which is an order of magnitude greater than that obtained in most other biological processes. Manipulating the volume of media added to a system, the fluidization velocity and the point in the reactor at which the bed height is controlled allows the average biofilm thickness and mean cell retention time to be designed for maximum performance.

The granular activated carbon (GAC) adsorption process employs a fixed-bed column that is used as a means of contacting the produced water with the carbon media. Produced water with dissolved organic compounds enters the inlet to the granular activated carbon container. Soluble organics are adsorbed on the surface of the carbon and the treated produced water exits the GAC container. The GAC must be reactivated when it can no longer absorb organics. The carbon can be reactivated in the canister or removed and reactivated off-site.

Figure 3.17 is a plot of the unit cost curves for dissolved organic treatment using bio-oxidation (GAC–FBR), GAC–FBR with a sand filter and GAC alone [137]. The GAC–FBR unit cost curve is a function of the flow rate and an influent chemical oxygen demand (COD) concentration of 34 mg/l. The GAC unit cost curve is a function of the flow rate and influent organic concentrations of 12 mg/l benzene, 1 mg/l naphthalene and 1 mg/l phenol. A sand filter would be needed to remove biosolids in certain situations, such as when total suspended solids (TSS) would be above

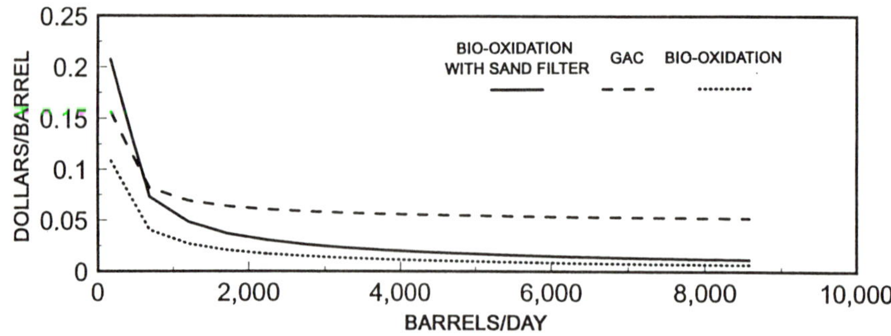

Figure 3.17 Cost of various techniques for removal of dissolved organics from produced water (1994 dollars; 20 year project life; 5% discount rate) [137].

permit limits or prior to electrodialysis, reverse osmosis or vapor compression, forced evaporation and solar evaporation. A brief descrip-significantly affect the unit cost of using a GAC–FBR. The cost for removing dissolved organics ranges from less than $0.01 to $0.25/bbl of produced water, depending on the process selected.

3.5.3 Salinity reduction

Demineralization technologies are electrodialysis, reverse osmosis, vapor compression, forced evaporation, and solar evaporation. A brief description of each of these processes is given below [137].

Electrodialysis accomplishes a selective separation of ionic compounds from produced water using semi-permeable, ion-selective membranes and electricity. Application of an electric potential between two electrodes causes cations to move toward the negative electrode and anions toward the positive electrode. Alternate spacing of cationic- and anionic-permeable membranes results in the formation of diluted (product) and concentrated (reject brine) salt solutions between the alternate membranes.

Reverse osmosis is a process in which produced water is partially demineralized by being forced through a semi-permeable membrane at a pressure greater than the osmotic pressure caused by the dissolved salts in the produced water. A partially demineralized water stream and a concentrated brine solution are produced.

Vapor compression is a process in which steam is used to heat the produced water above the boiling point. The vaporized produced water is compressed and also used to heat the incoming produced water in a heat exchanger. The condensate from the heat exchanger is the treated, demineralized, produced water.

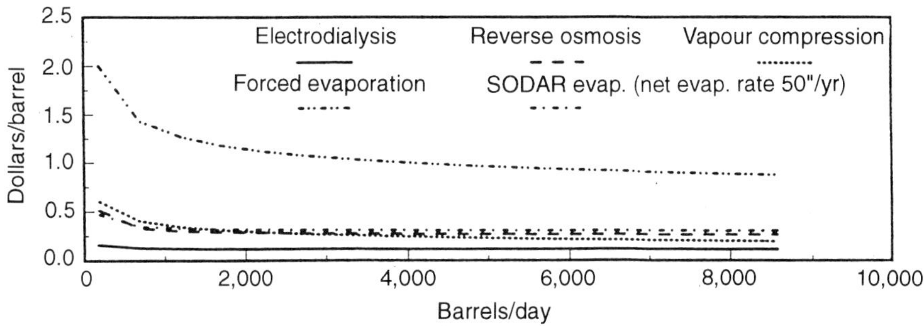

Figure 3.18 Cost of demineralization technologies for produced water (1994 dollars; 20 year project life; 5% discount rate) (50″ = 1.27 m) [137].

Forced evaporation uses a spray dryer into which the produced water is flashed at temperatures above boiling point, resulting in the production of steam and solid salt. The steam is then emitted to the atmosphere or recondensed.

Solar evaporation is accomplished in ponds and can be used in arid regions. Produced water evaporates from the surface of the pond, resulting in the build-up of solid salt in the pond.

Figure 3.18 is a plot of the unit cost curves for the five demineralization treatment options discussed above [137]. The cost ranges from \$0.10 to \$2.00/bbl of produced water. These unit costs are related to flow rate and have been calculated assuming an influent TDS concentration of 50 000 mg/l and an effluent TDS concentration of 500 mg/l. Electrodialysis is the least expensive technology for partial demineralization of produced water and ranges from \$0.11 to \$0.16 over a produced water flow rate range of 8570–170 bbl, respectively. Disposal cost of the rejected stream has not been included in the given unit costs. Forced evaporation is the most expensive technology for managing inorganic salts in produced water. The unit cost ranges from \$0.88 to \$2.00 over a produced water flow rate range of 8570–170 bbl, respectively. These unit costs do not include solids disposal or recovery of water. The solar pond unit costs were based on a 50 in/year net evaporation rate.

3.6 Subsurface injection of oilfield waste slurries

Technically, the term 'waste slurries' includes suspensions in fluids having various concentrations of solids, from less than 1% to over 20% by volume. All waste liquids from oilfield pits, contaminated produced water, drilling muds and slurrified (fluidized) drill cuttings fall into the category of oilfield waste slurries. Also, subsurface injection includes injection through the annular space between two strings of oilfield casing (annular injection) and injection well technology (tubular injection).

Subsurface disposal of solid waste has evolved from downhole injection of solids-free liquids combined with the well stimulation technique of hydraulic fracturing to the new technology of subsurface injection of slurrified solids. Conventional injection of solids-free liquids such as water flooding or deep-well disposal of the cleaned produced water is based upon mechanisms of flow and displacement in continuous porous media. On the other hand, injection of the waste slurry implies fracturing of the disposal zones, even for cases when these zones display very high permeabilities of the order of several darcies ($1\ \text{D} = 0.9868 \times 10^{12}\ \text{m}^2$), and low pore pressures. In high permeability zones, fracturing may occur at later stages of injection as a result of plugging off the disposal zone adjacent to the wellbore. For the purpose of this chapter, we shall call this technology high-permeability injection in contrast to slurry fracture injection, the technology of slurry disposal in artificial fractures that have been created in impermeable rocks.

In the early 1980s, high-permeability annular injection of drill cuttings became an environmentally sound alternative for one-site disposal of drilling waste, particularly in the Gulf Coast area [139–142]. Later, slurry fracture injection technology was developed for disposal of drill cuttings from oil-based muds in Alaska and the North Sea [143–145]. Recently, slurry fracture injection has been used for disposal of oilfield wastes other than drilling mud and cuttings. These other wastes include produced sand, sediment from tank bottoms, naturally occurring radioactive materials (NORM), unset cement and unused fracture sand [146–148].

3.6.1 *High-permeability annular injection of muds and cuttings*

Annular disposal of waste fluids from drilling mud reserve pits has been practiced for many years [139]. As shown in Fig. 3.19, annular injection is the injection of fluids between the annulus created by the space between the

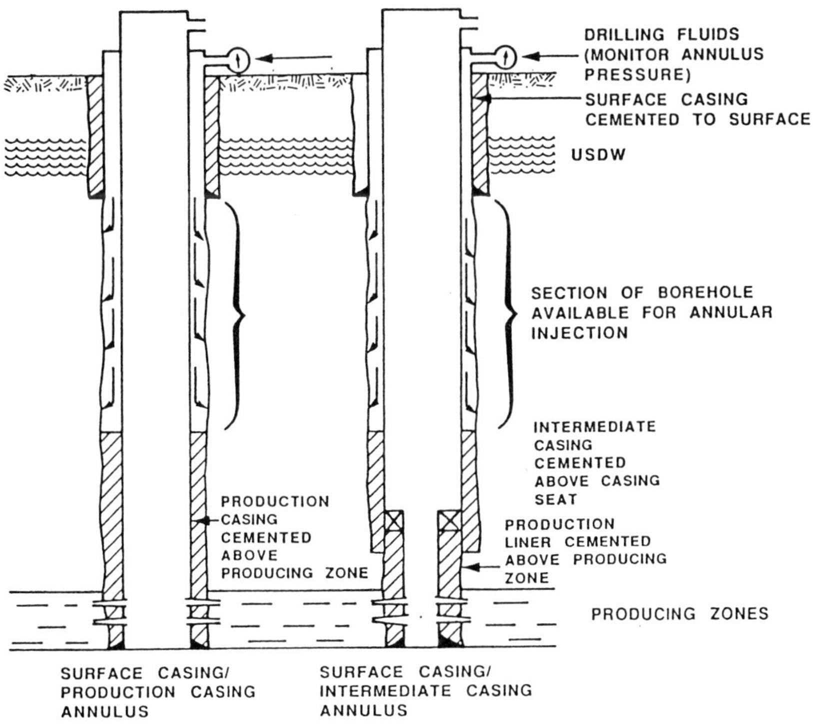

Figure 3.19 Well configurations for annular injection. USDW = underground source of drinking water.

surface and intermediate casings or between the surface and production casings. The surface casing is cemented all the way to the surface to protect fresh waters, and its setting depth may range from approximately 300 to 2000 feet. The intermediate casing is cemented below the depth at which the surface casing is set so there is an open hole annulus below the surface casing shoe. The annular space that has an open hole exposure enables the fluids to go down between the surface casing and the intermediate casing and out into the permeable formation. In wells with no intermediate casing strings, the fluid will go down below the surface casing and above the top of the cement on the production casing and out into the zones of least resistance. Usually, these zones of least resistance are low-pressure non-productive sands.

In the mid-1980s, the typical application of annular injection followed a fairly routine procedure. The pit fluid injection contractor connected the injection pump discharge line to the proper value at the wellhead that led to the annulus. Then, the waste drilling mud from the pit was pumped into the annulus to fill it up. (Some void space in the annulus, which was caused by settling of the mud, sometimes occurred.) Next, the pumping pressure was increased to 'break the formation down'. This breakdown pressure was usually higher than the average pumping pressure by 200–500 psi (~1360–3400 kPa). The process of formation breakdown is believed to have been in fact a fracturing treatment because gelled and thick mud was pushed out of the annulus and into the permeable rock.

After pumping for a few minutes, the pumping pressures were returned to normal. In most cases, the pumping was begun with water and was gradually changed from water to pit slurry, often with a corresponding increase in pressure. Most contractors injected the entire contents of the pit; therefore, at the end of injection, the pit was usually almost empty. Crowding (pushing) the pit levee with dozers ensured that most of the slurry was removed from the pit. By the time the pumping was finished, the dozers would have covered and closed the pit, grading the surface back to its original elevation. During the reserve pit injection, the wellhead pressure typically ranged from 500 to 1000 psi in most areas. For shallow wells, such as those in the Canadian counties of McClain or Kingfisher, for example, the average injection pressure ranged from 500 to 700 psi. In the Anadarko Basin, on the other hand, the deep-drilled wells usually required injection pressures ranging from 1000 to 5000 psi. The waste volume injected from a well depended upon the well's depth and pit volume and ranged from 15 000 to 60 000 barrels. The rates of injection, from two to ten barrels per minute, varied depending on the contractor's equipment. The equipment used in this technology was a type of centrifugal pump, known as a 'trash' pump, which homogenized the contents of the pit by circulating and stirring the pit and mixing the mud, cuttings and water together.

Since the late 1980s, the technology of annular injection of pit slurry has been greatly improved and is now applied to disposal of drill cuttings from oil-based muds used offshore in the Gulf of Mexico [140–142]. Specific for this technology is a lack of concern for hydraulic fracturing of the disposal zones. These zones are shallow (3600–4600 ft) unconsolidated sand strata with extremely high permeabilities due to the presence of shell deposits. Table 3.18 shows an example of the rock strata in the disposal zone. The high permeability of these formations allowed the successful disposal of materials such as slurrified, drilled-out cement, shredded paper waste (mud sacks and carboard boxes), shredded industrial plastic foil and ground wood with plastics (shredded wooden pallets and crates) [141]. Lack of concern for fracturing is based on the assumption that in highly permeable rocks fractures cannot be propagated far because most of the liquid phase of the injected slurry is lost from the fracture into the rock structure due to the 'screen-out'.

As shown in Fig. 3.20, screen-out can occur when the fluid phase of a solid–liquid mixture is lost into the fractured formation. As the liquid phase fraction diminishes, the solids fraction can increase in the fracture tip until there is no longer enough liquid phase to continue conveying the

Table 3.18 Description of subsurface disposal zone: Gulf of Mexico

Depth range (ft)	Rock	Per cent	Description
3810–3960	Sand	40–90	Clear, white, translucent, loose, very fine grained, well sorted
	Shale	10–50	Light gray, soft (occasionally firm), flaky, sticky, calcareous
	Shells	10	Loose fragments, macro fossils, micro fossils
3960–4080	Sand	70–90	Clear, white, moderately well consolidated, fine grained, well sorted, calcareous cement
	Shale	0–10	Gray, moderately firm, blocky, platy
	Shells	0–20	As above
4080–4280	Sand	30–70	Clear, translucent, unconsolidated, fine grained, moderately sorted, spherical
	Shale	10	Firm, blocky, platy, calcareous
	Shells	20–60	As above

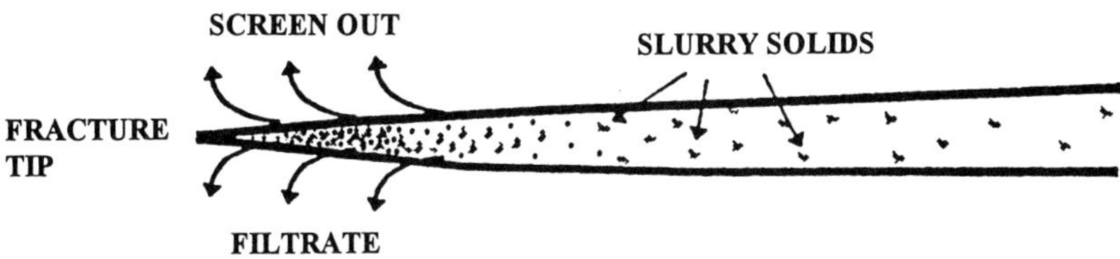

Figure 3.20 Fracture screen-out during high-permeability injection of slurrified solid waste.

solids. Cuttings slurries typically have a high potential for rapid screen-out across fracture walls since they tend to exhibit excessive fluid loss properties. However, data from various cuttings injection operations show that cuttings slurry can be successfully injected into formations with high permeability [142].

Figure 3.21 is a schematic diagram of the basic surface slurrification equipment and the downhole cuttings injection process. Cuttings generated

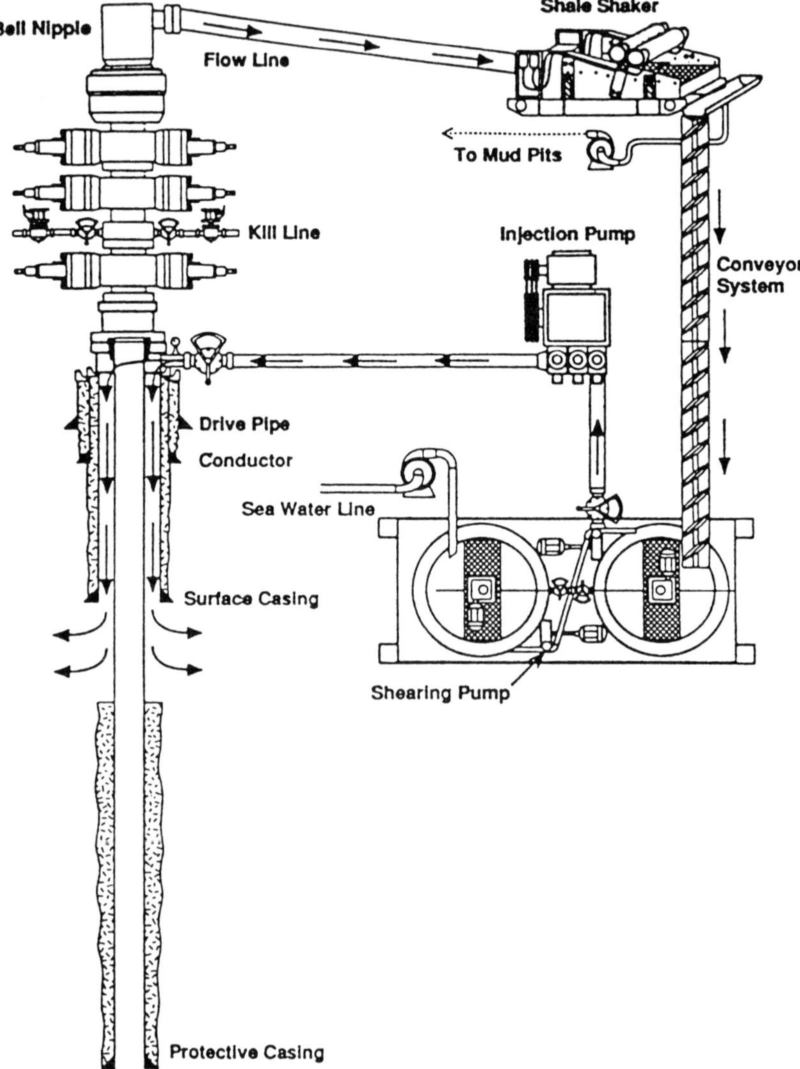

Figure 3.21 Schematics of slurrification and annular injection process for OBM cuttings in the Gulf of Mexico [140]

by drilling operations are removed from the drilling fluid using conventional solids control equipment and then transported to the cuttings slurrification system using conveying equipment. When the cuttings reach the system, they are transformed into a pumpable slurry by mixing water with the drilled cuttings at approximately a 3:1 ratio. Once the cuttings and water are blended into a homogeneous mixture, the cuttings are reduced to an acceptable particle size distribution by shearing them with specially modified centrifugal pumps and/or by grinding them using mechanical grinding equipment. Injection pumps are modified to enhance cavitation. Also, the pump impellers are hard faced so that erosion of the blades is minimized.

In the Gulf of Mexico area, drilled cuttings are so soft that the dispersion of the cuttings and the preparation of the slurry generally require only one pass through the centrifugal pump. Then, a small triplex pump takes the slurry from the slurrification pods and pumps it down the well's annulus. The slurry is kept at an optimum viscosity by adding sea water, dispersant, caustic or gel and is pumped at a specified rate. Typical properties of the slurry are shown in Table 3.19. When the pressure increase resulting from the pumping operation exceeds the strength of the exposed formation, the rock fractures and the cuttings slurry flows into the created fissure. The pumping operation continues until all slurry is injected into the formation. Table 3.20 gives the maximum injection parameters for four wells in the Gulf of Mexico. Maximum pumping pressures evidently exceeded the fracturing pressures of the disposal zones at times.

The high-permeability annular injection process has not yet been standardized. However, some basic guidelines have been developed from experience gained mostly in the Gulf of Mexico [142]. In the presence of a high-permeability disposal zone overlaid by a continuous sealing shale formation, the surface casing should be set and cemented at the bottom of the sealing zone. It has been proved by radioactive tracer surveys that the injected slurry would enter the high-permeability zone immediately below the surface casing shoe. Hydraulic fractures initiated in these zones are

Table 3.19 Properties of slurrified drill cuttings injected in the Gulf of Mexico[a]

Property	From	To
Density (lb/gal)	9.9	12.7
Funnel viscosity (s/qt)	41	92
Retort solids (vol %)	4	25
Retort liquid (vol %):		
Water	64	85
Oil	4	24

[a]After Ref. 140.

Table 3.20 Injection parameters for four wells in the Gulf of Mexico[a]

Well location	Surface casing		Intermediate casing		Leak-off test; equivalent mud weight (lb/gal)	Maximum injection parameters		
	TVD[b] (ft)	Size (in)	Size (in)	TOC[c] (ft)		Volume (bbl)	Rate (bbl/min)	Pressure (psi)
East Cameron	4724	10.75	7.625	5230	14.4	1 270	0.5	1500
Matagorda Island	4490	13.375	9.625	5800	14.3	9 560	4	1800
Galveston	3566	13.375	9.625	5200	14.1	19 579	2	2000
Galveston	3495	13.375	9.625	5890	14.3	9 990	3.5	1200

[a]After Ref. 140.
[b]TVD = true vertical depth.
[c]TOC = top of cement.

short and wide and do not propagate very far. Also modelling studies indicate that the amount of open hole below the surface casing shoe and the top of the cement controls the direction of fracture propagation [142]. As the length of the open hole section increases, the propagating fracture will tend to grow in the downward direction.

Since fracturing is not of much concern in high-permeability injection, the limiting factors for injection pressure and rate design are casing resistance to collapse, burst and erosion. Typically, operational practices use maximum injection pressure limits based on 70% of the burst rating for surface casing and 50% of the collapse pressure for intermediate casing string. Protection from erosion involves installation of a steel collar that deflects the stream of slurry entering the casing head and protects the intermediate casing hanger from exposure to the stream.

3.6.2 Slurry fracture injection of muds and cuttings

The technology of disposal to artificial fractures has been developed in drilling areas that lack low-pressure/high-permeability disposal zones typical for the Gulf of Mexico. In the North Sea, for example, permeable shallow sands having a porosity of 35% and permeability in the range of a few darcies are underlaid by massive Tertiary mudstones, as shown in Fig. 3.22. Two options for annular disposal can be considered theoretically: high-permeability injection to the lowermost sandstone formation or slurry fracture injection into the mudstone. A numerical simulation study showed that the disposal fracture in the sandstone would be shorter owing to slurry dehydration and would tend to propagate upwards into the overlying (impermeable) shales and siltstones [145]. Also, the calculations showed a rapid increase in injection pressure due to early screen-out (dewatering) of the slurry, as shown in Fig. 3.23. High-permeability injection was concluded to result in a smaller disposal volumes, a rapid increase in injection pressure for any new fracture created and a tendency of the fracture to propagate upwards into the sealing zone.

The other alternative, slurry fracture injection into a massive mudstone overlaid by permeable sandstone, proved superior to the high-permeability injection in the North Sea area. The conclusion was initially based upon theoretical simulation studies of fracture initiation, propagation, fracture shape and slurry screen-out [145, 149]. Fractures made in practically impermeable rocks were concluded to have a favorable, circular shape, i.e. they will propagate uniformly in vertical and horizontal directions. This process is shown in Fig. 3.24. Initially the vertical fracture expands as a radial fracture until its top reaches the permeable sand. Then, the cuttings-laden slurry would start to dehydrate, plugging the portion of the fracture that is in contact with the sand. Additional lateral fracturing would then occur (probably at a slightly higher pressure), as illustrated by fracture '2',

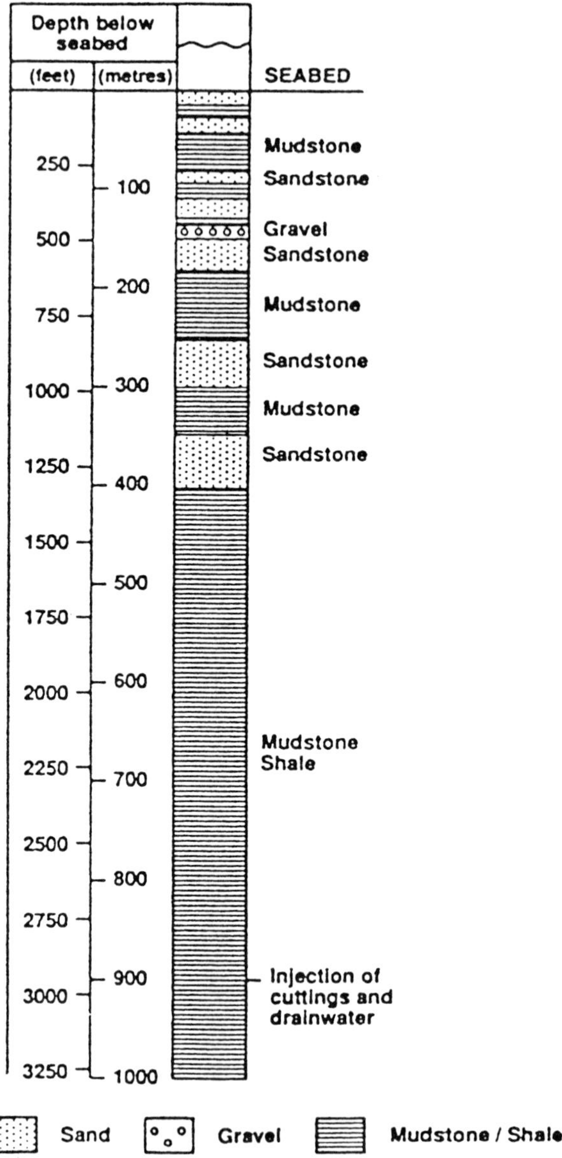

Figure 3.22 Example of shallow subsea stratigraphy in the North Sea area.

until again the fracture could grow vertically up into the permeable formation, where it would again screen-out, etc. Hence this mechanism of fracture propagation could conceivably allow significantly larger quantities of injection than might be possible for injection directly into a permeable formation.

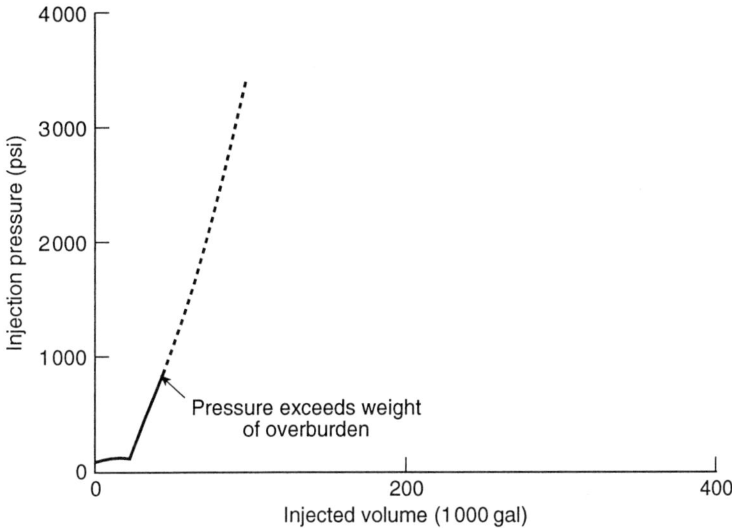

Figure 3.23 Computer-simulated trend of injection pressure during high-permeability injection to single fracture with early slurry screen-out [145].

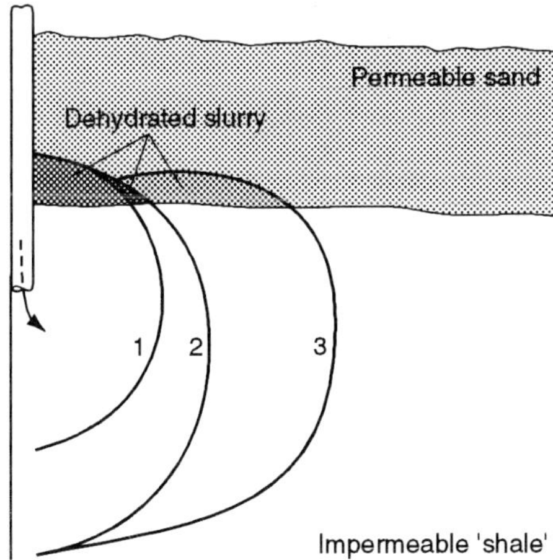

Figure 3.24 Propagation of disposal fracture during slurry fracture injection process [145].

The slurry fracture injection process for OBM cuttings has been fully implemented in the Gyda field [144, 150]. The BP Norway's Gyda was the first platform in the North Sea to dispose of all its drilling waste by downhole injection. The process is shown in Fig. 3.25 [150]. The oil-based mud is used to drill the three lower sections of 12¼, 8½ and 6 in holes.

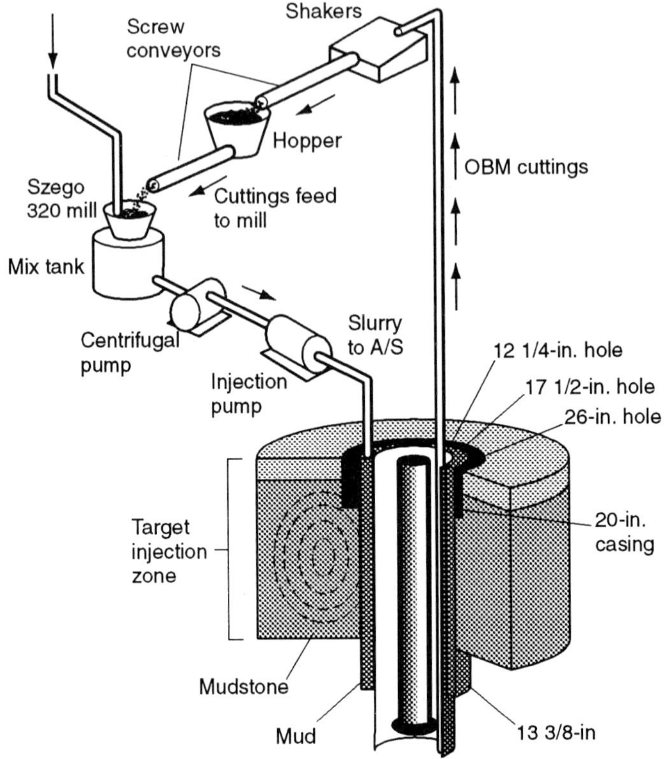

Figure 3.25 Slurry fracture injection process [151].

Approximately 500, 13 and 15 tonnes of rock and 35, 20 and 2 tonnes of oil are typically discharged from each of the respective hole sizes per well. As shown in Fig. 3.25, the surface installation for slurry-fracture injection is very similar to the high-permeability injection process used in the Gulf of Mexico. A simple centrifugal pump shearing system is used to grind and mix drill cuttings with sea water to produce a pumpable slurry. The slurry is pumped through the casing spool wing valve into the $9\frac{5}{8} \times 13\frac{3}{8}$ in casing annulus to fracture the massive Tertiary mudstones below the $13\frac{3}{8}$ in casing shoe, which is about 900 m below the sea-bed (Fig. 3.22). Several sand intervals with interbedded shales between 250 and 400 m below the sea-bed provide excellent geological barriers against fracture propagation and fluid migration to the sea-bed.

At Gyda, sequential annular injection, whereby cuttings from the well being drilled are injected into the annulus of the most recently completed well, has been adopted. On average, about 15 000 bbl of slurry per well were injected, including wash water and other watery drain-off wastes, with a maximum volume of 33 000 bbl in one well.

Table 3.21 Parameters of slurry fracture injection at Gyda[a]

Parameter	Well numbers: injection/drilled							
	A-23/A-09	A-09/A-22	A-22/A-16	A-16/A-19	A-19/A-27	A-27/A-15	A-15/A-26	A-26/A-24
Start injection	30/7/91	12/9/91	5/11/92	18/1/92	1/5/92	2/7/92	11/8/92	29/9/92
Duration (days)	42	31	47	41	42	21	30	Ongoing
Volume (bbl)	13 500	27 000	27 000	16 245	15 037	13 111	16 033	11 615
Injection rate (bbl/min)	8	3 8	7	7	7	7	9	11
Injection pressure (psi)	900	1000	1200	1100	1200	1400	1600	1450
Initial shut-in pressure (psi)	900	1100	700	NR[b]	NR	NR	NR	NR
Shut-in pressure (psi) (01/02/92)	700	150	700	NR	NR	NR	NR	NR
Shut-in presssure (psi) (10/10/92)	900	900	700	1100	1000	900	1100	950

[a]After Refs 149 and 150. Data as of 10 October 1992.
[b]NR = not recorded.

Performance of the fracture injection process is documented in Table 3.21 for the Gyda platform [149]. Note a sequential annular injection procedure in which cuttings from the well being drilled are injected into the annulus of the most recently completed well, etc. Also note in Table 3.21 that the annular shut-in pressure has not dropped over a 1 year period, which may become an environmentally significant fact regarding fracture disposal technology. This and other environmental considerations are discussed below.

3.6.3 Properties of injected slurries

Cutting slurry injection is similar to fracture stimulation technology in that both technologies inject liquids and solids into a fracture and both technologies rely on the ability to continue fracture propagation until the entire volume of materials has been injected. Still, there are differences between these two technologies, primarily because cuttings slurries exhibit fluid properties very different from those of fracture stimulation fluids.

During conventional fracture stimulation operations, a low-solids fluid with very low fluid loss properties is injected ahead of solids-laden (proppant) phase. This low fluid loss pad is essential to maximizing fracture propagation and to minimizing the chance of fracture screen-out. As shown above, screen-out can occur when the fluid phase of a solid–liquid mixture is lost into the fractured formation. As the liquid phase fraction filters out, the solids fraction can increase in the fracture tip until there is no longer enough liquid phase to continue conveying the solids.

In slurry injection technology the particle size distribution of solids in the slurry can be designed such that it controls the rate of the screen-out. If the selected injection zone is impermeable, the particle size of solids in the slurry should be increased to cause rapid fracture screen-out when the fracture propagates into a permeable formation. On the other hand, for high-permeability injection, the particle size of solids in the slurry should be reduced to minimize the rate of fracture screen out and to maintain fracture propagation into the permeable injection zone.

The size of particles in a slurrified suspension results from the type of grinding device used. These devices include a hard-faced centrifugal pump for weak cuttings (Gulf of Mexico), a vibrating ball-mill (Alaska [143]), an autogenous wet-crushing mill or a Szego ball-mill (North Sea [142, 145]). An example of the size distribution of solids in the slurry injected in the North Sea area is $d_{10} = 3$, $d_{50} = 9$ and $d_{90} = 120$ μm [149, 150]. With 50% of the particles smaller than 9 μm, the viscosity of the suspension is sufficient to prevent settling of larger solids in the fracture.

Rheological properties of injected slurries reported in the literature are plastic viscosity = 15 cP, yield point = 60 dyne/cm^2, flow behavior index = 0.26, consistency index = 0.148 lbf/ft^2/s$^{0.26}$, solids content ≈ 30% by volume and specific gravity = 1.68. Also reported was the use of polymeric viscosifiers with biocides [145], as well as thinners, bentonite and caustic, to control the rheology and biodegration of the slurries [140].

The filtration properties of injected slurries follow the theoretical mechanism of cake (or 'static') filtration, with filtrate volume directly proportional to the square root of time and with a proportionality constant equal to 0.004 ft/min$^{0.5}$ [145].

3.6.4 Environmental implications of subsurface slurry infection

The most important environmental concerns regarding fracture slurry injection are vertical propagation of the disposal fractures, loss of annular integrity of wellbore and the ultimate fate of the injected slurry. Typically, the risk of vertical propagation of fractures has been evaluated through mathematical modelling with the use of 3-D fracturing simulators. The simulator inputs include minimum *in situ* stresses, pore pressure gradients, Young's modulus and Poisson's ratio variations, slurry filtration (screen-out) and rheological properties, depth of injection and injection rate. The calculations typically show a relationship between the cumulative volume injected and the vertical height of the fracture for a given geological profile of sediments above the injection point. For example, simulation studies for the Gyda platform showed that, in the absence of any high-permeability sands above the massive mudstone (disposal zone), 90 000 bbl of slurry would be needed to propagate the fracture ot the sea-bed [149]. This study also showed that any shallow sand strata would become a barrier for fracture propagation. Similar studies were also reported for the Clyde platform in the North Sea [144].

In Alaska, field measurements of surface deformation were used to assess the potential for vertical propagation of disposal fractures under the permafrost in Prudhoe Bay field [146]. The fractures were initiated under the permafrost at 2000 ft. Then, a total of 2 million barrels of oilfield waste fluids were injected into three wells with injection rates averaging 1–2 bbl/min. Surface deformation of the permafrost was measured with an array of tiltmeters installed 25 ft into the permafrost. Analysis of the surface deformation was combined with transient pressure testing (step-rate and fall-off tests) of the injection wells. The analysis revealed the presence of horizontal fractures without discernible vertical fracturing.

Propagation of vertical disposal fractures in the highly permeable and thick (155 ft) Frio Sand at 4500 ft was effectively stopped by a 130 ft thick layer of shale overlaying the sand. This finding was documented by a recent field study involving computer simulation combined with a new method of real-time passive seismic monitoring and analysis [151].

Loss of external annular integrity of the borehole involves channeling outside the outer casing of the injection annulus and the flow of injected waste slurry to shallow aquifers or breaching the slurry to the surface. Verification of external integrity involves periodic additions of radioactive tracers to the slurry injected to the well's annulus while drilling the lower sections of the well. Typically, different types of short half-life tracers such as antimony, iridium and scandium are injected at the beginning, during and at the end of the annular injection process (upon reaching the total drilling depth). Upon completion of all drilling operations, a multiple isotope tracer log is run to determine actual injection points and flow behind the casing [141].

A long-term environmental risk results from the ultimate fate of injected slurry. When injecting wholly into shales, fluid screen-out is minimal. Here the fate of the solid waste slurry is dependent on chemical reaction with the surrounding shale. The hypothesis has been proposed that, since shales are usually reactive with water-based fluids, over time the sea water carrying the fluid reacts with the swelling clays to form an increasingly viscous, dehydrated slurry within the fracture, which will eventually seal the fracture over a long time period. The softened zone adjacent to the fracture would be relatively localized (a few feet at most, by virtue of the low permeability), thus posing little threat to subsequent well drilling, which may pass through the sealed fracture plane. In this new well the fracture will manifest itself as a localized tight-spot within the open hole without abnormally high pressure trapped in the fracture. Moreover, even if the pressure has been trapped, the high viscosity and gel strength of the remnant of dehydrated slurry preclude taking an unexpected kick. The above theory has never been verified experimentally. To date, field data indicate the continuing presence of pressurized fractures with no observed release of pressure in time, as shown in Table 3.21.

Significant fluid migration is also believed to be impossible, even in permeable strata. When disposal fractures intersect an unconsolidated sand of considerable thickness (10 m or so is usually sufficient), a rapid leak-off of the filtrate (screen-out), resulting in dehydration of the slurry, takes place. The dehydration assures permanent disposal of the solid particles, which remain trapped at the fracture–sand contact surface. Only the smallest clay particles may enter the sand formation. Also, the dehydrated solid cake will in time reduce the intrusion of the liquid phase into the sand. Because the pore volume of these laterally extensive shallow sands is large and because of their compressible nature, substantial volumes of slurry would be allowed to be injected without the risk of over-pressuring either the fracture or the sand formation.

3.6.5 Periodic fracture injection

A new concept of periodic fracture injection has been derived from the observation that for periodic injections a repetitive pattern of initial increase of injection pressure followed by pressure decrease and final stabilization occurs [152]. Also, the stabilized pressure level at the end of each injection tends to increase with the number of injections. This behavior contradicts the propagation of a single fracture, which would require a smaller propagation pressure due to the fracture size increase. This observation led to the conclusion that periodic injections may create multiple fractures in the same region of the formation around the injection borehole (disposal domain). The apparent environmental advantage of periodic fracturing is minimization of risk because of the small size of the disposal domain containing multiple fractures of controlled extent.

The new process has been pilot-tested in the Valhall field, North Sea, by comparing actual injection pressures with their calculated values from mathematical modelling of the disposal domain. The actual pressure response during injection of 460 000 bbl of slurrified cuttings matches the model predictions. These tests led to the conclusion that injection of slurrified cuttings by periodic fracturing may be a safe permanent solution to subsurface disposal of solid waste.

3.7 Integrity of petroleum wells: containment technology

Figure 3.26 shows the pollution mechanism due to the loss of external integrity of injection or production wells. The integrity loss causes upward migration of fluids outside cemented wellbores. Pollution of air, surface waters or groundwater aquifers may result from the migration of produced petroleum hydrocarbons, injected brines or other toxic waste fluids. The migration takes place in the annular space between the well casing string and borehole walls. This phenomenon has long been known in petroleum terminology as 'flow behind cement', 'gas migration', 'flow after cementing' or 'annular migration'. Most of these terms refer to the failure of well cements. In theory, at well completion, the subsurface isolation of aquifers and other strata is restored with annular seals (cement, grout, resin mixtures). Failure of the seals provides conduits for vertical transport of pollutants. The pollutants may originate from either wellbore fluids (drilling mud or injected wastewater) or formation fluids (oil, gas or brine).

3.7.1 Mechanism of cement seal failures

Typically, cement design specifications are based on the compressive strength of set cement; its tensile strength is assumed to be about 12 times smaller than the compressive strength. These properties have little effect on the quality of the annular seal. The failure of annular seals has been shown to be caused by poor bonding of cement or by the development of

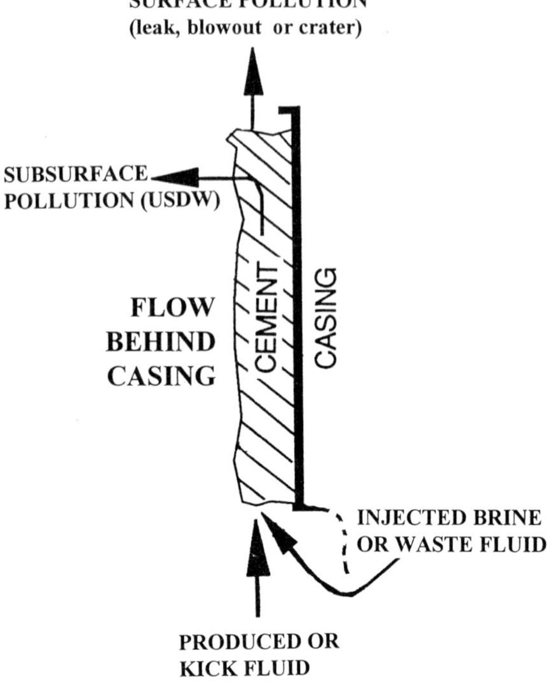

Figure 3.26 Pollution caused by lack of well integrity.

channeling during the cement setting process. The ability of set cement to isolate subsurface zones has been conventionally attributed to bonding of hardened cement to the pipe and borehole wall. Two magnitudes have been used to measure the quality of cement bond to the pipe (bond strength): shear bonding and hydraulic bonding. Shear bonding represents the force required to move pipe in a cement sheath [153]; hydraulic bonding represents the pressure required to initialize a leak between cement and pipe for liquid or gas [154]. Bond strength testing has been performed in laboratories for various pipe surfaces (rusty, sandblasted, resin-sand coated). This testing gave some basis for the actual design of cementing operations. The understanding of the cement–formation bond mechanism has been limited to the qualitative observations regarding the role of a mud cake and formation permeability [155] and the effect of mud displacement practices [156].

Channeling or development of secondary permeabilities in the cemented well annulus can be caused by either the annular gas migration during the cement thickening process [157, 158] or the sagging phenomenon (i.e.

formation of water channels in inclined wellbores caused by solids–water separation) [159]. Two causes of annular gas migration are the loss of hydrostatic pressure in the cement column and volumetric changes in the annulus. Annular pressure loss occurs during the transition of the cement slurry from the fluid state to the solid state due to fluid loss and development of static gel strength [160]. Simultaneously with the hydrostatic pressure, the pore pressure is reduced. The pore pressure loss mechanism results from the development of a matrix stress in the thickening cement so that the water pore pressure responds to the volumetric shrinkage, caused by dehydration of the matrix. The hydrostatic and pore pressure changes in cement are shown in Fig. 3.27.

Volumetric changes in the cemented well annulus may result from either a pressure drop inside the casing or volumetric shrinkage of the cement sheath. The casing pressure drop may create a microannulus between the casing and cement while cement shrinkage may cause the development of a microannulus between the formation and cement. Though the casing–cement microannulus is, by itself, too small to allow substantial flow, it is believed to be capable of initializing development of a flow channel and therefore must be prevented [161].

Shrinkage of cement, which is believed to be 3–4% by volume, is related to the concentration of calcium silicate crystals (which form during

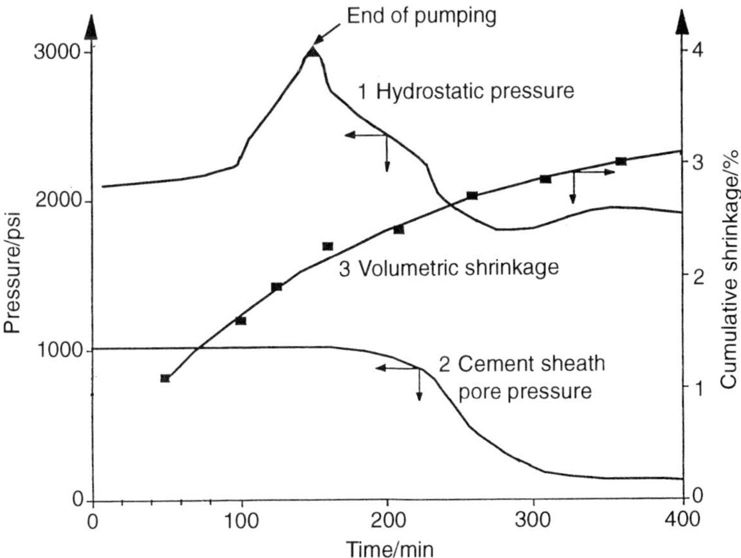

Figure 3.27 Loss of bottomhole pressure and shrinkage of cement slurry after cementing. (1) Field data at 4034 ft. (After Ref. 164.) (2) Laboratory gas flow simulator (pressure = 1000 psi). (After Ref. 158.) (3) Laboratory shrinkage cell at 250 °F (121 °C). (After Ref. 162.)

hardening) and the amount of available water during hardening [162]. An observation has also been made that 95% of volume shrinkage (up by 7% by volume) takes place after cement is in the solid state; therefore, the development of gas channeling through the bulk cement sheath when it is in a plastic state (transition state) is very unlikely [162, 163].

Sagging of cement slurries is an important mechanism of channeling in deviated wells. Settling of cement solids along the lower portion of the inclined well has been documented in well tests [164]. Also, the formation of a water channel along the upper portion of an inclined well, together with the resulting loss of the effective density, was observed in pilot-scale laboratory tests, as shown in Fig 3.28 [165].

3.7.2 *Improved cementing for annular integrity*

Annular seal integrity has been achieved through improvements in well cementing technology in three main areas: (1) steel–cement bonding techniques; (2) mud displacement practices; and (3) cement slurry design to prevent fluids from migrating after placement. The control of the steel–cement bond and mud displacement practices have long been incorporated into cementing technology [155, 156]. The most recent techniques have been developed to prevent the formation of channels due to gas migration in annuli after cementing [157, 161, 165–167].

Understanding the role of static gel strength in the mechanism of hydrostatic pressure loss has led to the development of delayed gel strength technology for oilwell cements. The technology was successfully demonstrated in the field when an addition of 0.4% of the delayed gel strength additive effectively stopped annular flow problems that had been traditionally experienced in the area [165].

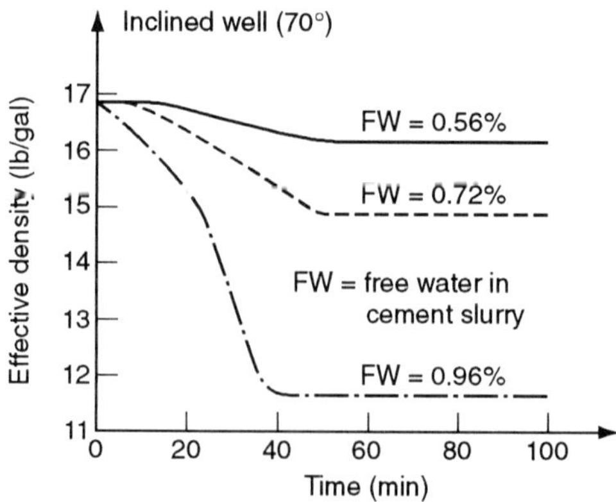

Figure 3.28 Loss of equivalent density in cement slurry column after cementing [159].

Another control measure, foam cementing technology, was derived from observation of the pore pressure drop in the annular cement column caused by shrinkage of the solids matrix and low compressibility of the matrix–water system. In typical applications of foam cements, gas is either added to the slurry at the surface or is generated by chemical reaction downhole. A recent improvement in this technology is to use foaming surfactants in an unfoamed cement slurry [157]. This new system employs a formation gas (invading the cement) to generate the foam.

A new laboratory procedure has been proposed to find an optimal composition of cement slurry for particular wellbore conditions. In this procedure, a sample of cement slurry is exposed to the expected gas invasion pressure in the gas flow cell simulating the downhole environment of the wellbore [166].

A more fundamental approach has been used in the slurry response number (SRN) method [161]. In principle, SRN is a ratio of static gel strength development rate to the fluid loss rate at a critical time. This critical time corresponds to the onset of a rapid increase in static gel strength. Fluid loss represents volumetric reduction of the slurry. The rate of fluid loss declines over time. At the critical time, the rate of fluid loss should be very small (high values of SRN). Otherwise, pressure at the bottom of the cement slurry could rapidly decline, causing gas migration.

SRN can be evaluated graphically from laboratory measurements of static gel strength and fluid loss versus time for a given cementing system. The optimal cement slurry selected is the one with the largest value of SRN. Recently, the SRN method was correlated with a conventional measure of gas migration tendency, i.e. gas flow potential (GFP) [167]. The analytical correlations, SRN versus GFP, in the form of two equations, constitute the first quantitative model of the annular seal integrity for a well.

3.7.3 *Integrity of injection wells*

The problem of hydraulic integrity of well annular seals has been addressed through both regulatory and technological measures. The two areas of regulatory initiatives to control annular integrity are drilling permit regulations and injection permit regulations. Drilling regulations focus mostly on the integrity of the surface casing. Typically, drilling permits require the surface pipe to be entirely cemented to protect freshwater sands from oil and gas zones. In addition, typical drilling regulations may specify minimum footage for surface pipe, minimum waiting-on-cement (WOC) time, minimum volume of cement slurry to be used, minimum length of cement sheath above the top producing zone and at the salt–fresh groundwater interfaces and the minimum testing requirements after completion [pressure test or cement-bond log tests (CBL)]. At present, no quantitative requirements exist to verify a potential annular flow between well casing and formations. For production casing, drilling permits are not very specific about the verification of annular integrity even though this integrity is most important in effectively isolating upper zones from produced hydrocarbons and brines.

Subsurface injection permits require an operator to provide evidence of the hydrodynamic integrity of the well's annular seal. However, no direct standardized tests for such integrity exist [165]. Usually, permit decisions are based upon indirect evidence of the well's integrity, such as CBL, electric logs, the driller's log and geological crossplots, which indicate to the regulatory agency that no unusual environmental risk is involved [168]. Typical generic criteria for wells injecting oilfield brines address the following issues: (1) the length of casing; (2) the mechanical integrity (pressure) test procedure (wellhead pressure, test duration, maximum pressure drop) and its frequency (usually before the operation, then every 5 years); and (3) the minimum distance to any abandoned well (usually 0.4–0.8 km). A permit is also required for the annular injection of solid drilling waste, the common method of on-site disposal during drilling operations (as discussed in the previous section).

In the area of subsurface brine injection, the permitting issue revolves around reliable techniques to prevent the stream of brine from migrating freely into the environment. The three main criteria are the 'internal' mechanical integrity of the borehole installation (IMI), the 'external' integrity of annular seals (EMI) and the integrity of the confining layer. The IMI practices of pressure testing casing as well as monitoring the annular pressure during injection are the most typical field technologies. However, since there are no standard procedures for IMI test analyses, the results of these tests are often left to the judgment of the permitting agency [169]. In addition, several factors may affect the result of pressure tests, such as the length and type of gas blanket, gas solubility in the annular liquid, temperature and the tubing–annulus pressure changes [170]. These effects should be included in quantitative interpretations of the tests.

A simple system to control continuously the internal integrity of an injection well has been developed by the chemical industry [171]. As shown in Fig. 3.29, the system does not use a packer at the bottom of the injection tubing or a surface pressurization system. Instead, it relies upon the laws of hydrostatics to separate the annular fluid from the injected fluid. A continuously recorded pressure differential between the injection and annular pressure is considered to be a sensitive indicator of tubing splits or casing leaks. Unlike the conventional 'packed' annular configuration, this system is believed to be insensitive to injection pressure variations and is unaffected by the packer leaks. Also, it has the unique ability to locate a point at which the mechanical integrity of a well is lost. Recently, the static fluid seal design was criticized for lack of precision,

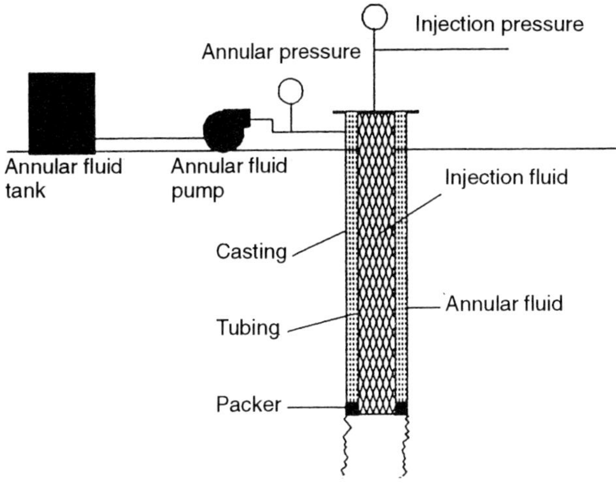

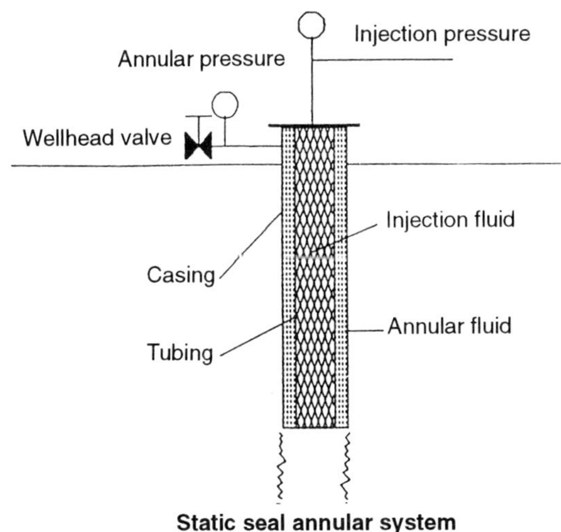

Figure 3.29 Two methods for continuous control of well integrity during subsurface injection [171].

which is caused both by slow mixing at the interface between the annular and the injected fluids and by the sensitivity of the design to injection fluid density/flow rate variations [172]. Therefore, unless the interface mixing problem is resolved (by placing a viscoelastic spacer, for example), conventional completions with packers will probably remain the accepted field practice.

Verification of the external (annular) mechanical integrity (EMI) of injection wells includes two groups of techniques: EMI tests and continuous monitoring systems. The most promising methods of EMI testing are radioactive tracer surveys [173], helium leak tests [174] and oxygen activation logging [also known as behind-casing water flow (BCWF) or neutron activation technique (NAT) [174–177]. None of these techniques, however, has been yet adopted as a single tool to demonstrate well integrity [178]. For hazardous waste injection wells, EMI is performed in a two-stage procedure using a combination of EMI tests. The first stage involves a demonstration of the absence of interzonal flow using noise, temperature or oxygen activation logs. In the second stage, the path of injected fluid as it exists in the wellbore is monitored, using the radioactive tracer survey to determine whether it is confined to the permitted injection zone. However, in the USA, for example, the use of the above procedure is not a required EMI test for oilfield brine injection but is considered the best achievable practice for oilfield injection wells [178]. In fact, the actually practiced requirements for EMI involve only reviews of cementing records; radioactive tracer surveys or temperature surveys are required infrequently [179–181].

NAT seems to be a particularly promising tool to detect flow in channels within annular seals. The wireline tool consists of the generator of neutrons and two gamma-ray detectors that are installed above and below the generator for detecting the upward and downward flow, respectively, The flowing water in the channel is irradiated with neutrons emitted by the generator. These neutrons interact with oxygen nuclei in the water to produce ^{16}N, which decays with a half-life of 7.13 s, emitting gamma radiation. Radiation energy and intensity is recorded by detectors and is used for computation of flow.

A concept of a continuous monitoring system installed in a single injection well is shown in Fig. 3.30. The suggested completion of a well with this system includes the following steps: (1) set a monitoring casing in the confining layer that overlays the injection zone and cement the monitoring casing inside the surface casing; (2) drill the well to the injection zone; (3) set a cement bridge plug and mill a short window in the monitoring casing opposite the permeable formation that is above the confining layer; (4) run the casing with a sophisticated packer (cement retainer) equipped with two (upper and lower) packing elements connected with two short tubing sections, one of which has been perforated; (5) install monitoring tubing in the annulus of the injection casing and land the monitoring tubing in the perforated section of the cement retainer; (6) cement the injection casing below and above the cement retainer; and (7) complete the well with injection tubing and a packer inside the injection casing [182].

During the injection operation, any change in pressure in the monitoring

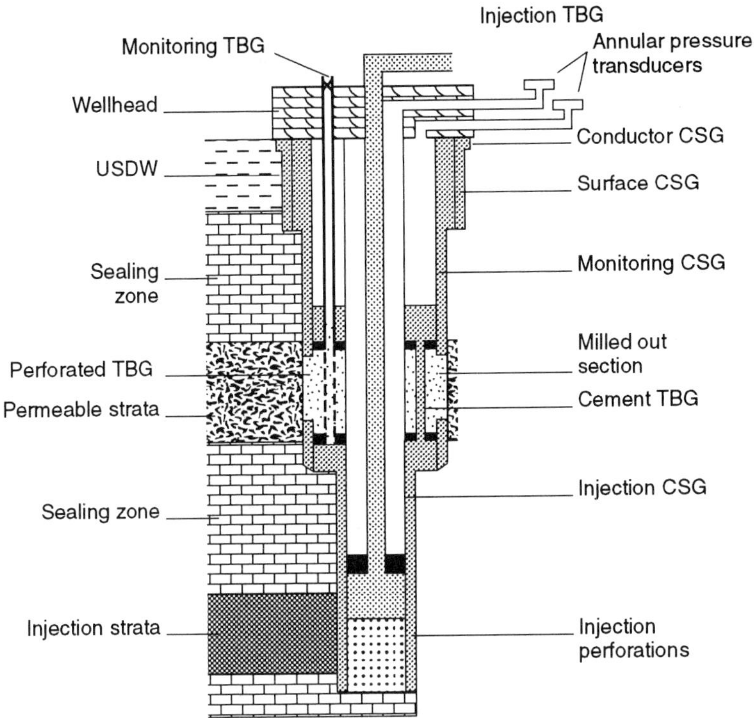

Figure 3.30 Dual completion for continuous monitoring of injection wells. (After Ref. 182.)

tubing becomes a sensitive indicator of fluid migration across the confining layer. Although theoretically sound, the system requires a complex well completion procedure, and its practical implementation still remains to be seen.

3.7.4 Field measurements of well integrity

In the early 1980s a systematic study was conducted in the USA to determine the state-of-the-art in EMI testing [183]. The first phase of the study was a survey of methods available for determining the mechanical integrity of oilfield brine injection wells. The second and third phases of the project involved experimental work using three research wells. The first two wells were used to evaluate the performance of CBL tools to detect channels in the cement sheaths behind the steel and fiberglass casings. The purpose of the third well was to evaluate the capability of various downhole tools to detect fluid movement behind the casing. The tested tools included an acoustic CBL tool, a noise logging tool and a neutron activation technique (NAT). In addition to the research well experiments, a 'real world' test was conducted in an abandoned 10 600 ft gas well using the NAT method. A known 100 ft long channel in the annular cement sheath of the well had been identified using a radioactive tracer survey.

The results of this study showed that most present commercial techniques do not provide sufficient information to determine the mechanical integrity of a well. With the acoustic CBL technique, the flow in channels behind the casing could only be detected when cement was not present. The noise logging tool showed great sensitivity to extraneous sources of sound to the extent that the log quality was low. Only the NAT method showed good detectability of flow in the annular channel when the tool was placed either in the casing or within the tubing.

In conclusion, the trend in permit regulations indicates that in the future more emphasis will be given to the verification of external integrity by test rather than by review of cementing records. NAT has great potential for testing EMI. NAT seems to be an excellent method for detecting flow in a channeled annular seal. Additional work needs to be done to determine specific applications of existing NAT tools. Also, since the cost of periodic EMI tests may be excessive, it seems possible that the oil industry can develop a new well completion system for injection wells that will permit continuous monitoring of pressures across confining zones.

3.8 Subsurface reduction of produced water

Recently, new technologies for subsurface management of produced water have rapidly developed, as shown in Fig. 3.31. These technologies represent attempts either to eliminate surface production of formation waters through injection *in situ* (bottomhole unloading, drainage injection) or to reduce the water inflow into the wellbore (water 'shut-off'), or to eliminate hydrocarbon contamination of the water by segregating inflows of petroleum and water. Several of these technologies improve the deliverability of petroleum wells and have been primarily developed as productivity measures having some environmental merit. For example, horizontal well completions are used for combating water coning problems in thin petroleum strata underlaid by strong aquifers. The environmental implication of this technology is that produced water to be disposed of is reduced. This implication has never been a main reason for the development of horizontal drilling. On the other hand, the technology of *in situ* injection of formation water has been solely developed for environmental reasons, but it also enhances well productivity by eliminating water coning (drainage-injection). Thus, in the ECT terminology each of these technologies shows both the upstream (productivity) and downstream (environmental) performances to some degree.

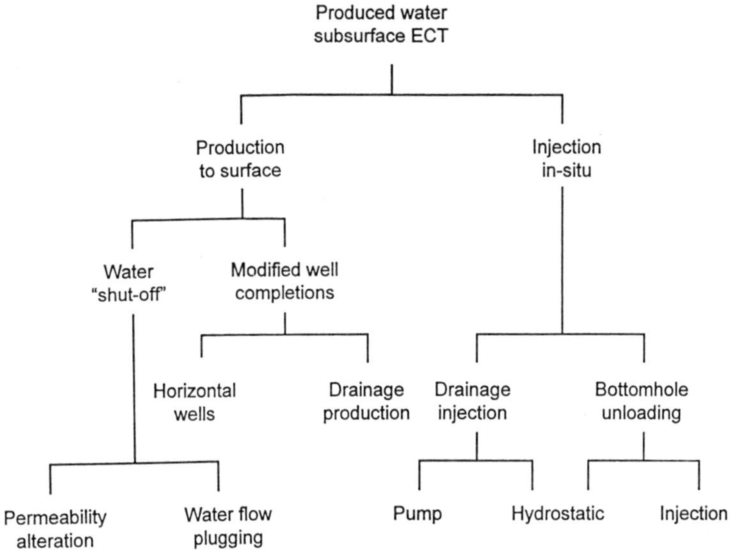

Figure 3.31 Subsurface environment-control technologies for produced water.

3.8.1 *Water shut-off technology*

Methods of water shut-off include techniques based on alteration of permeabilities or rock plugging. Alteration of relative permeability involves injection of a low-concentration polymer into the pay zone producing oil and water to create a selective near-well barrier with reduced permeability to water and unchanged permeability to oil. The selective effect has been evidenced in laboratory experiments with sandpacks [184] and rock cores [185, 186], as well as in field tests [187, 188]. The physical mechanism of this method is not very well known. Most researchers agree that the water permeability reduction is attributed to surface adsorption of the polymer and that the effect works only in small size pores. The effect is based on either selectively plugging the water-flowing pores [187] or, according to the other theory, altering the flow pattern in the two-phase flowing pores so that the annular flow of water is hindered while the central core flow of oil remains essentially unaffected [185].

The method of rock plugging is used to reduce brine flow when water and hydrocarbon flowpaths are clearly separated. This method requires selective placement of a reversible barrier into the water flowpath by injecting a gel slug. When the water flowpath consists of a system of high-conductivity fractures producing mostly water, the effect on oil production is small while the water flow is greatly reduced.

Three basic mechanisms of *in situ* gelation of the injected slug are polymer crosslinking, reversible gelation and sol stabilization. Crosslinking is accomplished with inorganic multivalent salts such as magnesium chloride, aluminum sulfate, aluminum nitrate or aluminum citrate [189]. These salts attract reactive sites on anionic polymer molecules so that they become larger and more rigid. At present, polyacrylamides crosslinked with solutions of inorganic Cr^{3+} are the most widely used gels. Their advantage stems from the ability to control the gelation time by selecting process parameters such as polymer and metal ion concentrations [190]. Also, crosslinking cationic polyacrylamide with organic crosslinking agents has recently been reported [188].

The biopolymer used in the reversible gelation process has the ability to change from the solution to the gel state by reducing the pH. This process has been proven to be reversible through an increase in pH. The proposed field procedure for this method, based upon laboratory tests [191], involves placing a biopolymer slug in the water zone and then displacing it with a solution of hydrochloric acid, which would create a barrier. To remove the barrier, an injection of sodium hydroxide would reverse the process and restore the initial permeability of the barrier zone.

The most environmentally attractive mechanisms of formation plugging, sol stabilization, is based on the gelling properties of colloidal silica suspensions. These suspensions are stable in fresh water, and their stability is sensitive to changes in pH and salinity. When destabilized, the suspensions form an impermeable gel structure. The time of destabilization and gelation can be controlled by pH and salinity changes. The field procedure involves pre-flushing the treated zone with fresh water to displace the *in situ* brine, followed by controlled on-line mixing and injection of the freshwater suspension of silica gel with a controlled volume of NaCl brine. The process has been field-tested with varying success and is considered a new alternative to polymer treatments [192].

A typical field example of successful gel treatments is shown in Table 3.22. Although the method reportedly works in the field, an actual outcome of the treatment is difficult to predict in the laboratory. Recent analysis of 57 field treatments with polymers and colloidal dispersion gels in water flood projects showed that 89% of these treatments were successful, despite laboratory predictions of a maximum 58% success rate [193]. The laboratory assessment of crosslinked gels' ability to build structure and resist shear rates in reservoir conditions was concluded to be inadequate, primarily owing to uncertainties regarding downhole flow variables. Another treatment design problem arises because the mechanism that triggers the disproportionate reduction in water permeability compared with oil permeability is poorly understood. Recent studies of this mechanism have suggested that segregation of oil and water pathways throughout a porous medium, which results in selective plugging of water flowing pores, may be a dominant effect of gel treatment [194]. The

Table 3.22 Example field performance of water 'shut-off'[a]

Well No.	Area	Oil production rate (bbl/day)		Water production rate (bbl/day)	
		Before treatment	After treatment	Before treatment	After treatment
1	Kansas	6	23	634	183
10	Kansas	7	10	384	96
4	Louisiana	33	12	440	0
7	Offshore Louisiana	30	30	720	370

[a] After Ref. 188.

conclusion was based on observations that the water-based gel reduced water permeability more than oil permeability, whereas the oil-based gel reduced oil permeability more than water permeability.

3.8.2 Downhole separation disposal: bottomhole unloading

As shown in Fig. 3.31, bottomhole unloading falls within the scope of technologies for *in situ* disposal of formation brines that do not involve producing these brines to the surface. In the bottomhole unloading technique, the process of separating water from petroleum has been moved from the surface to the bottom of the well. Moreover, the downhole separated brine is disposed of subsurface through the bottom (tail) section of the same well. Downhole separation can be accomplished either by using a gravity segregation mechanism for gas wells or by adding a liquid–liquid separator to the downhole completion installation for oil wells. Downhole brine disposal involves either hydrostatic drainage to a low-pressure disposal zone or *in situ* injection using a downhole pump and isolating packer.

Because of spontaneous and rapid separation of gas and water, these techniques were first used in producing natural gas, which requires unloading excessive water. For example, dewatering coalbed methane gas formations has been successfully applied in field operations to stimulate gas production. The Fruitland coal gas wells in the San Juan Basin in southwestern Colorado require artificial lift for dewatering. Most Fruitland wells produce 150–250 Mcf/day of gas, with flow rates improving gradually as dewatering continues. Dewatering is performed in this area using conventional plunger pumps that produce water concurrently with gas production. Similarly to the San Juan Basin, removing water from the Antrim shale gas reservoir in northern Michigan is necessary for efficient gas production in the area. Conventionally, the wells require continual dewatering to reduce the head of water. Therefore, several operators installed submersible pumps as a means of lowering the flowing pressure of the bottomhole water. In this application, water is pumped up through the tubing, and gas is produced from the annulus.

Although the concurrent water removal from the San Juan Basin coal seams and Antrim shales increases gas production rates, water pumping consumes energy, and the problem of brine disposal arises. In the Antrim shale wells a recent solution to this problem has been the waterless completion technique that employs downhole dumping of produced water to the Dundee limestone located about 1000 ft below the Antrim shale formations [195]. The completion is shown in Fig. 3.32. In this technique, the same well is used as both a production and disposal well. Its upper part, at about the gas–water contact (GWC), produces gas; the bottom part provides a conduit for the Antrim water downwards to the low-pressure

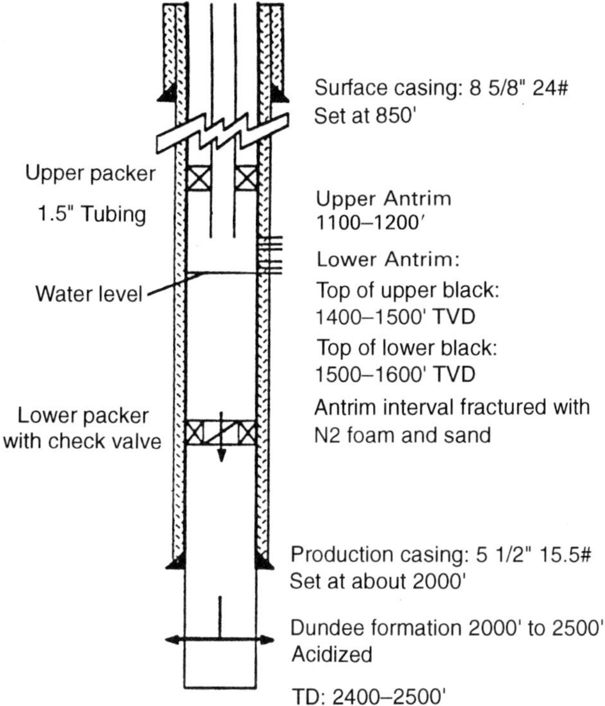

Figure 3.32 Schematic of 'waterless' completion in Antrim shales (1″ = 2.54 cm; 1′ = 0.3048 m) [195].

Dundee limestone. Since the water drainage is hydrostatic due only to the formation pressure difference between Antrim and Dundee, the water removal rate is limited and cannot be controlled. Despite this problem, the waterless completion has been successfully field tested and approved by both the US Environmental Protection Agency (EPA) and the Michigan Department of Natural Resources as a waste injection method [196, 197].

Another technique of downhole water disposal using hydrostatic drainage is to build a hydrostatic head of water inside the well to overcome the injection pressure of the disposal zone [198]. In this technique, the well is dually completed both in the gas reservoir and the deeper disposal zone. These two completions are separated by a packer. A mixed gas–water stream enters the well through the upper completion, where gravity separation takes place above the packer. The accumulated water is then picked up by a downhole pump and lifted inside the string of tubing, while the gas is produced to the surface through the tubing–casing annulus. When the hydrostatic head of water in the tubing exceeds the pressure in the disposal zone, the water flows down the tubing, bypasses the pump (through the seating nipple bypass valve) to the well section below the packer, moves to the bottom completion and then goes to the disposal zone. Because the underground injection is entirely controlled by the hydrostatic head of brine, this method can only be used in a specific geological area. Also, the disposal zone's pore pressure gradient must be substantially lower than its normal value. In addition, the permeability must be high enough to assure the minimum required injectivity index so that the water injection rate will match its inflow rate. In pilot tests conducted in southwestern Kansas and the Oklahoma panhandle, the required injection rates were from 50 to 300 bbl/day per well, with average inflow rates of 134 bbl/day of water and 105 000 scf/day of natural gas per well. One of the two reported failures of this method (out of the seven total wells tested) was attributed to the low injectivity of the disposal zone.

Development studies using mechanical downhole separators for oil and water have been reported in Canada and Norway [199, 200]. A downhole separation system developed in Canada is shown in Fig. 3.33. The system uses a dual-stream pump/hydrocyclone system to separate mechanically the produced water and oil. The bulk of the water is separated downhole

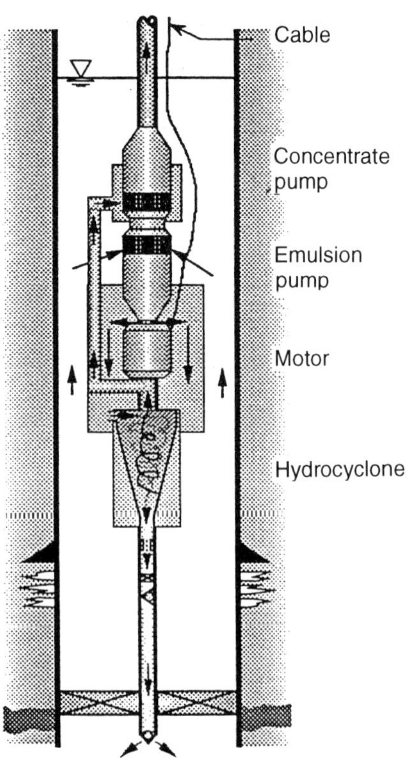

Figure 3.33 Downhole separation–disposal system [199].

(near the production zone) and reinjected into a disposal zone, while the oil-rich stream is pumped to the surface. The system includes a liquid–liquid hydrocyclone unit from Vortoil Separation Systems and standard artificial lift equipment modified to operate with the downhole separator. Systems have been tested in two separate field trials, the first with a Reda dual-stream electric submersible pump in a light crude application and the second with a progressive cavity pump in medium crude. In both cases, water production was reduced by 80–90% with no detrimental impact on oil production.

A prototype downhole separation system (DHS) developed in Norway has not yet been field tested [200]. The system is run on production tubing and temporarily connected with a polished bore receptacle to the permanent lower section of the tubing installed inside a 7 in liner string. The liner string goes all the way down through the oil reservoir, into the water disposal zone and is perforated in the oil zone. A packer at the top of the liner holds the lower tubing, while the second packer below the oil zone isolates the oil from the disposal zone below. A mixture of oil and water can enter the liner–tubing annulus and flow upwards, across and above the dual-bore top packer. Then, the mixture is segregated and oil is produced to the surface through the upper section of production tubing, while the separated water is pumped with the electrical submersible pump down the lower tubing and into the disposal zone. The DHS separation system consists of an integrated string with a bulk hydrocyclone, a dewatering hydrocyclone and a produced water hydrocyclone in series. This arrangement enables the oil to be dewatered down to 1% bottom sediment and water (BS&W) and the produced water to be deoiled down to 40 ppm.

3.8.3 In situ *water drainage technique*

The concept of *in situ* drainage of formation water (also known as the segregated production method or tailpipe water sink) draws on a hydrodynamic theory of water coning control that employs dual completion and segregated inflows of oil and water into the well. This theory postulates that, as petroleum and water are naturally segregated *in situ* in the reservoir rock, the water would not be contaminated with hydrocarbons if it was produced separately and independently from petroleum [201, 202]. The principles of the water drainage–production technique are shown in Fig. 3.34. The well is dually completed so that the lower perforations are placed in the water zone, and water can be produced both concurrently with, and independently of, oil production. These two producing streams are separated by a packer to prevent water from mixing with oil. Coning control is performed by adjusting the water production rate to the oil production rate so that the water cone does not break through the oil and enter the oil perforations. Physically, the water sink (water perforations) alters the flow

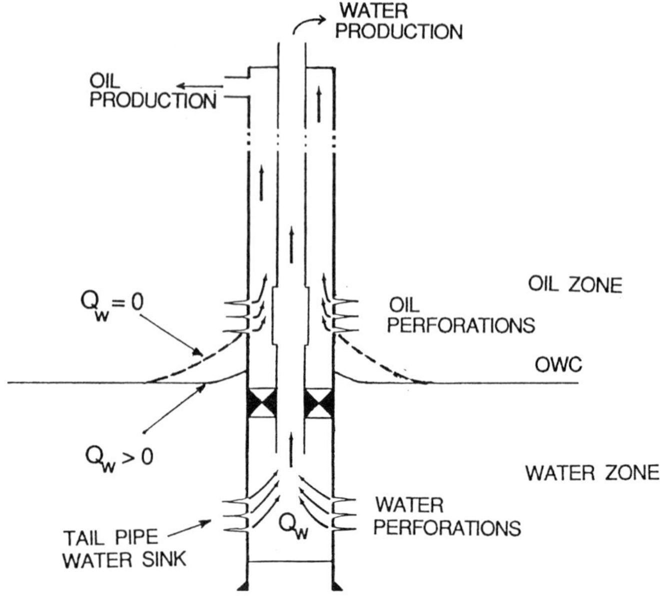

Figure 3.34 Schematics of water drainage–production system: tailpipe water sink [202].

potential field around the well so that the water cone is suppressed. Flow into the water sink generates a downward viscous force, which reduces the upwards viscous force that is generated by the flow into the upper (oil) perforations. At equilibrium, a stable water cone is 'held down' around and below the oil-producing perforations.

The segregated production method has several potential advantages:

- the oil production rate increases without water breakthrough;
- the well life extends beyond its value without coning control;
- the oil recovery per well (and for the whole reservoir) increases due to these mechanisms: (a) production can be continued with high levels of static OWC (caused by the bottom water-drive invasion), even when this level reaches the oil perforations; and (b) well productivity remains high because the near-well zone permeability to oil is not reduced by water encroachment;
- because produced water is not contaminated with crude oil, demulsifiers or other agents used in oil production, the method is more likely to meet effluent discharge limitations; and
- the water cut of the produced oil is minimized.

Theoretical simulation studies of *in-situ* water drainage revealed that, for each completion, a unique relationship exists between the oil production and water drainage rates, a performance window [203]. The window envelops the area of all possible combinations of oil and water rates that would provide stable operation of the drainage system. The window can be developed theoretically using data regarding reservoir and fluid properties in addition to well completion design. Also, the window can provide input for the economic analysis of the production project at hand.

Recently, the technique of water drainage–production has been used in the Wilcox sand oil reservoir of the Nebo Hemphil field in North Louisiana to resolve the problem of excessive water cuts experienced in conventional wells and to find out how clean the drainage water could be [42, 43]. Typically, for a conventional well in this area, a water problem would develop 60–90 days after the beginning of oil production. The excessive water cut would cause a reduction of the oil rate from 35 bbl/day initially to 12 bbl/day, with a 97% water cut.

This application was used in a new well that was drilled through the oil and water columns and dually completed in both zones. The water drainage completion (gravel packed) was isolated from the oil completion with a packer and 3½ in tubing. A downhole progressive cavity pump lifted the water in the tubing, while the formation pressure drove the water-free oil up the annulus between the tubing and 7 in casing.

The reported performance of drainage-production after 17 months of production was 57 bbl/day of almost water-free oil (0.2% BS&W), compared with 15 bbl/day using conventional completions [43]. The produced water/oil ratios were almost the same for both the new and conventional completions. However, the drainage water was free of hydrocarbons, as shown in Table 3.23 [42].

The analysis provided in Table 3.23 shows no detectable O&G contamination of the drainage water (below 2 mg/l). From a regulatory standpoint, this water would not require any clean-up to remove contamination with hydrocarbons before discharge or reuse. Also, additional analysis was made to compare concentrations of polyaromatic hydrocarbons (PAHs), the most toxic components of oil pollution in water.

Table 3.23 Hydrocarbon contamination of water produced from water drainage and conventional completions[a]

Contaminant	Conventional completion	Drainage production
Total dissolved solids (mg/l)	69 100	63 300
Oil and grease (mg/l)	484	UDL (2.0)[b]
PAHs (ppb)	592.6	ND[c]

[a]After Ref. 42.
[b]UDL (2.0) = under detection limit of 2 mg/l.
[c]ND = not detected.

Table 3.23 shows the results of the high-performance liquid chromatographic determination of PAHs in the drainage and conventional completion waters. These results clearly indicate that the drainage water is very clean relative to the conventionally produced water samples. Only 12 out of the total 55 PAHs determined this test were above the detection level of 0.005 ppb. Also, only a few of the most soluble aromatics, such as naphthalene and a few of its alkylated analogs, were detected, and these were found at very low levels. The total content of aromatics in the drainage water is approximately 11 ppb, almost one-fiftieth of that in water samples from conventional completions. The results of this test showed that, in view of present environmental regulations regarding hydrocarbon limits in produced water discharges, the drainage water does not need any treatment for hydrocarbon contamination prior to discharge.

3.8.4 In situ *water drainage disposal technique: downhole water loop*

As shown in Fig. 3.35, the technique of water drainage can be coupled in the same well with downhole injection of the drained water into a deeper disposal zone. Ideally, the disposal zone is isolated from the drainage zone both inside, with a packer, and outside the well, by an isolating stratum. Alternatively, when no outside isolating stratum exists between the disposal and drainage zones, the water will be drained from and pumped into the same aquifer, thus constituting a downhole water loop. This drainage disposal technology has not yet been implemented in petroleum wells. In gas wells, applications of this technology are often mistaken for downhole separation disposal techniques. However, the difference between the two is that separation disposal does not control water coning, whereas drainage disposal does.

For oil wells, the feasibility and design of drainage disposal systems were theoretically investigated in mathematical simulation studies [203–205]. Also, downhole installation for drainage injection was tested in the field [206]. In the field test, the pumping system was installed in an existing water flood well with one packer placed above the water drainage perforations and a second packer placed between these perforations and the injection perforations below (see Fig. 3.36). During the test, a sucker rod-driven, progressive cavity pump drained formation water from the upper water supply zone and pumped it into the injection perforations. The injection rate, measured with a downhole recording flow meter, was from 130 to 180 bbl/day at the differential pressure between the pump suction and discharge of 175 psi. The test proved that the drainage disposal system was functional. Also, the study resolved engineering problems regarding packing-off the system components inside the production casing and installing pressure gauges and a flow meter downhole. However, the test provided no information on annular isolation of the drainage and

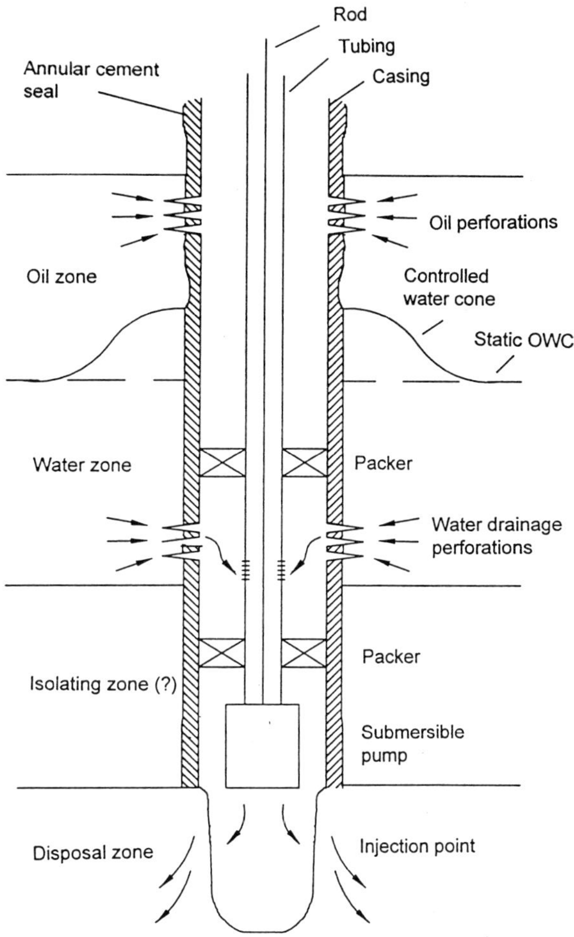

Figure 3.35 Downhole water drainage–disposal system [203].

injection zones because its objectives were limited to the installation and operation of the downhole tools inside the casing.

A single potential problem in using drainage disposal systems is hydraulic isolation of the system components. This problem is likely to be commonplace in practical applications and may be caused either by geological conditions or by installation failures. For example, the configuration of geological strata below the pay zone may lack an isolating zone between the aquifer and the water disposal strata. Also, some degree of leaking across the well's annular seal may develop as a result of the well completion operations. Therefore, actual field systems are likely to

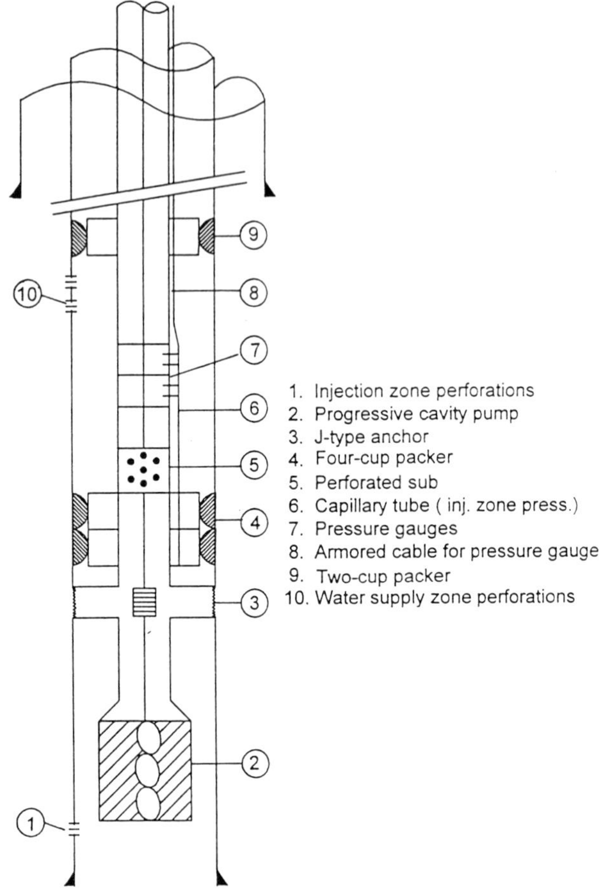

Figure 3.36 Field-tested downhole water loop [206].

operate under conditions of partial hydraulic communication between their components.

An analytical tool and computer program were developed and used to model dual well completions with downhole injection in a multilayered reservoir with crossflows and annular leaks [203, 205]. The analytical tool generates dynamic profiles of oil–water/gas–water contacts for a given geology, completions fluids and production/drainage injection rates. An example of a dynamic oil–water contact for a well with a deviated disposal section is shown in Fig. 3.37. The disposal section has been proven theoretically not to require being placed in a deviated section of a well, even for conditions of downhole water loop in the same aquifer [203]. The only requirement is to drill an adequate vertical rat hole and complete

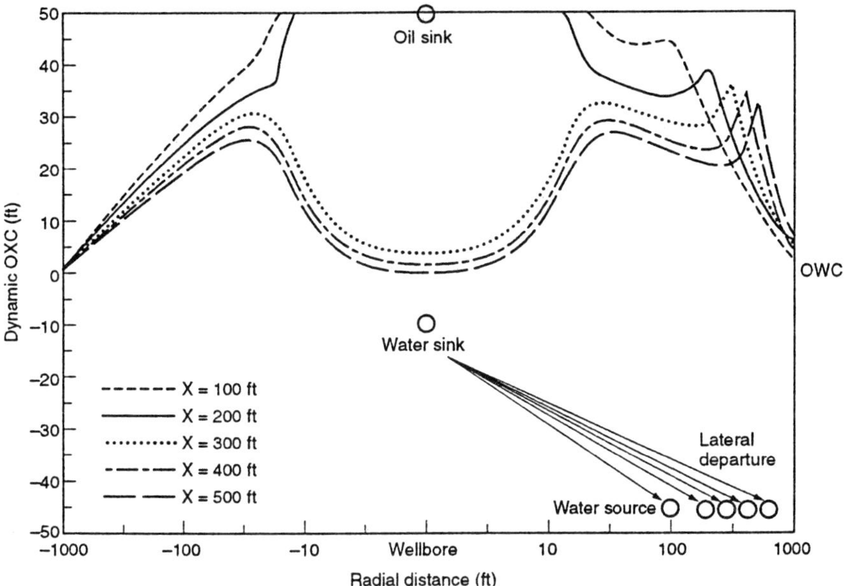

Figure 3.37 Dynamic oil–water contact (OWC) profiles for water drainage-disposal systems with deviated rat holes [201, 202].

disposal section deep enough so injection completion will have no effect upon the water cone. Drainage disposal systems have also been proven to be effectively operated with a leaking annulus outside the well.

When an annular leak develops around a well completed in isolated water zones, the amount of leaking water becomes proportional to the total water pumping rate. Therefore, a reduction in the system's performance caused by a leak depends only on the leak's conductivity. The reduced performance can be estimated using the predicted rate of leakage and the performance window plot. Thus, the performance window without the leak can be modified and used to predict the reduced performance with the leak.

3.9 Oilfield pit closure technology

On-site oilfield pits are surface impoundments usually excavated directly adjacent to the site of operation so that they can be used for temporary storage of waste generated from field operations prior to its final disposal. In the past, oilfield pits were typically used for both the temporary storage and final disposal. Such practices often resulted in surface damage due to excessive concentrations of buried hydrocarbons or permanent disposal of produced brines in pits. Modern technology of pit closure involves partial removal of waste from the pit, separation of liquids from solids and different treatment of these two phases prior to their final disposal on-site.

The petroleum industry has been using on-site pits in several different applications so the pits can be classified according to type of waste or function as follows:

- *Drilling reserve pits* are used to accumulate, store and, to a large extent, dispose of spent drilling fluids, cuttings and associated drill site wastes generated during drilling and completion operations.
- *Workover pits* typically contain workover fluids and are open only for the duration of workover operations. Workover fluids may contain total dissolved solids (TDS) in excess of 3000 ppm (approximately 4 mmho/cm conductivity) in addition to hydrocarbons or potentially toxic additives or compounds.
- *Produced water (collecting) pits* are used for storage of produced water prior to disposal to sea at a coastal (tidal) disposal facility or for storage of produced water or other oil and gas wastes prior to disposal at a fluid injection well.
- *Basic sediment pits*, also called burn pits, are used in conjunction with a tank battery for storage of basic sediment removed from a production vessel or from the bottom of an oil storage tank.
- *Blowdown/emergency pits* are used for storage of produced water for limited periods of time. They are not used for storage or disposal. Fluids diverted to emergency pits are removed as quickly as practical. After pit closure, contaminated soil should be remediated.
- *Skimming pits* are used for skimming oil off produced water prior to disposal of the water at a tidal disposal facility, disposal well or fluid injection well.
- *Percolation pits* allow liquid contents to drain or seep through the bottom and sides of the pit into surrounding soils. Percolation pits are unlined.
- *Evaporation pits*, defined as surface impoundments that are lined with clay or synthetics, are used in areas where small volumes of wastewaters are generated. Disposal of wastewater by evaporation results in the concentration of salts and residual hydrocarbons in the pit.

3.9.1 Impact of oilfield pit contaminants

Typical contaminants in oilfield pits are heavy metals, chloride salts and organics. Studies showed that soluble chloride salts and excess exchangeable sodium cause harmful effects on soil and plant growth [207, 208]. High levels of soluble salt lower the amount of water in the soil available to plants and reduce plant uptake of required nutrients [209, 210]. High levels of exchangeable sodium cause loss of soil structure, resulting in low water and air infiltration and excessive compaction of soil.

Heavy metals in soil can become incorporated and accumulated in the food chain or contaminate local sources of drinking water if leaching and migration occur from oilfield pits. Migration of metal ions from a pit site is usually limited by their attenuation in clay minerals and the formation of insoluble complexes in the soil. For drilling reserve pits, for example, researchers found little or no migration of metal ions from drilling muds because of clay attenuation and complexing [211, 212]. Attenuation and migration are affected by the type of soil; it is more extensive in porous soils than in clayey soils [209].

Incorporation of metals from oilfield pits into the food chain takes place through several possible pathways of exposure from soil to an individual. Research indicated that the exposure pathway may be different for each metal [213, 214]. In this research, a maximum soil concentration (MSC) (soil loading factor) was calculated using a so-called soil ingestion rate, i.e. the estimated amount of soil ingested by the individual per day. It was found out that of 14 possible exposure pathways for sewage sludge, four pathways have been identified as most likely to apply to oilfield pits. Maximum loading factors for 12 metals of concern in soils associated with oilfield pits are listed in Table 3.24. The table also shows the most likely exposure

Table 3.24 Maximum soil loading for oilfield pit metals[a,b]

Metal	Exposure pathway	API guidance	Louisiana 29-B[c]	Canadian agriculture	Maximum concentrations detected[d]
Arsenic	1	41	10	20	29/27.9/140
Barium[c]	1	180 000	20 000	750	56 200
			40 000		24 500
			100 000		10 700
Boron	3	2 mg/l	–	2 mg/l	290/73.6
Cadmium	4	26	10	3	14/1.5/3
Chromium	3	1500	500	750	368/145/54
Copper	3	750	–	150	82/124/210
Lead	1	300	500	375	446/302/970
Mercury	1	17	10	0.8	2.1/1.1/1.4
Molybdenum	2	–	–	5	16/9
Nickel	3	210	–	150	61/40.6/100
Selenium	1	–	10	2	3/0.6/1.4
Zinc	3	1400	500	600	823/413/400

[a]After Ref. 214.
[b]All concentrations in mg/kg unless otherwise specified.
[c]Louisiana 29-B barium values for wetlands, uplands and commercial landfarming facilities, respectively [216].
[d]Independent evaluations by American Petroleum Institute and US Environmental Protection Agency in 1987 and 1995.

pathway for each metal and its maximum concentration detected in oilfield waste.

The presence of organics in soil, typically measured as oil and grease (O&G) concentration, may severely limit revegetation efforts after oilfield pit closure (usually, the revegetation should be accomplished in one season). It has been established that, for most soils, an O&G concentration of 1% is an acceptable maximum [215, 216]. Surveys of oilfield pit content has indicated that 92.6% of the pits had organics concentrations below the soil loading level [217]. The remaining 7.4% of the pits required some dilution mixing of the waste with soil to reduce the O&G concentration to 1% by weight.

Table 3.24 gives a comparison of soil loading factors recommended by the API guidelines with those from Louisiana State Wide Order 29-B and Canadian Interim Soil Remediation Criteria for Agriculture [218]. The Louisiana 29-B criteria were developed primarily from early work on metals in sewage sludge (before 1980) (these early studies were later superseded by the research supporting the API guidelines). The Canadian Agriculture values for maximum loading have been adopted by the Canadian Council of Ministers for the Environment (CCME) from values that were currently in use in various jurisdictions across Canada. The API guidance criteria have resulted from a quantitative risk assessment, in combination with the best available data, which provided less conservative guidelines than those proposed by CCME.

3.9.2 Oilfield pit sampling and evaluation

The design of pit closure depends upon the degree of pit contamination. Oilfield pit samples must fully represent the concentration of pollutants in the pit waste material. Recent publications provide methodologies for representative sampling using grid networks and composite samples [219]. For example, sampling can be performed at the 50 ft × 50 ft (15 × 15 m) grid basis with subsamples collected over 2 ft (60 cm) intervals and the lowermost sample taken below the waste bottom. Then, at each of the sampling points (not necessarily a grid point), the subsamples are combined into a single composite for this point. Detailed testing procedures have been developed for environmental analysis of oilfield waste [216]. Particularly important in these procedures are the measurements of true total barium [220] and hot water-soluble boron [221].

Optimization of the sampling plan is an important issue because, theoretically, the cost of taking and analyzing samples at each grid point, multiplied by the number of grid points, is prohibitive. Usually, the number of sampling points can be much smaller than the number of grid points. An analytical method for determining a minimum required number of pit samples was developed using the variability of metals in the oilfield reserve pits [222].

In addition to oilfield pit content, sampling of the background soils is necessary on locations designated for pit closure by on-site land treatment. The land treatment area should be well drained and out of floodplains and wetlands. Background soil samples should be collected from the A soil horizon or upper 1 ft (30 cm), and be composited from a number of nearby locations. Details for designing and executing a soil sampling plan can be found in the relevant literature [219, 223, 224].

3.9.3 Oilfield pit closure: liquid phase

Oilfield pits are closed by segregating the liquid phase from the solid phase and disposing of each phase separately. The liquid phase can be broadly defined as an aqueous layer usually containing some suspended solids and situated above settled solids. The solid phase comprises the settled solids and significant amounts of liquids remaining in the pit after pumping the liquid phase out. Usually, the pumping continues until the remaining mixture becomes non-pumpable.

Three options for on-site disposal of the liquid phase are disposal to surface waters, land spreading or subsurface injection (annular injection or injection well). Disposal to surface waters requires dewatering the oilfield pit. The dewatering process can be accomplished *in situ* by chemical flocculation and settling or by using a portable process of chemically enhanced decanting [89, 225]. The principles of dewatering have been described earlier in this chapter. After dewatering, the pit liquid phase is practically solids free and may qualify for surface water disposal if it meets permit requirements for such disposal. An example requirement for disposal of oilfield pit liquids to surface waters is shown in Table 3.25.

If the liquid phase cannot meet requirements for surface water disposal, the only two options for disposal are subsurface injection or land spreading. The decision in this case is solely based upon electrical conductivity (EC) of pit liquids [226]. For an EC greater than 4 mmho/cm (4 Si/cm), liquids should be injected underground.

The design of land spreading of pit liquids requires calculation of the minimum land area for liquid application. Typically, water infiltration rates are used to determine the minimum required land spreading area that would not cause liquid phase run-off. Alternatively, the minimum land area can be calculated using the required values of ESP = 15% after the pit liquid phase infiltrates the soil to an assumed depth, usually 15 cm [226].

3.9.4 Oilfield pit closure: solid phase

The oldest and cheapest technique for pit closure is backfilling. This technique involves pushing the pit berm into the pit on top of waste, letting

Table 3.25 Effluent limitations (MAC) for reserve pit water discharge for Gulf of Mexico coast states[a]

Analysis[b]	Texas	Louisiana	Mississippi
pH	6–9	6–9	6–9
O&G (mg/l)	15.0	15.0	–
Chloride (mg/l)	500 (inland) 1000 (coast)	500	500
EC (μmho/cm)	–	–	1000
Total solids (mg/l)	–	–	–
TSS (mg/l)	50.0	50.0	100
TDS (mg/l)	3000	–	–
COD (mg/l)	200.0	125.0	250
TOC (mg/l)	–	–	–
Metals (mg/l):			
Arsenic	0.1	–	–
Barium	1.0	–	–
Cadmium	0.05	–	–
Chromium	0.5	0.5	0.5
Copper	0.5	–	–
Iron	–	–	–
Lead	0.5	–	–
Mercury	0.005	–	–
Nickel	1.0	–	–
Selenium	0.05	–	–
Zinc	1.0	5.0	5.0
Phenol (ppm)	–	–	0.1

[a]MAC = maximum allowable concentration for effluent discharge.
[b]COD = chemical oxygen demand; TOC = total organic carbon; TSS = total suspended solids; TDS = total dissolved solids.

pit fluids spread over the adjacent well and compacting the closure surface area. A potential environmental risk of this technique stems from the fact that waste is buried inside the pit in concentrated form, so it may become subject to leaching from periodic rainfalls. Also, hydrocarbon-contaminated waste may be buried too deep for biodegradation of organics due to insufficient supply of oxygen. At present, the method of backfilling meets regulatory approval only if the concentration of contaminants has been found to be below certain levels that render the waste harmless without dilutions [216]. Otherwise, land treatment techniques should be used for oilfield pit closure.

Land treatment technology which renders waste pit material harmless through soil incorporation employs dilution, chemical alteration and biodegradation mechanisms to reduce the concentrations of pollutants to acceptable levels consistent with intended land use [219]. The technique provides both treatment and final disposal of salts, petroleum hydrocarbons and metals. Land treatment of pit solids can be performed using techniques of land spreading, dilution burial (trenching or landfill) or solidification and burial. Laboratory analysis of waste composition must be made for

Table 3.26 Limits for oilfield pit closure and on-site disposal

Parameter (for waste material)	Units	Land treatment: Uplands (waste–soil mixtures)	Land treatment: Freshwater wetland (waste–soil mixtures)	Burial or trenching (waste–soil mixtures)	Solidification and burial (solidified material)
pH		6–9	6–9	6–9	6–12
EC (electrical conductivity)	mmho/cm	<8 mmho/cm sol. phase[a]	<4 mmho/cm sol. phase[a]	<12 mmho/cm sol. phase[a]	
SAR (sodium adsorption ratio)	ratio	<14 solution phase[a]	<12 solution phase[a]	–	–
ESP (Exchangeable sodium percentage)	%	<25% solid phase[a]	<15% solid phase[a]	–	–
CEC (cation-exchange capacity)	millieq.v/100g soil	–[b]	–[b]	–[b]	–[b]
O&G (oil and grease)		<1% by weight[a]	<1% by weight[a]	<3% by weight[a]	<10 mg/l[c]
Metals:	% dry weight				
As (arsenic)	ppm (or mg/l)	<10 ppm[a]	<10 ppm[a]	<10 ppm[a]	<0.5 mg/l[c]
Ba (barium)					<10 mg/l[c]
Elevated wetlands			<20 000 ppm[d]	<20 000 ppm[d]	
Uplands		<40 000 ppm[d]		<40 000 ppm[d]	
Cd (cadmium)		<10 ppm[a]	<10 ppm[a]	<10 ppm[a]	<0.1 mg/l[c]
Cr (chromium)		<500 ppm[a]	<500 ppm[a]	<500 ppm[a]	<0.5 mg/l[c]
Pb (lead)		<500 ppm[a]	<500 ppm[a]	<500 ppm[a]	<0.5 mg/l[c]
Hg (mercury)		<10 ppm[a]	<10 ppm[a]	<10 ppm[a]	<0.2 mg/l[c]
Se (selenium)		<10 ppm[a]	<10 ppm[a]	<10 ppm[a]	<0.1 mg/l[c]
Ag (silver)		<200 ppm[a]	<200 ppm[a]	<200 ppm[a]	<0.5 mg/l[c]
Zn (zinc)		<500 ppm[a]	<500 ppm[a]	<500 ppm[a]	<5.0 mg/l[c]
Soluble anions:					
Cl (chlorides)	ppm (or mg/l)	–	–	–	<500 mg/l[a]
Ratioisotopes:					
Coastal areas after 20 Oct. 90		–[b]	–[b]	–[b]	–[b]
Other requirements:					
Moisture content	% by weight	–	–	<50% by weight	–
Top of buried mixture	ft	–	–	<5 ft below ground level w/5 ft soil on top	<5 ft below ground level w/5 ft soil on top
Bottom of burial cell	ft	–	–	<5ft above high water table	<5 ft above high water table
Qu (unconfined compressive strength)	lb/in^2 (psi)	–	–	–	<20 psi[e]
Permeability	cm/s	–	–	–	<1 × 10^{-6}cm/s[e]
Wet/dry durability	cycles to failure	–	–	–	>10 cycles to failure

[a]Analyzed using 'standard soil' testing procedures [227].
[b]Mentioned as a parameter to analyze for, but no limitations are given.
[c]Analyzed using 'leachate' testing procedures [227].
[d]Analyzed using 'true total' testing procedures [227].
[e]Testing must be done according to ASTM.

GL = Ground level

each pit in order to evaluate levels of contamination [227]. Then, these levels are compared with their limiting values [loading factors or limiting constituents (LC)] to decide on the type of pit closure technique needed for successful land treatment design. Table 3.26 shows limiting constituents required for oilfield pit closures related to on-site disposal options in Louisiana [216].

The technique of land spreading involves addition of pit waste solids to the receiving soil, disking these solids to an appropriate depth such that the final waste–soil mixture meets the limiting constituent criteria.

The dilution burial technique involves both the mixing of soil with waste solids to reduce concentrations below LC values followed by burial of the mixture in trenches. The mixture is buried with at least 5 ft of soil cover above it and with at least 5 ft of undisturbed soil between the mixture and the highest level of groundwater table below. Management of waste in dilution burial is based on mechanisms of dilution and chemical alteration with little effect from the biodegradation mechanism due to lack of oxygen.

The technique of solidification and burial involves mixing solidifying agents, such as commercial cement, flash and lime kiln dust, with pit sediments to produce a relatively insoluble concrete matrix. Then, the solidified concrete is buried in the pit using the levee material, or in trenches using a protective liner. Solidification is a viable disposal option but is more expensive than land spreading or dilution burial. However, for highly contaminated waste and a small areas of available background soil for mixing, operators may find this option more cost effective than off-site disposal. Also, using the final solidified product the operator must demonstrate the integrity and strength of the product, as shown in Table 3.26 (compressibility, wet–dry cycling, permeability and leachate test).

References

1. Hirschhoth, J., Jackson, T. and Bass, L. (1993) Towards prevention – the emerging environmental management paradigm, in *Clean Production Strategies* (ed. T. Jackson), Stockholm Environmental Institute, Lewis, Chelsea, MI, pp. 130–3.
2. Baas, L.W. and Dieleman, H. (1990) Cleaner technology and the River Rhine: a systematic approach, in *Industrial Risk Management and Clean Technology* (eds S. Maltezou, A. Metry and W. Irwin), Verlag Orac, Vienna, p. 139.
3. Wojtanowicz, A.K. (1993) Oilfield environmental control technology: a synopsis. *Journal of Petroleum Technology*, February, 166–72.
4. Oil legacy in Louisiana. *The Wall Street Journal*, 23 October, 1984.
5. Wasicek, J.J. (1983) Federal underground control regulations and their impact on the oil industry. *Journal of Petroleum Technology*, August, 1409–11.
6. State moves to regulate wastes to hit industry with added costs. *Oil and Gas Journal*, 24 June 1985.
7. Modesitt, L.E., Jr (1987) Environmental regulations: leading the way towards restructuring the petroleum industry. *Journal of Petroleum Technology*, September, 1113–8.
8. Sullivan, J.N. (1990) Excellence and the environment. *Journal of Petroleum Technology*, February, 130–3.

9. EPA (1987) *Report to Congress: Management of Wastes from the Exploration, Development, and Production of Crude Oil, Natural Gas and Geothermal Energy – Volume 1: Oil and Gas*, US Environmental Protection Agency, Washington, DC, pp. 11.1–11.26 and 111.5–111.29.
10. API (1995) Characterization of exploration and production associated wastes. *Production Issue Group – API Report*, American Petroleum Institute.
11. Lawrence, A.W. and Miller, J.A. (1995) *A Regional Assessment of Produced Water Treatment and Disposal Practices and Research Needs*. GRI-95/0301, Gas Research Institute.
12. Schumaker, J.P. *et al.* (1991) Development of an Alaskan north slope soils database for drill cuttings reclamation. SPE 22094, *Proc. International Arctic Technology Conference*, Anchorage, AK 29–30 May, 321–2.
13. Ohara, S. and Wojtanowicz, A.K. (1995) A drilling mud management strategy using computer-aided life-cycle analysis. *Proc. Fourth International Conference on Application of Mathematics in Science and Technology*, Krakow, Poland, 20–21 June.
14. Brandon, D.M., Fillo, J.P., Morris, A.E. *et al.* (1995) Biocide and corrosion inhibition use in the oil and gas industry: effectiveness and potential environmental impacts. SPE 29735, *Proc. SPE/EPA E&P Environmental Conference*, Houston, Texas, 27–29 March, 431–44.
15. Wojtanowicz, A.K. (1991) Environmental control potential of drilling engineering: an overview of existing technologies. SPE/IADC 21954, *Proc. 1991 SPE/IADC Drilling Conference*, Amsterdam, 11–12 March, 499–516.
16. Walker, T.O. and Simpson, Y.P. (1989) Drilling mud selection for offshore operation, Part 3. *Ocean Industry*, October, 43–6.
17. Hou, J. and Luo, Z. (1986) The effect of rock cutting structure on rock breaking efficiency. SPE 14868, *SPE 1986 International Meeting on Petroleum Engineering*.
18. Hayatadavoudi, A., *et al.* (1987) Prediction of average cutting size while drilling shales. Proc. SPE/IADC Drilling Conf., New Orleans, LA, 15–18 March.
19. Fullerton, H.B., Jr (1973) *Constant-Energy Drilling System for Well Programming*, Smith Tool, Division of Smith International.
20. Mohnot, S.M. (1985) Characterization and control of fine particles involved in drilling. *Journal of Petroleum Technology*, September.
21. Chenevert, M.E. and Ossisanya, S.O. (1989) Shale/mud inhibition defined with rig-site methods. *SPE Drilling Engineering Journal*, September.
22. Ritter, A.J. and Gerant, R. (1985) New optimization drilling fluid program for reactive shale formations. SPE 14247, *Proc. 60th Annual Technical Conference and Exhibition of SPE*, Las Vegas, NV, 22–25 September.
23. Bol, G.M. (1986) The effect of various polymers and salts on borehole and cutting stability in water-base shale drilling fluids. IADC/SPE 14802, *Proc. 1986 IADC/SPE Drilling Conference*, Dallas, TX, 10–12 February.
24. Bourgoyne, A.T. *et al.* (1986) *Applied Drilling Engineering*, SPE, Richardson, TX, p. 78.
25. Lu, C.F. (1985) A new technique for evaluation of shale stability in the presence of polymeric drilling fluid. SPE 14249, *Proc. 60th Annual Technical Conference and Exhibition of SPE*, Las Vegas, NV, 22–25 September.
26. Baroid (1979) *Manual of Drilling Fluid Technology; Borehole Instability*. NL Baroid Industries, p. 4.
27. Lal, M. (1988) Economic and performance analysis models for solids control. SPE 18037, *Proc. 63rd Annual Conference and Exhibition of SPE*, Houston, TX, 2–5 October.
28. Rodt, G. (1987) Drilling contractor proposes new methods for mud solids control optimization on the rig. SPE 16527/1, *Proc. Offshore Europe 87*, Aberdeen, 8–11 September.
29. Hoberock, L.L. (1991) Fluid conductance and separation characteristics of oilfield screen cloths, in *Advances in Filtration and Separation Technology, Volume 3: Pollution Control Technology for Oil and Gas Drilling and Production Operations*, American Filtration Society, Houston, TX.
30. Wojtanowicz, A.K. *et al.* (1987) Comparison study of solid/liquid separation techniques for oil pit closure. *Journal of Petroleum Technology*, July.

31. Thoresen, K.M. and Hinds, A.A. (1983) A review of the environmental acceptability and the toxicity of diesel oil substitutes in drilling fluid systems. IADC/SPE 11401, *Proc. IADC/SPE 1983 Drilling Conference*, New Orleans, LA. 20–23 February.
32. Jackson, S.A. and Kwan, J.T. (1984) Evaluation of a centrifuge drill-cuttings disposal system with a mineral oil-based fluid on Gulf Coast offshore drilling vessels. SPE 14157, *Proc. SPE Annual Technical Conference and Exhibition*, Houston, TX, 16–19 September.
33. Bennett, R.B. (1983) New drilling fluid technology – mineral oil mud. IADC/SPE 11355, *Proc. 1983 Drilling Conference*, New Orleans, LA, 20–23 February.
34. Boyd, P.A. *et al.* (1983) New base oil used in low-toxicity oil muds. SPE 12119, *Proc. 58th Annual Technical Conference and Exhibition*, San Francisco, CA, 5–8 October.
35. Hoiland, H. *et al.* (1986) The nature of bonding of oil to drill cuttings, in *Oil-Based Drilling Fluids*, Norwegian State Pollution Control Authority, Trondheim, pp. 24–6.
36. Cline, J.T. *et al.* (1989) Wettability preferences of minerals used in oil-based drilling fluids. SPE 188476, *Proc. SPE Int. Symp. on Oilfield Chemistry*, Houston, TX, 8–10 February.
37. Candler, J. *et al.* (1992) Sources of mercury and cadmium in offshore drilling discharges. *SPE Drilling Engineering*, December, 279–83.
38. Jacobs, R.P.W.M. *et al.* (1992) The composition of produced water from Shell operated oil and gas production in the North Sea, in *Produced Water* (ed. J.P. Ray), Plenum Press, New York, pp. 13–21.
39. Reilly, W.K., O'Farrell, T. and Rubin, M.R. (1991) *Development Document for 1991 Proposed Effluent Limitation Guidelines and New Source Performance Standards for the Offshore Subcategory of the Oil and Gas Extraction Point Source Category*. US Environmental Protection Agency, Washington, DC.
40. Daly, D.J. and Mesing, G. (1993) *Gas Industry-Related Produced-Water Management Demographics*, Gas Research Institute, GRI Contract No. 5090–253–1988 and 5090–253–1930 (Draft Report).
41. Matovitch, M.A. (1978) *The Existence of Effects of Water Soluble Organic Components in Produced Brine*, Report to EPA Region IV, Shell, 26 May.
42. Swisher, M.D. and Wojtanowicz, A.K. (1995) In situ-segregated production of oil and water – a production method with environmental merit: field application. SPE 29693. *Proc. SPE/EPA Exploration & Production Environmental Conference*, Houston, TX, 27–29 March.
43. Swisher, M.D. and Wojtanowicz, A.K. (1995) New dual completion method eliminates bottom water coning. SPE 30697, *Proc. SPE Annual Technical Conference and Exhibition, Volume: Production*, Dallas, TX, 22–25 October, 549–55.
44. Otto, G.H. (1990) *Oil and Grease Discharge Characteristics of Methods 413.1 and 503E (42 Platform Study)*, Report to Offshore Operators Committee, University of Houston, Houston, TX.
45. Stephenson, M.T. (1992) A survey of produced water studies, in *Produced Water* (eds J.P. Ray and F.R. Engelhart), Plenum Press, New York, pp. 1–11.
46. Brown, J.S., Neff, J.M. and Williams, J.W. (1990) *The Chemical and Toxicological Characterization of Freon Extracts of Produced Water*. Report to Offshore Operators Committee, Battelle Memorial Institute, Duxbury, MA.
47. Ayers, R.C., Sauer, T.C. and Anderson, P.W. (1985) The generic mud concept for NPDES permitting of offshore drilling discharges. *Journal of Petroleum Technology*, March, 475.
48. Hudgins, C.M. Jr (1992) Chemical treatment and usage in offshore oil and gas production systems. *Journal of Petroleum Technology*, May, 604–11.
49. Dunn, H.E. and Beardmore, D.H. (1989) Status report: Gulf of Mexico, mud toxicity limitations. *Petroleum Engineer International*, October.
50. Clark, R.K. (1994) Impact of environmental regulations on drilling-fluid technology. *Journal of Petroleum Technology*, September, 804–9.
51. Jones, R.D. *et al.* (1983) The development and performance of a nontoxic oil mud. *Drilling*, April.
52. Clark, R.K. *et al.* (1976) Polyacrylamide/potassium chloride mud for drilling water-sensitive shales. *Journal of Petroleum Technology*, June, 719; also in *Transactions AIME*, 261.

53. Steiger, R.P. (1982) Fundamentals and use of potassium/polymer drilling fluids to minimize drilling and completion problems associated with hydratable clays. *Journal of Petroleum Technology*, August, 1661.
54. Reid, P.I. *et al.* (1993) Reduced environmental impact and improved drilling performance with water-base muds containing glycols. SPE 25989, *Proc. 1993 SPE/EPA E&P Environmental Conference*, San Antonio, TX, 7–10 March.
55. Downs, J.D. *et al.* (1993) TAMME: a new concept in water-base drilling fluids for shales. SPE 26699, *Proc. 1993 SPE Offshore European Conference*, Aberdeen, UK, 7–10 September.
56. Reid, P.I., Minton, R.C. and Twynam, A. (1992) Field evaluation of a novel inhibitive water-based drilling fluid for Tertiary shales. SPE 24979, *Proc. 1992 SPE European Petroleum Conference*, Cannes, France, 16–18 November.
57. Elsen, J.M. *et al.* (1991) Application of a lime-based drilling fluid in a high-temperature/high-pressure environment. *SPE Drilling Engineering*, March, 51.
58. Hale, A.H. and Mody, F.K. (1993) Mechanism for wellbore stabilization with lime-based muds. SPE 25706, *Proc. 1993 SPE/IADC Drilling Conference*, Amsterdam, Netherlands, 23–25 February.
59. Beihoffer, T.W. *et al.* (1992) Field testing of a cationic polymer/brine drilling fluid in the North Sea. SPE 24588, *Proc. 1992 SPE Annual Technical Conference and Exhibition*, Washington, DC, 4–7 October.
60. Hemphill, T. *et al.* (1992) Cationic drilling fluid improves ROP in reactive formations. *Oil and Gas Journal*, 8 June, 60.
61. Welch, O. and Lee, L. (1992) Cationic polymer mud solves gumbo problems in North Sea. *Oil and Gas Journal*, 13 July, 53.
62. Burba, J.L., *et al.* (1988) Laboratory and field evaluation of novel inorganic drilling fluid additive. IADC/SPE 17198, *Proc. 1988 IADC/SPE Drilling Conference*, Dallas, TX, 28 February–2 March.
63. Burba, J.L. *et al.* (1990) Field evaluations confirm superior benefits of MMLHC fluid system on hole cleaning borehole stability, and rate of penetration. IADC/SPE 19956, *Proc. 1990 IADC/SPE Drilling Conference*, Houston, TX, 27 February–2 March.
64. IDF (1990) Visplex system. *Technical Bulletin*, International Drilling Fluids, Houston, TX.
65. Simpson, J.P., Walker, T.O. and Jiang, G.Z. (1994) Environmentally acceptable water-base mud can prevent shale hydration and maintain borehole stability. SPE 27496, *Proc. 1994 IADC/SPE Drilling Conference*, Dallas, TX, 15–18 February.
66. API (1974) Drilling fluids processing equipment. *API Bulletin*, 13C.
67. Goldsmith, R. and Hare, M. (1982) Solids analysis improves mud cost control. *World Oil*, June.
68. Dearing, H.L. (1990) Material balance concepts aid in solids control and mud system evaluation. IADC/SPE 19957, *Proc. 1990 IADC/SPE Drilling Conference*, Houston, TX, 27 February–2 March.
69. Beasley, R.D. and Dear, S.F., III (1989) A process engineering approach to drilling fluids management. SPE 19532, *Proc. 64th Annual Technical Conference and Exhibition of SPE*, San Antonio, TX, 8–11 October.
70. Lal, M. (1988) Economic and performance analysis models for solids control. SPE 18037, *Proc. 63rd Annual Conference and Exhibition of SPE*, Houston, TX, 2–5 October.
71. Sharples, V.M. and Nance, G.W. (1979) New computerized model optimizes solids control. SPE 8227, *Proc. 54th Annual Conference and Exhibition of SPE/AIME*, Las Vegas, NV, 23–26 September.
72. Ardrey, W.E. *et al.* (1986) Automation of solids control system. IADC/SPE 14751, *Proc. 1986 IADC/SPE Drilling Conference*, Dallas, TX, 10–12 February.
73. Sharma, S.P. (1984) Experimental investigation of the performance characteristics of hydrocyclones. SPE 13209, *Proc. 59th Annual Technical Conference and Exhibition of SPE*, Houston, TX, 16–19 September.
74. Young, G.A. (1987) An experimental investigation of the performance of a 3-in. hydrocyclone. SPE/IADC 16175, *Proc. 1987 SPE/IADC Drilling Conference*, New Orleans, 15–18 March.

75. Hoberock, L.L. (1980) A study of vibratory screening of drilling fluids. *Journal of Petroleum Technology*, November.
76. Hoberock, L.L. (1990) Fluid conductance and separation characteristics of oilfield screen cloths. *Proc. Fall Meeting of American Filtration Society*, Baton Rouge, LA, 29–30 October.
77. Thurber, N.E. (1988) Decanting centrifuge performance study. *MS Thesis*, Univ. of Tulsa, OK.
78. Field, J. and Anderson, B. (1972) An analytical approach to removing mud solids. *Journal of Petroleum Technology*, June.
79. Patel, J. and Steinhauser, J. (1979). A material balance method to evaluate drill-fluid solids removal equipment. *Petroleum Engineer International*, March.
80. Froment, T.D. *et al.* (1986) A drilling contractor tests solids control equipment. IADC/SPE 14753, *Proc. IADC/SPE Drilling Conference*, Dallas, TX, 10–12 February.
81. Wojtanowicz, A.K. (1993) Dry drilling location: an ultimate source reduction challenge: theory, design and economics. SPE 26013, *Proc. SPE/EPA Exploration and Production Environmental Conference*, San Antonio, TX, 7–10 March.
82. Mongomery, M.S. and Love, W.W. (1993) Improve your solids control. *World Oil*, October, 59.
83. Thurber, N.E. (1992) Waste minimization for land-based drilling operations. *Journal of Petroleum Technology*, May, 542.
84. Lal, M. and Thurber, N. (1989) Drilling wastes management and closed-loop systems, in *Drilling Wastes* (eds F.R. Englehardt, J.P. Ray and A.H. Gillam), Elsevier, London, pp. 213–28.
85. Halliday, W.S. *et al.* (1993) Closed-loop operation with alternative dewatering technology. *SPE Drilling and Completion*, March, 45.
86. Wojtanowicz, A.K. (1994) Optimization of drilling mud conditioning for chemically-enhanced centifuging. *Proc. 45th Annual Technical Meeting of CIM*, Calgary, Canada, 12–15 June, 18.1–18.12.
87. Wojtanowicz, A.K. (1994) Testing protocol and economic analysis of mud dewatering for closed-loop drilling systems. *Proc. 45th Annual Technical Meeting of CIM*, Calgary, Canada, 12–15 June, 19.1–19.21.
88. Wojtanowicz, A.K. (1988) Modern solids control: a centrifuge dewatering process study. *SPE Drilling Engineering*, September, 315–23.
89. Wojtanowicz, A.K. *et al.* (1987) Comparison study of solid/liquid separation techniques for oilfield pit closures. *Journal of Petroleum Technology*, July, 845–56.
90. Malachosky, E. *et al.* (1989) The impact of the use of dewatering technology on the cost of drilling waste disposal. SPE 19528, *Proc. 64th Annual Technical Conference and Exhibition of SPE*, San Antonio, TX, 8–11 October.
91. Sanders, J.M. (1989) Minimized hauloff while drilling in a zero-discharge area. SPE 19529, *Proc. 64th Annual Technical Conference and Exhibition of SPE*, San Antonio, TX, 8–11 October.
92. Wojtanowicz, A.K. and Griffin, J.M. (1990) Drilling fluids dewatering: application survey and case history economics, in *Advances in Filtration and Separation Technology: Volume 1: Filtration and Separation in Oil and Gas Drilling and Production Operations*, American Filtration Society, Houston, TX.
93. Sirgo, M.A. Jr (1986) Regulatory changes impacting drilling and producing operations. *Journal of Petroleum Technology*, September.
94. How to choose a bioassay laboratory. *Offshore*, April 1989.
95. Ray, J.P. *et al.* (1989) Drilling fluid toxicity test: variability in US comemrcial laboratories, in *Drilling Wastes* (eds F.R. Englelhardt, J.P. Ray and A.H. Gillam), Elsevier, London, pp. 731–55.
96. Bland, R.G. *et al.* (1987) Toxicity of drilling fluids using photobacteria and Mysid shrimp bioassays. SPE 16689, *Proc. 62nd Annual Technical Conference and Exhibition of SPE*, Dallas, TX, 27–30 September.
97. API (1989) *Rapid Bioassay Procedures for Drilling Fluids*, Publication No. 4481, API.
98. Dorn, P.B. *et al.* (1990) Rapid methods for predicting compliance to drilling mud toxicity tests with Mysids for offshore NPDES general permits. *Environmental Auditor*, **1**, 221–7.

99. Hoskin, S.J. and Strohl, A.W. (1993) On-site monitoring of drilling fluids toxicity. SPE 26005, *Proc. 1993 SPE/EPA E&P Environmental Conference*, San Antonio, TX, 7–10 March.
100. Wojtanowicz, A.K. *et al.* (1991) Cumulative bioluminescence – a potential rapid test of drilling fluid toxicity: development study. SPE/IADC 21938, *Proc. 1991 SPE/IADC Drilling Conference*, Amsterdam, The Netherlands, 11–14 March.
101. Jones, F.V. and Jardiolin, R.A. (1991) Field evaluation of toxicity of drilling fluids using computer applications. *Proc. 1991 Petro-Safe Conference*, Houston, TX, 6 February.
102. Boyd, P.A. *et al.* (1983) New base oil used in low-toxicity oil muds. SPE 12119, *Proc. 58th Annual Technical Conference and Exhibition*, San Francisco, CA, 5–8 October.
103. Thoresen, K.M. and Hinds, A.A. (1983) A review of the environmental acceptability and the toxicity of diesel oil substitutes in drilling fluid systems. IADC/SPE 11401, *Proc. IADC/SPE 1983 Drilling Conference*, New Orleans, LA, 20–23 February.
104. Park, L.S. (1988) A new chrome-free lignosulfonate thinner performance without environmental concerns. *SPE Drilling Engineering*, September.
105. James, R.W. and Helland, B. (1992) The Greater Ekofisk area: addressing drilling fluid challenges with environmental justifications. SPE 25044, *Proc. 1992 SPE European Petroleum Conference*, Cannes, November.
106. Park, S., Cullum, D., and McLean, A.D. (1993) The success of synthetic-based fluids offshore Gulf of Mexico: a field comparison to conventional systems. SPE 26354, *Proc. 1993 SPE Annual Technical Conference and Exhibition*, Houston, TX, 3–6 October.
107. Hinds, A.A., Carlson, T. and Peresich, R. (1991) Industry searching for alternative to transport of drill cuttings: biodegradable esters substitute for oil in drilling performance. *Offshore*, October, 30.
108. Growcock, F.B. and Frederick, T.P. (1994) Operational limits of synthetic drilling fluids. SPE 29071, *Proc. 1994 Offshore Technology Conference*, Houston, TX, 2–5 May.
109. Halliday, W.S. and Clapper, D.K. (1989) Toxicity and performance testing of non-oil spotting fluid for differentially stuck pipe. SPE 18684, *Proc. 1989 SPE/IADC Drilling Conference*, New Orleans, LA, 28 February–3 March.
110. Clark, R.K. and Almquist, S.G. (1992) Evaluation of spotting fluids in a full-scale differential-pressure sticking apparatus. *SPE Drilling Engineering*, June, 121.
111. Jones, M. and Evangelist, R. (1986). A review of drill cuttings discharge technology and regulations. *Proc. National Conference on Drilling Muds*, EGWI, Norman, OK, 29–30 May, 188–211.
112. Davies, S. (1986) An assessment of the development of cuttings cleaning systems within the context of North Sea offshore drilling. *Conference Proceedings, Oil Based Drilling Fluids*, Norwegian State Pollution Control Authority, Trondheim, 24–26 February, pp. 71–8.
113. Kristensen, A. *et al.* (1986) Ultrasonic cleaning of oil rock particles. *Conference Proceedings, Oil Based Drilling Fluids*, Norwegian State Pollution Control Authority, Trondheim, 24–26 February, pp. 63–70.
114. Hughes creates environmentally safe mud cleaning unit (1982) *Offshore*, April, 147.
115. Warnheim, T. and Sjoblom, J. (1986) A theoretical study of surfactant and surfactant systems suitable for washing of drill cuttings contaminated by oily mud. *Conference Proceedings, Oil Based Drilling Fluids*, Norwegian State Pollution Control Authority, Trondheim, 24–26 February, 47–54.
116. Wiig, P.O. (1986) Cleaning of drilling cuttings in seawater/surfactant wash systems. *Conference Proceedings, Oil Based Drilling Fluids*, Norwegian State Pollution Control Authority, Trondheim, 24–26 February, pp. 55–62.
117. Ayers, R.C. *et al.* (1989) The EPA/API diesel pill monitoring program, in *Drilling Wastes* (eds F.R. Englelhardt, J.P. Ray and A.H. Gillam), Elsevier, London, pp. 757–72.
118. Cornitius, T. (1988) Advances in water treating solving production problems. *Offshore*, March.
119. Meldrum, N. (1988) Hydrocyclones: a solution to produced-water treatment. *SPE Production Engineering*, November, 669–76.
120. Yong, G.A. *et al.* (1991) Oil–water separation using hydrocyclones – an experimental search for optimum dimensions in environmental-control technology in petroleum

engineering. *Journal of Petroleum Science and Engineering*, Special Issue (ed. A.K. Wojtanowicz), **11**(1), April, 37–50; *Advances in Filtration and Separation Technology*, Volume 3, American Filtration Society, Gulf, pp. 102–23.

121. LADEQ (1994) *Water Quality Regulations*, 3rd edn, LAC 33:1X.708. Louisiana Department of Environmental Quality, Baton Rouge, LA.
122. Broek, W.M.G.T. and Plat, R. (1991) Characteristics and possibilities of some techniques for de-oiling of production water. SPE 23315, *Proc. First International Conference on Health and Safety and Environment*, The Hague, The Netherlands, 10–14 November, 47–54.
123. Brown and Root Inc. (1974) *Determination of Best Practicable Control Technology Currently Available to Remove Oil from Water Produced With Oil and Gas*, Report prepared for the Offshore Operations Committee, Sheen Subcommittee, New Orleans, LA.
124. EPA (1981) *Oil Content in Produced Brine on Ten Louisiana Production Platforms*, EPA-600/2-81-209. EPA, Cincinnati, OH.
125. Jackson, G.F. and Humes, E.E. (1983) Oil content in produced brine on 10 production platforms. SPE 12195, *Proc. 58th Annual Technical Conference and Exhibition*, San Francisco, CA, 5–8 October.
126. Strickland, W.T., Jr (1980) Laboratory results of cleaning produced water by gas flotation. *Society of Petroleum Engineers Journal*, June, 175–90.
127. Leech, C.A. *et al.* (1978) Performance evaluation of induced gas flotation (IGF) machine through math modeling, OTC 3342, *Proc. of the 10th Annual Offshore Technology Conference*, Houston, TX, 8–11 May, 2513–22.
128. Gay, J.C. *et al.* (1987) Rotary cyclone will improve oily water treatment and reduce space/weight requirement on offshore platforms. SPE 16571, *Proc. Offshore Europe '87*, Aberdeen, UK, 8–11 September.
129. Alfa-Laval Industry (1989) *Oily Water Treatment in Oilfield Industry*, Alfa Laval, Maarssen, The Netherlands.
130. Chen, A.S.C. *et al.* (1990) Removal of oil, grease, and suspended solids from produced water using ceramic crossflow microfiltration, in *Advances in Filtration and Separation Technology, Volume 3: Pollution Control Technology for Oil and Gas Drilling and Production Operations*, American Filtration Society, Kingwood, TX, 292–317.
131. Hagen, R.D. and Lorne, C.G. (1988) Membralox crossflow microfiltration of heavy oil produced water. *Proc. Quarterly Meeting of Canadian Heavy Oil Association*, Calgary, Alberta, 25 October.
132. Petrakis, L. and Ahner, P.F. (1976) Use of high-gradient magnetic separation techniques for the removal of oil and solids from water effluents. *IEEE Transactions on Magnetics*, **MAG-12**(5), September, 486–8.
133. Weintraub, M.H. *et al.* (1983) Development of electrolytic treatment of oily wastewater. *Environmental Progress*, **2**(1), February, 32–7.
134. Balmer, L.M. and Foulds, A.W. (1986) Separating oil from oil-in-water emulsions by electroflocculation/electroflotation. *Filtration and Separation*, November/December, 366–70.
135. Burke, R.G. (1985) Cleanup of oily water poses a challenge. *Offshore*, September, 100–6.
136. Clean water, clear choices. *Journal of Petroleum Technology*, January 1994, 24–6.
137. GRI (1995) *A Regional Assessment of Produced Water Treatment and Disposal Practices and Research Needs*. GRI-95/0301, Gas Research Institute, Chicago.
138. GRI (1995) *GRI-Pro W Calc™, Gas Research Institute – Produced Water Calculation Cost Model User's Manual*. GRI-95/0302, Gas Research Institute, Chicago.
139. McCaskill, C. (1985) Well annulus disposal of drilling waste. *Proc. National Conference on Disposal of Drilling Waste*, University of Oklahoma, Environmental and Ground-water Institute, Norman, OK, 30–31 May, 28–34.
140. Malachosky, E., Shannon, B.E. and Jackson, J.E. (1993) Offshore disposal of oil-based drilling fluid waste: an environmentally acceptable solution. SPE 23373, *Proc. First International Conference on Health Safety and Environment*, The Hague, The Netherlands, 10–14 November, 465–73; also *SPE Drilling and Completion*, December, 283–7.

141. Louviere, R.J. and Reddoch, J.A. (1993) Onsite disposal of rig-generated waste via slurrification and annular injection. SPE/IADC 25755, *Proc. 1993 SPE/IADC Drilling Conference*, Amsterdam, The Netherlands, 23–25 February, 737–51.
142. Anderson, E.E., Louviere, R.J. and Witt, D.E. (1993) Guidelines for designing safe, environmentally acceptable downhole injection operations. SPE 25964, *Proc. SPE/EPA Exploration and Production Environmental Conference*, San Antonio, TX, 7–10 March, 279–86.
143. Smith, R.I. (1991) The cuttings grinder. SPE 22092, *Proc. SPE International Arctic Technology Conference*, Anchorage, AL, 28–31 May.
144. Minton, R.C. (1992) Annular reinjection of drilling waste. SPE 25042, *Proc. European Petroleum Conference*, Cannes, France, 16–18 November, 247–53.
145. Sirevag, G. and Bale, A. (1993). An improved method for grinding and reinjection of drill cuttings. SPE/IADC 25758, *Proc. 1993 SPE/IADC Drilling Conference*, Amsterdam, The Netherlands, 23–25 February, 773–88.
146. Abou-Sayed, A.S., Andrews, D.E. and Buhidma, I.M. (1989) Evaluation of oily waste injection below the permafrost in Prudhoe Bay field. SPE 18757, *Proc. SPE California Regional Meeting*, Bakersfield, CA, 5–7 April, 129–42.
147. Bruno, M.S. *et al.* (1995) Economic disposal of solid oilfield wastes through slurry fracture injection. SPE 29646, *Proc. 1995 SPE Western Regional Meeting*, Bakersfield, CA, 8–10 March.
148. Bruno, M.S. and Qian, H.X. (1995) Economic disposal of solid oilfield wastes, *Journal of Petroleum Technology*, September, 775.
149. Willson, S.M., Rylance, M. and Last, N.C. (1993) Fracture mechanics issues relating to cuttings reinjection at shallow depth. SPE/IADC 25756, *Proc. 1993 SPE/IADC Drilling Conference*, Amsterdam, The Netherlands, 23–25 February, 753–62.
150. Kech, R.G. and Withers, R.J. (1994) A field demonstration of hydraulic fracturing for solids waste injection with real-time passive seismic monitoring. SPE 28495, *Proc. SPE 69th Annual Technical Conference and Exhibition*, New Orleans, LA, 28–28 September, 319–34.
151. Minton, R.C., Meader, A. and Willson, S.M. (1992) Downhole cuttings injection allows use of oil-base muds. *World Oil*, 47–52.
152. Moschovidis, Z.A. *et al.* (1994) Disposal of oily cuttings by downhole periodic fracturing injections, Valhall, North Sea: case study and modelling concepts. *SPE Drilling and Completion*, December, 256–62.
153. Carter, L.G. and Evans, G.W. (1964) A study of cement–pipe bonding. *Journal of Petroleum Technology*, February.
154. Becker, H. and Peterson, G. (1963) Bond of cement composition for cementing wells. *Proceedings of Sixth World Petroleum Congress*, Frankfurt, 19–26 June.
155. Smith, K.D. (1976) *Cementing*. SPE Monograph, Vol. 4, Society of Petroleum Engineers, New York, Dallas.
156. Hartog, J.J. *et al.* (1983) An integrated approach for successful primary cementation. *Journal of Petroleum Technology*, September.
157. Stewart, R.B. (1986) Gas invasion and migration in cemented annuli: causes and cures. IADC/SPE 14779, *Proc. 1986 Drilling Conference*, Dallas, TX, 10–12 February.
158. Cheung, P.R. (1982) Gas flow in cements. SPE 11207, *Proc. 57th Annual Fall Technical Conference and Exhibition of SPE*, New Orleans, LA, 26–29 September.
159. Webster, W.W. and Eikerts, J.V. (1979) Flow after cementing – a field and laboratory study. SPE 8259, *Proc. 54th Annual Technical Conference and Exhibition*, Las Vegas, NV, 23–26 September.
160. Sutton, D.L. and Sabins, F.L. (1990) Interrelationship between critical cement properties and volume changes during cement setting. SPE 20451, *Proc. 65th Annual Technical Conference and Exhibition of SPE*, New Orleans, LA, 23–26 September.
161. Sutton, D.L. and Ravi, K.M. (1989) New method for determining downhole properties that affect gas migration and annular sealing. SPE 19520, *Proc. 64th Annual Technical Conference and Exhibition of SPE*, San Antonio, TX, 8–11 October.
162. Chenevert, M.E. and Shrestha, B. (1987) Shrinkage properties of cement. SPE 16654, *Proc. 62nd Annual Technical Conference and Exhibition of SPE*, Dallas, TX, 27–30 September.

163. Sabins, F.L. (1990) An investigation of factors contributing to the deposition of cement sheaths in casing under highly deviated well conditions. IADC/SPE 29934, *Proc. 1990 IADC/SPE Drilling Conference*, Houston, TX, 27 February–2 March.
164. Cooke, C.J. Jr *et al.* (1982) Field measurements of annular pressure and temperature during primary cementing. SPE 11206, *Proc. 57th Annual Fall Technical Conference and Exhibition of SPE*, New Orleans, LA, 26–29 September.
165. Sykes, R.L. and Logan, J.L. (1987) New technology in gas migration control. SPE 16653, *Proc. 62nd Annual Technical Conference and Exhibition of SPE*, Dallas, TX, 27–30 September.
166. Beirute, R.M. and Cheung, P.R. (1989) A scale-down laboratory test procedure for tailoring to specific well conditions: the selection of cement recipes to control formation fluids migration after cementing. SPE 19522, *Proc. 64th Annual Technical Conference and Exhibition of SPE*, San Antonio, TX, 8–11 October.
167. Harris, K.L. *et al.* (1990) Verification of slurry response number evaluation method for gas migration control. SPE 20450, *Proc. 65th Annual Technical Conference and Exhibition of SPE*, New Orleans, LA, 23–26 September.
168. Thornhill, J.P. and Benefield, B.G. (1986) Mechanical integrity research. *Proc. International Symposium on Subsurface Injection of Liquid Wastes*, New Orleans, LA, 3–5 March.
169. Kamath, K.I. (1989) Testing injection wells for casing leaks: theory, practice and regulatory requirements. *Proc. International Symposium on Class I & II Injection Well Technology*, Underground Injection Practices Council, Dallas, TX, 8–11 May, 505–19.
170. Langlinais, J.L. (1981) *Waste Disposal Well Integrity Testing and Formation Pressure Buildup Study*. Report for Louisiana Department of Natural Resources, September.
171. McKay, J.E. *et al.* (1989) Static fluid sealed injection wells – an acceptable alternative. *Proc. International Symposium on Class I & II Injection Well Technology*, Underground Injection Practices Council, Dallas, TX, 8–11 May, 141–9.
172. Kamath, K.I. (1988) Significance of regulatory constraints on the operation of packerless waste injection well. *Journal of Petroleum Technology*, November, 1501–5.
173. UIPC (1986) Radioactive tracer surveys. *Mechanical Integrity Testing Seminar*, Underground Injection Practices Council, Oklahoma City, OK, 113–57.
174. Dewan, J.P. (1983) *Mechanical Integrity Tests – Class II Wells: Review and Recommendations*, Report for Environmental Protection Agency Regions II and III.
175. Williams, T.M. (1987) Measuring behind casing water flow. *Proc. International Symposium on Subsurface Injection of Oilfield Brines*. Underground Injection Practices Council, New Orleans, LA, 4–6 May, 467–84.
176. DeRossett, W.H. (1986) Examples of detection of water flow by oxygen activation on pulsed neutron logs. *Proc. SPWLA 27th Annual Logging Symposium*, 9–13 June.
177. Wieseneck, J.B. (1989) Practical experience with oxygen activation logging in South Mississippi. *Proc. International Symposium on Class I and II Injection Well Technology*, Underground Injection Practices Council, Dallas, TX, 8–11 May, 7–24.
178. Lyle, R. (1989) Demonstration of mechanical integrity utilizing radioactive tracer surveys. *Proc. International Symposium on Class II Injection Well Technology*, Underground Injection Practices Council, Dallas, TX, 8–11 May, 65–73.
179. UIPC (1989) *The Texas Railroad Commission, Oil and Gas Division UIC Program: a Peer Review*, Report of Underground Injection Practices Council, 21–7.
180. UIPC (1989) *The Louisiana Department of Natural Resources, Office of Conservation, Injection and Mining Division UIC Program: a Peer Review*, Report of Underground Injection Practices Council, 19–22.
181. Gould, L.A. and Landry, D.A. (1987) Status of mechanical integrity testing in Mississippi. *Proc. International Symposium on Subsurface Injection of Oilfield Brines*, Underground Injection Practices Council, New Orleans, LA, 4–6 May, 435–7.
182. Poimboeuf, W.W. (1989) Injection well with monitoring system installed in a single wellbore. *Proc. International Symposium on Class I and II Injection Well Technology*, Underground Injection Practices Council, Dallas, TX, 8–11 May, 461–88.
183. Tornhill, J.T. and Kerr, R.S. (1990) *Injection Well Mechanical Integrity*. EPA/625/9–89/007, US Environmental Protection Agency, Washington, DC.
184. Sparlin, D.D. (1976) An evaluation of polyacrylamides for reducing water production. *Journal of Petroleum Technology*, August, 906–13.

185. Zaitoun, A. and Kohler, N. (1989) Improved polyacrylamide treatments for water control in producing wells. SPE 18501, *Proc. SPE International Symposium on Oilfield Chemistry*, Houston, TX., 8–10 February, 379–88.
186. Kohler, N. and Zaitoun, A. (1991) Polymer treatment for water control in high-temperature production wells. SPE 21000, *Proc. SPE International Symposium on Oilfield Chemistry*, Anaheim, CA, 20–22, February, 37–48.
187. White, J.L. *et al.* (1973) Use of polymers to control water production in oil wells. *Journal of Petroleum Technology*, February, 143–50.
188. Avery, M.R. (1988) Field evaluation of a new gelant for water control in production wells. SPE 18201, *Proc. 63rd Annual Technical Conference and Exhibition of the Society of Petroleum Engineers*, Houston, TX, 2–5 October, pp. 203–14.
189. Needham, R.B. *et al.* (1974) Control of water mobility using polymers and multivalent cations. SPE 4747, *Proc. Third Symposium on Improved Oil Recovery*, Tulsa, OK, 22–24 April.
190. Terry, R.E. *et al.* (1981) Correlation of gelation times for polymer solutions used on mobility control agents. *Society of Petroleum Engineers Journal*, April, 229–35.
191. Vossoughi, S. and Putz, A. (1991) Reversible in-situ gelation by the change of pH within the rock. SPE 20997, *Proc. SPE International Symposium on Oilfield Chemistry*, Anaheim, CA, 20–22 February, 7–12.
192. Jurinak, J.J. *et al.* (1989) Oilfield application of colloidal silica gel. SPE 18505, *Proc. SPE International Symposium on Oilfield Chemistry*, Houston, TX, 8–10 February, 425–54.
193. Smith, J.E. (1994) Closing the lab-field gap: a look at near-wellbore flow regimes and performance of 57 field projects. SPE 27774, *Proc. SPE/DOE Ninth Symposium on Improved Oil Recovery*, Tulsa, OK, 17–20 April, 445–59.
194. Liang, L., Sun, H. and Seright, R.S. (1995) *Why Do Gels Reduce Water Pearmability More Than Oil Permeability*? SPE 27829 (peer reviewed).
195. Lang, K.R., Oliver, S.J. and Scheper, R.J. (1991) Optimization of gas production in the Antrim shale. SPE 21694, *Proc. SPE Production Operation Symposium*, Oklahoma City, OK, 7–9 April.
196. EPA reclassifies waterless Antrim wells. Texas inventor holds patent on use of pump. *Michigan's Oil and Gas News*, **96**(45), 9 November, 1990.
197. Lintemuth, R. (1990) Antrim play proves strong catalyst in boosting Michigan drilling permit totals. *Michigan's Oil and Gas News*, **96**(51–52), 21 December.
198. Grubb, A.D. and Duball, D.K. (1992) Disposal tool technology extends gas well life and enhances profits. SPE 24796, *Proc. 67th Annual Technical Conference and Exhibition of SPE*, Washington, DC, 309–16.
199. Bell, S. (1995) Downhole separation technology tested. *Hart's Show Daily, Suppl. to Petroleum Engineer International*, 23 October, 3.
200. Kjos, T. *et al.* (1995) Downhole water–oil separation and water reinjection through well branches. SPE 030518, *Proc. 1995 SPE Annual Technical Conference and Exhibition, Volume: Drilling*, Dallas, TX, 22–25 October, 689–701.
201. Wojtanowicz, A.K. *et al.* (1991) Oil well coning control using dual completion and tailpipe winter sink. SPE 21654, *Proc. SPE Production Operations Symposium*, Oklahoma City, OK, 7–9 April.
202. Wojtanowicz, A.K., Xu, H. and Bassiouni, Z. (1994) Segregated production method for oil wells with active water coning. *Journal of Petroleum Science and Engineering, Special Issue: Environmental Control Technology in Petroleum Engineering*, **11**(1), April.
203. Wojtanowicz, A.K. and Shirman, E.I. (1996) An in-situ method for downhole drainage-injection of formation brine in a single oil producing well, in *Deep Injection Disposal of Hazardous and Industrial Waste* (eds. J.A. Apps and C. Tsang), Academic Press, New York, pp. 403–20.
204. Wojtanowicz, A.K. and Xu, H. (1992) A new in-situ method to minimize oilwell production watercut using downhole water loop. CIM 92-13, *Proc. 43rd Annual Technical Meeting of CIM*, Calgary, Canada, 7–10 June.
205. Wojtanowicz, A.K. and Xu, H. (1995) Downhole water loop – a new completion method to minimize oilwell production watercut in bottom-water-drive reservoirs. *Journal of Canadian Petroleum Technology*, **3**(8), October, 56–62.
206. Klein, S.T. and Thompson, S. (1992) Field study: utilizing a progressive cavity pump for

a closed-loop downhole injection system. SPE 24795, *Proc. 67th Annual Technical Conference of SPE*, Washington, DC 301–8.
207. Moseley, H.R. (1983) Summary and analysis of API onshore drilling mud and produced water environmental studies. *American Petroleum Institute Bulletin D19.*
208. Miller, R.W. and Honarvar, S. (1975) Effect of drilling fluid component mixtures on plants and soil. EPA 560/1–75–004, *Conf. Proc. Environmental Aspects of Chemical Use in Well Drilling Operations*, Houston, TX, May, 125–43.
209. Murphy, E.C. and Kehew, A.E. (1984) The effect of oil and gas well drilling fluids on shallow groundwater in Western North Dakota. *Report No. 82*, North Dakota Geological Survey.
210. Miller, R.W., Honarvar, S. and Hunsaker, B. (1980) Effects of drilling fluids on soils and plants: 1. Individual fluid components. *Journal of Environmental Quality*, 547–52.
211. Whitmore, J. (1982) Water base drilling mud land spreading and use as a site reclamation and revegetation medium. *Fosgren-Perkins Engineering Report*, API Production Department, Dallas, TX.
212. Henderson, G. (1982) Analysis of hydrologic and environmental effects of drilling mud pits and produced water impoundments. *Dames and Moore Report*, API Production Department, Dallas, TX.
213. EPA (1993) *Standards for the Use or Disposal of Sewage Sludge*, 40 Code Federal Register, Parts 257, 403 and 503, US Environmental Protection Agency, Washington, DC.
214. API (1995) *Metals Criteria for Land Management of Exploration and Production Wastes: Technical Support Document for API Recommended Guidance Values.* Publication 4600, American Petroleum Institute, Washington, DC.
215. Pal. D and Overcash, M.R. (1978) Plant oil assimilation capacity for oils. *Proc. 85th National Meeting of American Institute of Chemical Engineers*, Philadelphia, PA.
216. LADNR (1990) *Amendment to Statewide Order No. 29-B*, Office of Conservation, Louisiana Department of Natural Resources, Baton Rouge, LA.
217. Freeman, B.D. and Deuel, L.E. (1984) Guidelines for closing drilling waste fluid pits in wetland and upland areas. *Proc. 7th Annual Energy Source Technology Conference and Exhibition*, Petroleum Division of ASME, New Orleans, 12–16 February.
218. API (1995) *Criteria for pH in Onshore Solid Waste Management in Exploration and Production Operations.* Publication 4595, American Petroleum Institute, Washington, DC.
219. Deuel, L.E. Jr and Holliday, G.E. (1994) *Soil Remediation for the Petroleum Extraction Industry*, Pennwell, Tulsa, OK.
220. Deuel, L.E., Jr and Freeman, B.D. (1989) Amendment to Louisiana Statewide Order 29–B: suggested modifications for barium criteria. *Proc. SPE/IADC Drilling Conference*, New Orleans, LA, 28 February–3 March, 481–6.
221. Carter, M.R. (1993) *Soil Sampling and Methods of Analysis*, Lewis, Boca Raton, FL. 91–3.
222. Wojtanowicz, A.K., Field, S.D., Krilov, Z. and Spencer, F.L. Jr (1989) Statistical assessment and sampling of drilling fluid reserve pits. *SPE Drilling Engineering*, June.
223. Peterson, R.G. and Calvin, L.D. (1986) Sampling, in *Methods of Soil Analysis, Part 1: Physical and Mineralogical Methods* (eds A. Klute *et al.*), American Society of Agronomy Monograph, **9** (1), 33–50.
224. Crepin, J. and Johnson, R.L. (1993) Soil sampling for environmental assessment, in *Soil Sampling and Methods for Analysis* (ed. M.R. Carter), Canadian Society of Soil Science, Lewis, Ann Arbor, MI.
225. Osterman, M.C., Wojtanowicz, A.K. and Field, S.D. (1987) The development study of a centrifuging technique for oilfield production pit closure. *Transactions of the ASME, Energy Resources Technology*, December.
226. API (1993) *Evaluation of Limiting Constituents Suggested for the Land Disposal of Exploration and Production Wastes*, Publication 4527, American Petroleum Institute, Washington, DC.
227. LADNR (1989) *Laboratory Procedures for Analysis of Oilfield Waste*, Office of Conservation, Louisiana Department of Natural Resources, Baton Rouge, LA.

4 Drilling and production discharges in the marine environment

A.B. DOYLE, F.V. JONES, S.S.R. PAPPWORTH
and D. CAUDLE

4.1 Introduction

Drilling and production discharges to the marine environment present different environmental concerns to those in onshore areas. Determining the potential environmental impacts in this area is a complex task. Such potential impacts have historically not been well understood and this has led to various misperceptions of the importance of these impacts. There are three areas that cause environmental concerns about offshore oil and gas operations: those occurring because of the physical presence of the facilities, those occurring as a result of the waste discharges from facilities and those occurring as a result of accidental discharges. These concerns may also be heightened by the sensitivity of the particular environment. Areas containing breeding grounds for fish and other sea life or adjacent to salt marshes or mangroves or areas containing sea floor communities such as coral are thought to be particularly vulnerable.

Regulatory control of the offshore oil industry operations is under a variety of agencies, international, regional and national. In addition, in some areas such as the USA some portion of the offshore is regulated by the states or other local bodies. Although the objectives of these regulatory bodies are often similar, for example protection of the marine environment, frequently specific concerns differ from region to region and regulatory approaches vary. For example, the USA uses end-of-pipe controls and the countries around the North Sea tend to control the processes involved and limit the materials that can be used in processes that produce waste discharges. These variations have arisen for many different reasons, including the attitude and perceptions of governments and citizenry toward business, past experiences with large pollution events, competition with other users of the marine environment, impact of oil and gas operations on the local economy and the sensitivity of the local environment.

The type of regulatory scheme that the platform is governed by its dependent on the location of the platform and the method(s) of waste disposal. The location, for example near shore or outer Continental Shelf (OCS), can determine whose jurisdiction the platform is under. Disposal method(s) include discharge to the environment and/or hauling the waste back to shore for onshore disposal. Some countries, such as the USA, regulate the end-of-pipe discharge, whereas other countries, such as the UK, require materials to be pre-approved prior to usage. Still other countries, such as Venezuela, issue permits for entire projects.

Marine environments differ from place to place. Some of the factors that distinguish them are water depth, temperature, type of bottom, distance from shore or wetlands, the biological communities supported and significant bottom features such as coral reefs and sea grasses. In addition, shore lines bordering marine areas may be affected by oil and gas operations in the marine environment. People living near the shore might also perceive an effect on the alteration of the view.

Drilling and activities associated with the production of crude oil and natural gas typically produce three distinct waste streams: drilling waste, produced water and associated wastes. Drilling waste includes drilling fluids or muds and cuttings. Produced water is the aqueous phase separated from the recovered hydrocarbons produced during oil and gas operations and, besides water, can include chemicals used to enhance or perform production work, gas processing, stimulation and/or workovers. Associated wastes include such materials as tank bottoms, sludges, contaminated soil, well treatment fluids, produced sand, spent filters, used oil, unused and damaged chemicals and a wide variety of other wastes. Usually associated wastes are transported to shore for disposal. This chapter will focus predominantly on drilling waste and produced water, as these are the wastes which typically are discharged to the offshore environment.

As the term indicates, drilling waste is generated primarily during the drilling of the wells and is composed of both cuttings which retain a thin film of the drilling fluids on them and, in the case of water-based muds, excess mud. The volume of drilling waste is greatest in the early life of an operation, and drops off as development is completed. The produced water, as the name implies, is water that is produced with the oil and/or gas. Over the life of a field the volumes and composition of the produced water vary greatly. Accidental discharges are discharges that are not permitted or planned. These discharges are usually shorter in duration, but can produce a more intense impact.

4.2 Nature of offshore discharges

4.2.1 Introduction

The API estimated that in 1985 in the USA the oil and gas industry (both onshore and offshore) generated 361 million barrels of drilling wastes

(1.5% of total waste) and 20.9 billion barrels of produced water (98%). Another 118 million barrels of associated wastes (0.5%) were generated (Freeman and Wakin, 1988). From this, it is clear that the majority of waste generated by oil and gas operations is in the form of produced water. The E&P forum have estimated that in 1991, oil and gas platforms in the North Sea discharged 160 million m^3 (~1 billion barrels) of produced water, with about 5% of the total volume coming from gas platforms (E&P Forum, 1994).

4.2.2 Drilling waste

Drilling wastes are composed of drilling fluids and cuttings produced during the drilling of oil and gas wells. Drilling fluids, also referred to as muds, are suspensions of solids and other materials in a liquid base. Three of the primary functions that drilling muds perform during drilling are lubricating and cooling the drilling bit, maintaining downhole hydrostatic pressure and cleaning out the hole by bringing cuttings to the surface. Cuttings are small pieces of formation rock that are generated by the crushing action of the drill bit. Drill cuttings are carried out of the borehole by the drilling fluids. Drill cuttings themselves are inert solids from the formation. However, drill cuttings discharges also contain drilling fluids that adhere to the cuttings. The volume of the mud that adheres to the cuttings can vary considerably depending on the formation being drilled and the cutting's particle size distribution (EPA, 1993a). A general rule of thumb is that 5% mud, by volume, is associated with the cuttings (Ray, 1979).

Based on the definition and nature of the drilling fluids and cuttings, the material is generated only while the well is being drilled and for a very short time after completion of the well. The volume of material peaks very quickly while the well is drilling and then drops off once the well is completed. For water-based muds, drilling discharges can be of two types: excess mud and drill cuttings coated with mud. Excess mud typically is discharged in discrete events intermittently over the life of the well and at completion of the well when mud tanks may be emptied. Cuttings are discharged more or less continuously while the hole is being made; however, many drilling operations, such as running pipe in and out of the hole, do not result in the generation of cuttings. For non-water-based muds (oil- and synthetic-based muds), bulk mud is not discharged and the pattern for generation of cuttings is similar to that of water-based mud. The greatest volume of drill cuttings is generated during the initial stages of drilling when the borehole diameter is greatest.

Depending on the formations and the drilling characteristics, the total volume of solids generated will vary. The volume generated can be several times more than the borehole volume due to excessive caving of formations into the borehole, equal to the borehole volume, or be up to 50% less than the borehole volume, owing to the dispersion of very small cuttings in the drilling fluid.

Drilling fluids fall into one of three classes: water-based muds, oil-based muds and synthetic muds. The water in water-based muds can be fresh water or salt water. Bulking components and other chemical additives are added to mud systems to create the desired characteristics. The types and compositions of drilling fluid may vary during the drilling.

A water-based drilling fluid or mud is the conventional drilling fluid in which water is the continuous phase and the suspending medium for solids, whether or not oil is present (EPA, 1993a). Water-based drilling muds are relatively inexpensive. They are generally non-toxic. Discharged cuttings will disperse in the water column.

An oil-based drilling fluid or mud has diesel, mineral or some other oil as its continuous phase with water as the dispersed phase (EPA, 1993a). Oil-based muds are currently used only for specific drilling conditions because of limitations on discharge. They are more expensive to use than water-based muds. Industry has indicated that oil-based muds continue to be the drilling fluid of choice for certain drilling conditions. Conditions warranting the use of oil-based muds include required thermal stability when drilling high-temperature wells, require specific lubricating characteristics when drilling deviated wells, the ability to reduce stuck pipe or hole wash-out problems when drilling thick, water-sensitive formations and drilling through water-soluble formations such as salt.

A primary concern when using conventional, oil-based mud systems is their potential for adverse environmental impact as a result of either a permitted or accidental discharge. Because of the relatively high toxicity of diesel oil used in early oil-based systems, they were replaced by mineral oil-based mud systems.

Restrictions on oil-based (both diesel and mineral) mud use in the North Sea led to the development of synthetic mud bases (Friedheim, 1994). Synthetic mud-based compositions range from saturated (non-aromatic) hydrocarbons to more complex molecules such as acetals, ethers and esters (Candler, Rushing and Leuterman, 1993). Synthetic-based muds are less toxic than oil-based muds (Candler, Rushing and Leuterman, 1993).

The use of synthetic mud bases has become of interest in the USA after the present US New Source Performance Standards (NSPS) guidelines were published (Park, Cullum and McLean, 1993; Burke and Veil, 1995). Since the US regulations did not specifically address limitations for discharges from the use of synthetic muds, some uncertainty has developed around the use of these materials and the subsequent discharge of waste. These uncertainties are over such issues as 'are synthetic materials oil?' and 'how can the EPA tell if operators have added diesel or mineral oil to their synthetic mud?'

Since synthetic mud bases are water insoluble and also low in toxicity, none of them has a significant potential for toxic impacts on the marine environment. Synthetic mud bases have no potential to cause a priority pollutant impact on the environment, since they do not contain priority pollutants. The potential for a bioaccumulation impact from synthetic mud bases has been found to be very low. Any organic material discharged to the sea will impart oxygen demand. All synthetic mud bases share this characteristic with diesels and mineral oils.

4.2.3 *Produced water*

Produced water is the water generated from the oil and gas extraction process. Produced water includes the formation water brought to the surface with the oil and gas, the injection water used for secondary recovery that has broken through the formation and various well treatment solutions and chemicals added during production and the oil–water separation process.

Formation water, which comprises the bulk of the produced water, is found in the same rock formation as the crude oil and gas or an adjoining level of the same formation (i.e. below the oil/gas cap). Formation water is classified as meteoric, connate or mixed. Meteoric water comes from rain water that percolates through bedding planes and permeable layers. Connate water (sea water in which marine sediments were originally deposited) contains chlorides, mainly sodium chloride (NaCl), and dissolved solids in concentrations many times greater than common sea water. Mixed water is characterized by both a high chloride and sulfate–carbonate–hydrogencarbonate content, which suggests multiple origins.

Produced water is the highest volume waste source in the offshore oil and gas industry. Produced water can constitute from 2 to 98% of the gross fluid production at a given platform. In general, produced water volume is small during the initial production phase when hydrocarbon production is the greatest, and increases as the formation approaches hydrocarbon depletion.

Produced water volumes are much greater for structures producing oil or a combination of both oil and gas as compared with gas-only platforms. Although the gas-only platforms generate less produced water, the concentration of the chemical constituents of the water is considerably higher than those from oil-producing platforms (op ten Noort *et al.*, 1994). The volume of produced water at a given platform is a site-specific phenomenon. In some instances, no formation water is encountered whereas in others an excessive amount of formation water is encountered at the start of production.

According to Walk, Haydel and Associates (1984), the average produced water discharge rate from an offshore platform off the USA is usually less than 1800 barrels per day (bpd), whereas discharges from large treatment facilities handling water from many platforms may be as high as 157 000 bpd.

Disposal options for offshore produced water is basically either discharge to the sea or reinjection. Prior to these disposal methods the water may need to be treated to remove oil. Owing to the changing volumes and compositions, designing treatment systems for constantly changing production rates presents a challenge and results in many of the problems encountered in such treatment operations.

4.2.4 Accidental discharges

The accidental discharges which are covered by the media and brought to the attention of the public are large tanker spills, and not production facility spills. However, production facilities can have spills. During drilling, causes of unauthorized releases include well blowouts, treatment/storage vessel ruptures and gathering line ruptures. Accidental discharges from offshore oil and gas operations are typically instantaneous events consisting predominantly of crude oil. If the crude oil was left alone, it would eventually degrade naturally to harmless by-products, and the impact on the marine environment would be negligible.

The first line of defense against accidental discharges is prevention. Even with the most diligent preventative methods, however, accidents will still occur. There are three steps to be taken after an accidental discharge has occurred. The first step is to stop the leak. This will minimize the amount of material that comes into the environment to begin with. Next, the spill should be contained so that further environmental damage does not occur. Finally, the spill should be removed or treated quickly. This could, but is not likely to, reduce the volume of spill material back to the pre-spill level.

4.2.5 Site abandonment

Although platform disposal is discussed in Chapter 5, there is an aspect of site abandonment in the North Sea that belongs with the oil impacts discussed here. Production in the North Sea is beginning to mature and has begun to decline. Some of the platforms are older and are reaching the point where they will have to be removed. Under the older platforms in the North Sea large volumes of oily cuttings are still in place, which have settled in piles around the base of the platforms. The piles contain oil from the discharge of oily cuttings during the early drilling programs, and are usually covered with a bacterial crust. One of the major issues discussed recently is whether these cutting piles should be removed once the platform is decommissioned. Estimates of the life of these piles could be as high as 50–100 years if left undisturbed. Removal of the cuttings piles would be very expensive and the most environmentally sound removal procedure is uncertain.

4.3 Potential impacts on the environment

4.3.1 Introduction

'Environmental impacts' mean different things to different people and, once defined, are still difficult to measure. In this chapter, 'environmental impacts' will be interpreted as any issues that raise concern in public or regulatory bodies, whether or not actual lasting effects occur.

4.3.2 Potential impacts from drilling waste

Major impacts to the marine environment from drilling waste generated by oil and gas operations include:

- toxicity;
- bioaccumulation and fish tainting;
- biochemical oxygen demand (BOD); and
- persistence.

The major impacts from drilling waste can be caused by either organic or inorganic components. The organic load is predominantly composed of the associated oil. The oil can be a component of the drilling fluid or a residual component of the cuttings. Inorganic components consist mainly of inorganic salts, with trace metals and nutrients.

Toxicity is a measure of the power to poison an organism. Acute toxicity is a measure of immediate danger of poisoning while chronic toxicity is a measure of long-term effects. Chronic impacts include such things as growth inhibition and reproductive interference. Toxic impacts require:

- a minimum concentration;
- a minimum exposure time; and
- a recovery rate that is sufficiently slow for the effect to be detected.

Regulations in most areas ensure that toxicity is not a serious problem.

A more subtle effect is tainting of sea creatures. The taste of seafood can be influenced by the fish consuming food that has been exposed to discharges associated with the exploration and exploitation of oil and gas. Even though the fish are still alive, healthy and catchable, the market for them is reduced. Fish taint is really a special case of bioaccumulation, in which the materials that accumulate affect the taste of the fish, and therefore its suitability as food for human beings. Bioaccumulation is a process whereby a material is absorbed into the body of an organism. For example, polynuclear aromatic hydrocarbons (PAHs) can be taken into the bodies of fish through contact with water containing PAHs or through ingestion of contaminated sediment. This process not only affects the organism involved, but can affect species that prey on them and ultimately humanity. Fish tainting is a local issue that has rarely proved to be a problem. Where it is a concern, regulations effectively control it.

Organic materials are removed from the aquatic environment through either aerobic or anaerobic biodegradation. Organic materials in both the water column and sediment are consumed by bacteria and converted into simpler material and ultimately into carbon dioxide and water. Aerobic biodegradation requires an oxygen source in the affected environment. The oxygen necessary for biodegradation is termed the biochemical oxygen demand (BOD). Neither the water column nor the sediment contains much oxygen, and a high concentration of organic materials will consume available oxygen rapidly making that environment unable to support aerobic life. Oxygen is easily replaced in the water column, however, because wind, waves and currents act to replace the oxygen at a rate higher than most degradation depletes it. On the other hand, oxygen in the sediment is easily depleted by biodegradation. In anoxic (oxygen-free) sediments anaerobic (non-oxygen) biodegradation takes place. Since drill cuttings usually end up on the sediment, if they have an oxygen demand impact it is in the sediment, not in the water column (Davies *et al.*, 1988). However, it should be noted that the floor of the ocean in deep water, such as the northern North Sea, is sparsely populated, and so the impact is small and the areal extent limited. This concern is recognized and addressed by most regulatory bodies.

The persistence of the contaminant in the environment also plays a role in determining the overall impact to the environment. Persistence is the ability to remain in the environment in a detrimental form and not be broken down into more innocuous materials. The only materials in oil and gas operations which would persist in the environment are present in very low concentrations and the threat of build-up is low.

4.3.3 Potential impacts from produced water

The major impacts and the discharge components of produced water are similar to those from drilling waste. The major impacts to the marine environment from produced water generated by oil and gas operations include:

- bioaccumulation and fish tainting;
- BOD;
- persistence; and
- sedimentation of constituents.

Laboratory tests have demonstrated that produced water has an intrinsically low toxicity level (E&P Forum, 1994). Therefore, toxicity is not an issue in the produced water discussion.

However, bioaccumulation (fish tainting), BOD and persistence are potential impacts that should be considered. In addition, produced water discharge plumes that touch the ocean floor affect the communities living in the area. Produced water discharges do not affect dissolved oxygen concentrations in the immediate vicinity of the discharge, as demonstrated by field measurements.

The biodegradation of organic compounds in produced water will create oxygen demand, but initial dilution following discharge will ensure that this is readily compensated for by natural aeration. The oxidation of inorganic compounds will not create significant oxygen demand (E&P Forum, 1994).

4.3.4 Potential impacts from treating chemicals

Chemicals used to perform specific functions during oil and gas exploration and production operations can also impact the marine environment. Treating chemicals can be classified into four categories:

- production treating chemicals;
- gas processing chemicals;
- stimulation and workover chemicals; and
- drilling chemicals.

Production treating chemicals include scale and corrosion inhibitors, biocides, emulsion breakers, water-treating chemicals (reverse emulsion breakers, coagulants and flocculants), anti-foams, and paraffin/asphaltene-treating chemicals. Gas processing chemicals are limited to hydrate inhibition and dehydration chemicals. Stimulation and workover chemicals include acids, brines and appropriate additives (Stephenson, 1992). Drilling chemicals perform many of the same functions as production chemicals and in addition they are used to adjust mud properties. The first three classes can affect produced water discharges. Drilling chemical impacts are covered under drilling wastes.

If production treating chemicals have an impact on the environment, it would be a toxic effect. These potential effects are controlled in several ways. In the USA, end-of-pipe toxicity limits ensure that they are controlled. In Europe, limitations on the characteristics of chemicals that can be used to control these potential effects. For example, there are both toxicity limits and concentration limits on chemicals used in the North Sea.

The addition of treatment chemicals, however, may increase the efficiency of production operations. This may in turn, therefore, actually decrease the overall environmental impacts. For example, if scale and corrosion of the piping can be controlled, the amount of waste generated by removing scale and corrosion from piping may be reduced.

4.3.5 Potential impacts from accidental discharges

Accidental discharges will consist primarily of crude oil. The environmental impacts of crude oil are basically the same as the impacts from drilling fluids and produced water, namely:

- fouling;
- tainting;
- BOD; and
- persistence.

These impacts have been discussed above and the general discussion can be applied to accidental discharges.

It should be kept in mind that accidental discharges that occur near or in sensitive coastal environments, although causing minimal long-term impacts, may have significant short-term impacts which must be dealt with. Typically, production facility accidental discharges are composed of smaller volumes than the more publicized tanker spills. This, in conjunction with the location of the facilities, many of which are some distance offshore, means that immediate impacts to the surrounding marine environment are less likely to occur. In fact, the accidental release may be allowed to degrade naturally without any noticeable impact on the environment.

The UK Royal Commission (1981), after reviewing a substantial body of information on the environmental effects of actual oil spills, concluded that there is no evidence to substantiate claims for long-term irreversible impact on the marine environment. On the other hand, the short-term consequences in relation to amenity loss, interruption of fishing activities and impact on individual sea birds (although not on bird populations) are sufficiently serious to justify efforts to develop and implement effective means of oil spill clean-up.

4.4 Regulatory approaches

Typically, regulatory controls over permitted and accidental discharges differ in their aims. The aim of the regulations for permitted discharges is to control the impact on the natural environment to an acceptable level, in order to gain the benefits of the production activity. There are basically two objectives for accidental discharges: prevention of the discharge and mitigation or minimization of the impact if an accidental discharge does occur.

Regulatory schemes can differ in many ways: the target of the regulations can be different; the grouping of the activities can be different; and the regulatory attitude can be different. These differences are illustrated by the following examples.

Some national bodies tend to target only the waste disposal or only the impacting activity. The USA, for example, tends to target the end-of-pipe discharge. Limits are put on the actual discharge and it is the responsibility of the discharger to control the process. Dischargers can use whatever means are at their disposal as long as the limits on the discharge are met. Some authorities tend to control what goes into a process, assuming that the waste from it will thereby be controlled. The countries around the North Sea, for example, put limits on chemicals that can be used and controls on their use and assume that the resulting discharges will be within acceptable toxicity limits.

Some regulatory bodies tend to look at oil and gas production facilities as a unified whole. An assessment is made of the entire activity and limits are set for each of its potential impacts. For example, this approach is used in Venezuela. Other regulatory bodies tend to look at specific discharge streams and put limits on them regardless of what operating unit they are attached to. For example, the USA regulates produced water discharges for all facilities in the same way, regardless of what facility they are discharged from.

Regulatory attitudes are also frequently different. Some countries view industry, including oil and gas production, as a national asset and attempt to help their industry protect the environment. Other countries tend to take an adversarial approach and attempt to protect the citizenry against uncaring industrialists.

These factors should be kept in mind when considering the regulatory histories discussed below.

4.4.1 Regulation of drilling waste in the USA

Waste disposal in the offshore areas of the USA is regulated under the Clean Water Act. The National Pollutant Discharge Elimination System (NPDES) is the permitting program that controls the discharge of waste to the sea (Hanson *et al.*, 1986). The NPDES program allows the EPA to issue general permits to cover similar discharges within specified geographic areas. The general permit specifies limits that must be met in order to discharge waste. Under NPDES regulations, guidelines are written by the EPA in Washington for the use of the various EPA regions in actually issuing permits (EPA, 1993b).

In a 1987 report to Congress, the EPA determined that oil and gas wastes were exempt from Hazardous Waste Subtitle C regulations under RCRA (Lyon, 1994). However, under Subtitle D these wastes were still regulated.

The EPA also regulated injection of wastes and discharges through the Safe Drinking Water Act and the Clean Water Act. In the 1987 determination, the EPA stated their intent to promulgate tailored criteria for non-hazardous waste management of exploration and production wastes.

As NPDES regulations developed, several features of the program emerged. One was that the regulations were implemented in phases. Another feature was the categorization of the waste into conventional pollutants, non-conventional pollutants and priority pollutants. A third feature for the offshore was categorization of the receiving environment as coastal, territorial seas or outer Continental Shelf (OCS). The first phase of the program was the development of Best Practicable Control Technology Currently Available and Economically Achievable (BPT). In general, this phase covered conventional pollutants (oil and grease, BOD, TSS, etc.) and required the use of the best technology that had already been applied in the industry. BPT guidelines were formally or informally used for all offshore areas. Waste drilling mud and cuttings had limits on 'free oil', which was interpreted as 'no sheen on the surface of the receiving waters'.

After the BPT guidelines had been published, the EPA was required by law to have Best Available Technology (BAT) guidelines for offshore areas by 1 July 1984. The task proved too ambitious for the EPA to accomplish by that deadline with the resources available to them. This brought litigation against the EPA to speed up the development of the guidelines. During the period from the late 1970s to 1993, BAT guidelines for the OCS were developed and implemented into general permits for several areas. BAT guidelines cover all classes of pollutants: conventional, non-conventional and priority. Separate guidelines have been developed or are under development for the different receiving environments: coastal, territorial seas and OCS.

Under BAT guidelines, discharges are not permitted in all area categories. For example, discharges are not permitted in the coastal subcategory in most locations. Mud and cuttings cannot be discharged into the territorial seas, but produced water can still be discharged under permit limitations. In the OCS, discharges of both drilling wastes and produced water are still permitted, but under increasingly stringent restrictions. Drilling fluids have limitations on free oil, toxicity, discharge rate, cadmium and mercury in the barite used in drilling mud and the types of oil that may be added to mud. Cuttings discharges are under similar restrictions.

Restrictions on oils in drilling discharges are as follows: no oil-based or inverse emulsion muds, or cuttings generated using them, can be discharged; no oil-contaminated mud can be discharged; no mud containing diesel oil or cuttings generated using a diesel oil-based mud can be discharged; and mineral oil can only be added to a drilling mud as a carrier for additives.

If it is allowed, in an offshore environment, drilling fluids and cuttings are usually discharged overboard. If the drilling fluids and cuttings cannot be discharged, they are hauled to shore for disposal. Water-based muds that meet regulatory limits and have no free oil are usually discharged along with the associated cuttings (Ray, 1989). Oil-based muds are recycled whereas the oily cuttings are hauled to shore. Synthetic-based muds have not been specifically addressed by the regulatory agencies, but synthetic muds must meet all limitations imposed on oil- and water-based muds.

Some EPA staff have communicated a desire to prohibit all discharges to the sea from oil and gas operations. Economics have been a major factor in preventing this, and will probably continue to bear a significant impact on preserving overboard discharge as a waste disposal option. In recent years the Clinton administration has tried to foster a more supportive philosophy for industry at the EPA. The Department of Energy has been taking a more active role in representing the oil industry on environmental issues. Solutions to waste disposal problems based on sound, scientific research and reasoning are still a possibility, but not yet a certainty.

4.4.2 Regulation of drilling waste in the North Sea

The majority of oil produced in Europe is from the North Sea. Development of the northern North Sea began in earnest in the 1960s. This oil province is characterized by very deep waters compared with the Gulf of Mexico. The typical development is from very large self-contained platforms from which a number of directional wells are drilled. Originally wells were drilled with water-based muds, but the geology of the area is such that water-insoluble muds are much more efficient. When this was discovered, many operators switched to diesel-based muds. The use of diesel-based muds caused environmental concerns and these concerns led to the development of low-toxicity mineral oils for use as mud base materials. Drilling with water-insoluble muds caused the cuttings to fall into a pile beneath the platform instead of dispersing into the water column as they do when water-based muds are used. At first the discharge of cuttings from the use of oil-based muds was justified on the basis that the impact of the oily cuttings could be contained within a small radius around the platform. As more and more experience was gained, however, it was discovered that materials leaching from the cuttings piles did have impacts, however slight, at distances from the platform greater than the authorities found acceptable (Hartley and Watson, 1993). This led to the development of synthetic-based muds that theoretically could degrade anaerobically much faster than oil-based muds. However, by this time several older platforms had large existing cuttings piles beneath them.

Each of the North Sea countries has separate regulations dealing with the discharge of drilling fluids and cuttings. The Paris Commission (PARCOM) and Oslo Commission, which have now been combined into OSPARCOM, are international agreements or standards which have set the pattern of testing and evaluation of discharges into marine waters and are used by the North Sea countries.

Regulations dealing with waste management and disposal in the North Sea area are controlled by the Paris/Oslo Commission. This treaty organization sets standards for environmental regulations which are then adopted by the member countries. The adoption of such regulations is not always swift or uniform. In Europe, environmental regulatory development is much more interactive than in the USA and industry has more immediate input in the regulatory development process. European regulations differ from US regulations in another important way. In Europe an attempt is made to control pollution at the source, whereas in the USA control is exercised on the discharge stream and the EPA has little control over the process generating the waste (Futsaeter, 1991; Henriquez, 1991).

Historically, the development of international treaty law for offshore oil and gas activities can be traced to the UN Geneva Conference of 1958, sometimes referred to as UNCLOS I. This conference adopted four conventions on the law of the sea. One of these, the Convention on the Continental Shelf (1958), provided coastal states with sovereign rights to explore and exploit the mineral resources of their continental shelves, thus providing for the development of the offshore oil industry.

In 1972, the UN Conference on the Human Environmental and its Declaration of Principles laid down in general terms requirements for the protection of the marine environment from toxic and other wastes, for the development of liability arrangements and for the conservation of flora and fauna. The conference established an action plan and instituted the United Nations Environment Program (UNEP), part of which has currently resulted in the development of framework conventions for 11 regional sea areas. The North Sea legal regime commenced prior to the regional seas program and is similar to that provided by the regional sea conventions.

The North Sea legal regime derives principally from the Oslo Convention on the Prevention of Marine Pollution by Dumping from Ships and Aircraft (1972), the Paris Convention on the Prevention of Marine Pollution from Land-Based Sources (1974) and the Bonn Agreement for Co-operation in Dealing with Pollution of the North Sea by Oil and Other Harmful Substances (1983). Under the Paris Convention, offshore platforms were considered land-based sources. The Oslo and Paris Conventions apply to the northeast Atlantic area as a whole, excluding the

Baltic and Mediterranean Sea areas. Both the Oslo and Paris Conventions established commissions, now combined in a joint Oslo and Paris Commission (OSPARCOM), which additionally hosts the meetings held under the Bonn Agreement.

A new Convention on the Protection of the Marine Environment of the North East Atlantic signed in Paris in 1992 will replace the earlier Oslo (1972) and Paris (1974) Conventions when it comes into effect. While this will provide a greater degree of unification, it will also introduce some significant changes from the historical approach of the prevention of pollution of the sea towards a regime concerned more generally with the protection of the marine environment.

Controlling discharges at the source means that all components of drilling fluids are approved for use before they are actually used. Systems are set up to test mud additives and treating chemicals for such characteristics as toxicity, oxygen demand, bioaccumulation and persistence in the environment. In addition, limits are placed on the oil content of cuttings generated using oil-based muds. At first the limit was 10% oil, then 1% oil, and for some locations a limit of 0.1% oil is being considered. In the North Sea, oily cuttings with an oil content of less than 1% can still be discharged if the base oil is approved for overboard discharge (i.e. it is not diesel based). Limits of 0.1% oil are not easily achieved, if at all, and push operators toward the use of synthetic muds (Davies *et al.*, 1988).

One future trend of international initiatives can be identified as perhaps at least having been started at the United Nations Conference on Environment and Development held in Rio in June 1992 and the Declaration of Principles adopted by it. In this, the world's nations embraced the concept of sustainable development and recognized the need to adopt sustainable approaches to both resource and environmental conservation and management.

In the North Sea, the development of agreed upon regulations and conventions under the European Community (EC) will be a long and drawn out affair. However, eventually standards will be developed that will control North Sea operations. The critical discharge issues under debate are removal of the oily cuttings piles, removal of platforms and the best method of disposal and produced water discharges and controls. The issue of testing and banning chemicals for use in drilling fluids and produced water will continue to develop and restrictions on discharges will increase. The use of synthetic drilling fluids over oil-based drilling fluids will increase and eventually the issue of discharges of these pseudo-oils will become a major issue. The Dutch have already banned the discharge of oily cuttings and operators are hauling the cuttings to shore for treatment and disposal (Hendriks and Henriquez, 1995). This may not be a realistic solution for the UK and Norway owing to the long distances from the shore and the lack of locations for treatment and disposal; however, the trend to limit and/or ban oily cuttings discharges and other toxic chemicals will increase.

4.4.3 Regulation of produced water in the USA

Under the old BPT guidelines, produced water had limits on oil and grease and pH. Now where discharges are permitted, such as the territorial seas and OCS, produced water discharges currently have limits on oil and grease, toxicity, discharge rate and data gathering requirements for radium and bioaccumulation.

Produced water disposal regulations covering the US offshore have increasingly become more and more stringent about what can be discharged. Once, produced water could be discharged in all subcategories with limitations on oil and grease content and little else. Now produced water cannot be discharged in the coastal subcategory, except in a few special areas, and discharges may not be permitted in these special areas for much longer. Produced water can still be discharged in the OCS, but the oil and grease limits have been sharply reduced and limitations on toxicity have been added. Guidelines are not yet final for the territorial seas subcategory and so there is uncertainty about produced water discharges in that subcategory.

4.4.4 Regulation of produced water in the North Sea

As fields in the North Sea mature, the volume of produced water being discharged is constantly increasing. For the past few years there has been rising concern about the amount of oil involved. There are two ways of controlling the amount of oil: control the volume discharged or control the amount of oil allowed in the discharge. In the past, regulators were satisfied to control the concentration of oil discharged. At present the allowable oil content of produced water is 40 mg/l; however, some discussions are being undertaken to reduce this to 30 mg/l. Various industry organizations, such as the E&P Forum, are working to find an acceptable resolution of these produced water concerns (Kingston, 1991). A recent attempt to control oil discharges to the North Sea involves schemes to control the total amount of oil, not just the concentrations.

4.5 Accidental discharges

4.5.1 Behavior of oil spilled at sea

When oil is spilled in the marine environment, the first thing that usually happens is that the oil tends to spread evenly over the top of the water, unless the crude oil has a pour point that is higher than the water temperature, in which case it will quickly solidify. The lighter ends of the crude that has been spilled will evaporate, and the water-soluble fractions will dissolve, leaving behind the more viscous residue which forms the slick.

Typically an oil slick spreads fairly quickly until it reaches a thickness of about 3 mm, at which point the spreading slows until a sheen has formed. Eventually all the oil is dispersed, or evaporates, and the slick disappears.

Gravity and surface tension are currently thought to be the principle physical forces that affect the rate of spreading of the spilled oil. In addition, many oils contain surface-active agents (e.g. long-chained carboxylic acids) that help the oil spread. Initially the main forces influencing the horizontal movement of the oil are gravity and the surface tension of the water. Over time, as the slick spreads, the effect of gravity lessens, and the principle driving mechanism becomes the surface tension of the water, which is typically greater than that of the floating oil. Inertia and viscosity act to retard the spreading of the slick.

A 'chocolate mousse' is formed when a heavier oil stays in a relatively thick layer for a long enough time that a water-in-oil emulsion is formed by wave action. The viscosity then increases rapidly, and the spreading rate consequently is reduced. A 'chocolate mousse' slick tends to break up into wind rows. Wave action therefore tends to retard the spreading of the oil.

After the natural spreading tendency of the oil, the wind is the second most important factor in causing the slick to expand. The movement of a slick in the Northern Hemisphere is marginally to the right of the wind, whereas in the Southern Hemisphere it is to the left. If there is no wind, tidal or river currents act to move the oil. In the event that the wind and the currents are in the same direction, their effects are additive, but if they are in opposite directions, the wind will prevail. It should be noted, however, that the presence of a large amount of surface debris can affect the movement of a spill.

When oil is spilled under ice, it will usually stay in close contact with the underside of the ice as it seeks the highest level as its specific gravity is less than water.

4.5.2 Regulatory programs

Much of the regulatory emphasis has been on reducing and responding to accidental releases from transportation-related incidents. As production of oil and gas has expanded throughout much of the world, a concerted effort to address how to respond to accidental releases has been made. The initial steps in this direction tended to come as a direct response to a specific incident.

The first such incident to attract massive public attention was the grounding of the *Torrey Canyon* off the southwest coast of England in April 1967, which resulted in pollution of the English and French beaches. As a result of the *Torrey Canyon*, a number of individual governments began to study the situation urgently and to look for remedies. However, they quickly realized that oil spills do not recognize or respect international boundaries and, as such, unilateral action would be of very little use. It was clear that there was a need to handle these issues internationally, and so the governments went to what was then called the Inter-Governmental Maritime Consultative Organization (IMCO), a specialized organization of the United Nations, and asked for help. IMCO has since changed its name to the International Maritime Organization (IMO), but it still continues to take the lead in this area.

In the meantime, during the late 1960s, while IMCO began its work, the tanker and oil industries decided to move ahead with their own plans to address the problem of accidental releases. The objective of the work was to develop a scheme that would ensure that governments and people adversely impacted by oil spills anywhere in the world would be promptly and fairly compensated for any damage that they had suffered. Industry also endeavoured to come up with a scheme that would help ensure that cargo and tanker owners would take immediate steps to prevent or mitigate any environmental damage.

In order to meet their objectives, the tanker and oil industries entered into two voluntary agreements:

- the Tanker Owners Voluntary Agreement Concerning Liability for Oil Pollution (TOVALOP); and
- the Contract Regarding an Interim Supplement to Tanker Liability for Oil Pollution (CRISTAL).

The biggest change that TOVALOP brought about was that instead of the burden of proof being on the government to prove liability by a ship owner for a pollution incident before any monies could be collected, the tanker owner had to prove that they were not liable. This greatly speeds up the system, and allows for much quicker responses. The other big difference is that it allowed for the taking of all reasonable precautions to prevent pollution from occurring. Before the agreement, an accident had actually to happen before any action could be taken or monies could be collected. In addition, previously the emphasis in dealing with an accident was on the saving of life and recovery of property. In fact, if the value of the vessel was less than the cost of clean-up, the clean-up was not carried out. TOVALOP changed the system and allowed for clean-up operations, the cost of which could far exceed the value of the vessel.

TOVALOP is only an agreement, and as such has no funds. It is administered by the International Tanker Owner Pollution Federation

(ITOPF). The monies for the fund actually come from the traditional 'Protection and Liability Clubs' (P&I Clubs). These are professionally managed, non-profit protection and indemnity associations which exist to share their collective expenses on a mutual basis. The P&I Clubs agreed to fund TOVALOP as a natural extension of their existing role.

In November 1969, the International Legal Conference on Marine Pollution met in Brussels under IMCO. The majority of the Governments attending signed the Civil Liability for Oil Pollution Damage Convention (CLC), which closely matched TOVALOP. The CLC did not actually come into effect until June 1975, at which time it had been ratified by 18 countries.

In December 1971, the Convention on the Establishment of an International Fund for Compensation for Oil Pollution Damage (Fund Convention) was signed. The Fund Convention is in addition to CLC and was adopted with the purpose of providing additional compensation to those who could not obtain full and adequate compensation for oil pollution damage under the CLC. The Fund Convention set up the International Oil Pollution Compensation Fund (IOPC Fund), an intergovernmental organization, to administer the Fund. The Fund is financed by companies who receive crude oil and heavy fuel oil in member states after transport by sea. The Fund Convention came into force in October 1978, at which time the IOPC Fund was established.

CRISTAL was an additional significant change to the previously existing scheme as it established a mechanism and fund whereby the owners of the oil cargo would contribute monies to compensate individuals as well as governments who had sustained pollution damage as the result of an oil spill.

The first international convention on the prevention of oil pollution at sea, the International Convention for the Prevention of Pollution of the Sea by Oil (OILPOL 1954), was signed in 1954. It specifically controlled oily water discharges from general shipping and oil tanker transportation operations. OILPOL has now been largely superseded by MARPOL 73/78, the International Convention for the Prevention of Pollution from Ships 1973, as Modified by the Protocol of 1978. MARPOL 73/78 defines a ship to include 'floating craft and fixed or floating platforms' and as such oil production platforms are covered by the Convention. This means, for example, that drainage discharges must not exceed 15 ppm, and so, in the UK, offshore installations are required to maintain an oil record book of all such discharges.

The latest convention concerning oil pollution at sea is the International Convention on Oil Pollution Preparedness, Response and Cooperation 1990 (OPRC). It was adopted in November 1990, and will enter into force 12 months after its ratification by 15 of its signatories. The objective of OPRC is to improve the level of preparations and preparedness to respond to an oil pollution incident, and to increase and promote international co-operation. OPRC seeks to build on the regional agreements (such as the Bonn Agreement for the North Sea area) to establish an interlocking series of plans that will ensure that all affected countries can adequately respond to any oil pollution incident in a co-ordinated, effective and rapid manner.

The impetus for the development of the OPRC was the much publicized *Exxon Valdez* spill in Prince William Sound, Alaska. The incident pointed out that to some extent governments and industry, having developed spill prevention and response plans, had become complacent, and some of the plans had become merely paperwork exercises to meet a regulatory requirement, rather than working documents. The Oil Pollution Act of 1990 (OPA 90) was passed in the USA in response to the same incident.

4.5.3 Government and industry initiatives

The previous section addressed the conventions and agreements that govern the response to an accidental release. This section will discuss some of the initiatives that have been taken to prevent accidental releases, and to minimize the impact of any releases that might still occur. Obviously, as stated elsewhere, the best method of avoiding environmental damage from an accidental release of oil is to prevent the release from ever occurring. To this end, industry groups and governments have developed voluntary and regulatory requirements to ensure that plans are in place with the objective of prevention, control and clean-up of any release. The plans range from individual facility prevention and response plans to regional inter-governmental and industry plans, as oil spills do not recognize or respect international boundaries.

To be effective, spill prevention planning needs to be done on a site specific, local and regional basis. This is because successful planning has to start with prevention at the source, but then must address the potential regional impact of a spill, and how best to respond quickly and decisively to minimize any potential negative impact.

The first generation of facility spill plans were fairly rudimentary. They covered a description of the facilities involved, discussed the possible type and size of releases that could occur, identified appropriate control measures that would be employed to prevent a release, addressed what to do in the event of a release and listed both the internal and external notifications that must be made in the event of a reportable spill, as well as some of the contractors who could help in a clean-up. A good example of such a plan is the Spill Prevention, Control and Countermeasure (SPCC) Plan that is required in the USA under the Clean Water Act. The regulations also require that all personnel are adequately trained to respond appropriately in the event of a release.

Although the SPCC type of plans were an excellent start to good spill prevention planning, over the years they have had the tendency to become merely paperwork exercises. This was graphically illustrated with the *Exxon Valdez* spill in Prince William Sound, Alaska. The contingency planning that had been done, when tested, did not perform as had been anticipated. Consequently, the new breed of spill planning not only requires extensive reviews of the potential impact of any release, but also requires detailed planning that ensures that responders will know exactly how to respond to all types of releases. Equipment has to be either on-site, or available on-site within specified time limits. In order to do this, operators have to enter into binding contracts with equipment providers who will guarantee a certain level of response within a specific time. The equipment has to be regularly inspected for operability, and the equipment has to be actually used in drills or actual responses on a specified schedule. Company and agency personnel who would be responsible for responding to a release have to receive regularly scheduled training that must include classroom and field segments. A good example of this type of plan is the Facility Response Plan required under the Oil Pollution Act of 1990 (OPA 90) in the USA.

On a regional basis, industry groups and governments have recognized the need for a co-operative effort to pool resources so that spill response can be as quick and effective as possible. The initial thrust came from industry, who formed regional equipment co-operatives which allowed each company to have access to a stockpile of equipment usually stored at strategic locations. Examples of these co-operatives include the United Kingdom Offshore Operators Association (UKOOA) equipment stockpiles and those of Clean Gulf and Clean Seas in the USA. On a world-wide basis, groups such as the Marine Spill Response Corporation (MSRC) stockpile equipment at strategic locations throughout the world.

Again, in response to a series of usually tanker spills, although there were also a few exploration and production releases (Ixtoc blowout, Ekofisk and the Santa Barbara release), individual governments began to set up their own response groups. Each country has established a program that meets its individual needs, and as such they vary from country to country.

As the programs are developed to meet specific needs, there is a wide variation in the nature and type of system that is established and how it operates. However, their objective is to be as prepared as possible to respond to any oil pollution incident.

For example, the Marine Pollution Control Unit (MPCU) was established in the UK in 1978 to replace the previous non-dedicated central government organization for dealing with oil and chemical pollution at sea, with a small dedicated unit. This change came about as a result of the work done by the United Kingdom Royal Commission on Environmental Pollution (which was set up after some major tanker incidents that impacted the UK coast). Amongst other things, the Royal Commission stated that they considered it essential that the response to a major spill should be a single coordinated operation overseeing the response at sea, inshore and on the land, hence the MPCU.

Incidentally, it should be noted that the Royal Commission also reported that, having reviewed the substantial body of information on the environmental effects of actual oil spills, it concluded that there is no evidence to substantiate claims for long-term irreversible impact on the marine environment. On the other hand, the short-term consequences in relation to amenity loss, interruption of fishing activities and impact on individual sea birds (although not on bird populations) are sufficiently serious to justify efforts to develop and implement effective means of oil spill clean-up.

The UK system is based on the idea of providing a central co-ordinating group, the MPCU, which oversees and co-ordinates the spill contingency planning and training of the local authorities who have jurisdiction from the beach to 1 mile (1.6 km) out; owns and maintains strategically placed stockpiles of equipment (in particular, aircraft and dispersant stocks); liaises with industry to dovetail equipment stockpiles, contingency planning and training; directs government research and development groups to help ensure that research is practical and useful; and finally is the point of liaison with the Bonn Agreement partners. The contingency planning, including the choice and location of stockpiled equipment, can be particularly effective in the UK because the country has the luxury of having developed detailed sensitivity maps that identify each area's particular ecosystem, including all the sensitive organisms and systems. Preplanning allows for a rapid response, for example, permission to use dispersants can be obtained in a matter of minutes, as opposed to the weeks, or even months, that it might take in, for example, the USA.

The UK has studied carefully the short- and long-term impacts an accidental release could have on the environment and leisure activities, and established its resources within financial limits set by the level of impact anticipated. Generally, the UK endeavors to achieve maximum response through the pooling of resources, for example having government-owned, strategically located stockpiles of equipment, co-ordinating the government-owned stockpiles with the industry co-operative stockpiles and the Bonn signatory government ones.

In contrast to the UK, which has a well established program that has developed over many years, China has taken a different approach, which more closely meets its specific needs. Unlike the UK, China is a vast country which has only recently been opened up to oil exploration and production. Consequently, its initial program is based on requiring the operator to do the spill contingency planning and to maintain any equipment necessary to provide an initial response until the international spill response community could get equipment and expertise into the area, if needed. The China National Offshore Oil Company (CNOC) is charged with reviewing the contingency planning and equipment to ensure that it is adequate. The current law requires that an environmental impact statement must be completed, submitted and approved by the National Environmental Protection Agency prior to a company being able to begin exploration and production activities.

The information collected in the environmental impact statement is used in the contingency planning phase. The contingency plan must include at a minimum the following elements: a general description of the project; the environmental conditions of the area, including the oceanography, meteorology and the sensitive environmental zones; risk analysis; response organization and responsibilities; oil spill response procedures; and how spilled oil will be handled (in particular taking into account that most of the offshore discoveries have been of high density, high pour point, waxy crudes, which means that standard skimmers and dispersants might not be effective).

The Ivory Coast, West Africa, has developed a co-ordinated approach to responding to oil spills. In the early 1990s, the government teamed with the Danish International Development Agency (DANDIA), which sponsored a study to determine the current situation and to propose and implement any needed changes, and to purchase any necessary equipment. The Centre Ivoirien Antipollution (CIAPOL) under the Ministry of Environment is the organization that deals with marine pollution problems. CIAPOL has three divisions: an administrative division; a division that deals with combating oil and chemical spills at sea, known as the Centre Ivoirien de Lutte contre les Pollutions Marines et Lagunaires (CIPOMAR); and the Central Laboratory for the Environment (LCE), which carries out most types of water analyses, including analyses for total and individual hydrocarbons.

The national oil spill plan, Plan Pollumar, was originally developed in the early 1980s, and has now been completely revised. The government has decided that CIAPOL will act as the national responsible authority, and so is responsible for all matters related to marine oil and chemical spill contingency planning in the Ivory Coast. The day-to-day running of the program, and the implementation of Plan Pollumar, have been delegated to the CIPOMAR division. CIPOMAR has been organized into three sections, namely Operations, Maintenance and Administration. The Operation Section has set up a national communications center which receives the reports of spills in the Ivory Coast response area, as well as pollution reports from neighboring countries within the West and Central Africa region. The duty officer at the communications center evaluates the report, and decides on the appropriate response, including, for example, enacting Plan Pollumar. The Maintenance Section is responsible for maintaining the spill response equipment. The Administrative Section is responsible for the financial, accounting and personnel functions. In the event that Plan Pollumar is enacted, the Administrative Section is responsible for creating all the documentation that will be used for the claim and compensation procedures. Employees from all three sections have been trained to perform the functions of the Incident Commander and On-Scene Coordinators.

4.5.4 Should the release be remediated?

Since the first oil spill and resultant clean-up, the question has been raised as to how clean is clean. Over the years, considerable effort and resources have been expended to determine not only the impact of spilled crude oil on the environment, but also the impact of the clean-up. In the early days the cure was often worse than the original incident. For example, the dispersants used on the *Torrey Canyon* spill were several orders of magnitude more toxic than the oil that they were trying to disperse. Eventually the results of the scientific studies began to be implemented into both individual company and country response planning. There began to be a feeling that the net impact on the environment should be an important factor in deciding on the appropriate response to an accidental release.

However, it is important to remember that the political reality will not always allow the responders to a spill to base their decisions solely on what is best for the environment. For example, natural biodegradation, and bioremediation of a beach may be the best ecological solution; however, the company responsible for the spill and clean-up, and the agency overseeing the response may have to attempt to clean the area in order to be seen as responsive!

In spite of political pressures, it is important to try to always make the minimum net environmental impact the objective of a response plan. Exactly how to do this will depend on the nature of the crude oil spilled, the location of the oil and the systems that are, or may be, impacted. For example, it is now generally accepted that crude oil spilled in a salt marsh is best left to degrade naturally, as any attempt to remove the oil mechanically will result in a much greater negative impact on the system.

In addition, in the how clean is clean debate, it is important for the parties to agree on the appropriate end-point, beyond which the cost of remediation far exceeds the net benefit to the environment.

4.6 Summary and conclusions

Discharges of oil and gas drilling and production wastes to the marine environment have been studied extensively in the last 30 years. Regulations have been developed in most areas that assure a high level of environmental protection.

Environmental compliance, including disposal of waste products, is a major economic factor in offshore exploration and production. Disposal of drilling waste and produced water, including required laboratory testing of discharges, pretreatment of materials prior to discharge and contingencies for accidental discharges, are a major factor in the planning and design of drilling and production operations.

References

Burke, C.J. and Veil, J.A. (1995) Synthetic drilling muds: environmental gain deserves regulatory confirmation. *Proceedings, SPE/EPA Exploration and Production Environmental Conference*, Houston, TX, 27–29 March, 457–68.

Candler, J.E., Rushing, J.H. and Leuterman, A.J.J. (1993) Synthetic-based mud systems offer environmental benefits over traditional mud systems. SPE 25993, *Proceedings, SPE/EPA Exploration and Production Environmental Conference*, San Antonio, TX, 7–10 March.

Davies, J.M., Bedborough, D.R., Blackman, R.A.A. *et al.* (1988) Environmental effect of oil-based mud drilling in the North Sea. *Proceedings, 1988 International Conference on Drilling Wastes*, Calgary, Canada, 5–8 April.

E&P Forum (1994) *North Sea Produced Water: Fate and Effects in the Marine Environment.* E&P Forum Report No. 2.62/204.

EPA (1993a) *Development Document For Effluent Limitation Guidelines and New Source Performance Standards for the Offshore Subcategory of the Oil and Gas Extraction Point Source Category, Final.* EPA 821-R-93-003, US Environmental Protection Agency, Washington, DC.

EPA (1993b) Clarification of the regulatory determination for wastes from the exploration, development, and production of crude oil, natural gas and geothermal energy. *Federal Register*, **58**, 15284–7.

Freeman, B.D. and Wakin, P.G. (1988) API survey results on 1985 onshore wastes volumes and disposal practices within the US petroleum extraction industry, in *Drilling Wastes* (eds F.R. Engelhardt, J.P. Ray and A.H. Gillam), Elsevier Applied Science, New York.

Friedheim, J.E. (1994) Drilling with synthetic fluids in the North Sea – an overview. Presented at the IBC Conference on Drilling Technology, Aberdeen, Scotland, 22–23 November.

Futsaeter, G. (1991) Environmental guidelines: environmental policy and regulations and supportive government actions regarding discharges from the offshore petroleum industry. *Proceedings, 5th Norwegian Petroleum Society Northern Europe Drilling Conference*, Kristiansand, Norway, 5–6 November, 14.

Hanson, P.M., Jones, F.V., Moffitt, C.M. and Rhea, M.R. (1986) A review of mud and cuttings disposal for offshore and land based operations. *Proceedings, Drilling Muds National Conference*, Norman, OK, 29–30 May, 109–22.

Hartley, J.P. and Watson, T.N. (1993) Investigation of a North Sea oil platform drill cuttings pile. *Proceedings, 25th Annual SPE et al. Offshore Technology Conference*, Houston, TX, 3–6 May, Volume 4, 749–56.

Hendricks, R. and Henriquez, L. (1995) Operators agree to further limits on use of oil-based muds. *Offshore*, August, 196.

Henriquez, L.R. (1991) Current regulations regarding the testing and evaluation of chemicals and drilling fluids discharged offshore The Netherlands. *Proceedings, 5th Norwegian Petroleum Society Northern Europe Drilling Conference*, Kristiansand, Norway, 5–6 November, 20.

Kingston, P.F. (1991) The North Sea oil and gas industry and the environment. *Proceedings,*

Financial Times North Sea Oil and Gas Conference, London, 2–3 July, *Oil Gas Europe Magazine*, **17**(4), 6–10.
Lyon, F.L. (1994) Nonhazardous oilfield waste (NOW) disposal, liability and the law. *Proceedings, International Petroleum Environmental Conference*, Houston, TX, 2–4 March, 287–95.
op ten Noort, F.J., Marquenie, J.M., Cofino, W.P. and van Hattum, B. (1994) Environmental aspects of produced water from offshore platforms on the Dutch Continental Shelf. *Proceedings, The Second International Conference on Health, Safety and Environment in Oil and Gas Exploration and Production*, Jakarta, Indonesia, Volume 1, 391–8.
Park, S., Cullum, D. and McLean, A.D. (1993) *The Success of Synthetic-Based Drilling Fluids Offshore Gulf of Mexico*, SPE 26354, SPE, Richardson, TX.
Ray, J.P. (1979) *Offshore Discharges of Drilling Cuttings, Proceedings, Outer Continental Shelf Frontier Technology*, National Academy of Sciences, Washington, DC, 6 December, (*Offshore Rulemaking Record*, Volume 18).
Ray, J.P. (1989) Offshore drilling waste issues, in *Drilling Wastes* (eds F.R. Engelhardt, J.P. Ray and A.H. Gillam), Elsevier Applied Science, New York, pp. 849–60.
Royal Commission on Environmental Pollution (1981) *Oil Pollution of the Sea*, Eighth Report. HMSO, London, 307 pp.
Stephenson, M.T. (1992) Components of produced water: a compilation of industry studies. *Journal of Petroleum Technology*, **44**(5), 548–603.
Walk, Haydel and Associates (1984) *Potential Impact of Proposed EPA BAT/NSPS Standards for Produced Water Discharges from Offshore Oil and Gas Extraction Industry*, Report prepared for Offshore Operators Committee, Walk, Haydel and Associates, New Orleans.

Further reading

Balkau, F. (1990) International aspects of waste management, and the role of the United Nations Environment Program (UNEP). *Proceedings, First International Symposium on Oil and Gas Exploration and Production Waste Management Practices*, New Orleans, LA, 10–13 September, 543–52.
Bender, K., Jensen, S.K., Ostergard, J. and Nogbou, P. (1993) Oil spill contingency planning in the Ivory Coast. *Proceedings, International Oil Spill Conference*, Tampa, FL, 29 March–1 April, 31–4.
Derkies, O.L. and Souders, S.H. (1993) Pollution prevention and waste minimization opportunities for exploration and production operations. *Proceedings, SPE/EPA Exploration and Production Environmental Conference*, San Antonio, TX, 71–6.
Dutta, S. and Alam, W. (1995) Oil and gas exploration and production wastes – RCRA exemptions and non-exempts. *Proceedings, SPE/EPA Exploration and Production Environmental Conference*, Houston, TX, 27–29 March, 295–304.
E&P Forum (1993) *Exploration and Production (E&P) Waste Management Guidelines*. E&P Forum Report No. 2.58/196.
EPA (1987) *Report to Congress: Management of Wastes from Exploration, Development, and Production of Crude Oil, Natural Gas, and Geothermal Energy*. NITS Publication No. PB 88-146212.
EPA (1988) Regulatory determination for oil and gas and geothermal exploration, development and production wastes. *Federal Register*, **53**, 25446–59.
EPA (1990) *Proceedings, First International Symposium on Oil and Gas Exploration and Production Waste Management Practices*, New Orleans, LA, 10–13 September.
EPA (1995) *Crude Oil and Natural Gas Exploration and Production Wastes: Exemption from RCRA Subtitle C Regulations*. EPA 530-K-95-003, US Environmental Protection Agency, Washington, DC.
Fillo, J.P. and Evans, J.M. (1995) Natural gas industry waste production and management practices. *Proceedings, SPE/EPA Exploration and Production Environmental Conference*, Houston, TX, 27–29 March, 267–82.

Fingas, M. (1995) Oil spills and their cleanup. *Chemistry and Industry*, 18 December, 1005–8.
Fitzpatrick, M. (1990) Common misconceptions about the RCRA Subtitle C exemption for wastes from crude oil and natural gas exploration, development and production. *Proceedings, First International Symposium on Oil and Gas Exploration and Production Waste Management Practices*, New Orleans, LA, 10–13 September, 169–78.
Holt, W.F. (1995) Implementing the OPRC – translating diplomatic concepts into reality. *Proceedings, 1993 Oil Spill Conference*, Tampa, FL, 29 March–1 April, 655–8.
Jacobsson, M. (1989) The International Oil Pollution Compensation Fund: ten years of claims settlement experience. *Proceedings, 1989 Oil Spill Conference*, San Antonio, TX, 13–16 February, 509–12.
Lu, M.Z. (1989) Oil spill prevention and treatment in the offshore oil industry of China. *Proceedings, Oil Spill Conference*, San Antonio, TX, 13–16 February, 235–8.
Mead, D.A. and Lillo, H. (1990) The Alberta Drilling Waste Review Committee – a cooperative approach to development of environmental regulations. *Proceedings, First International Symposium on Oil and Gas Exploration and Production Waste Management Practices*, New Orleans, LA, 10–13 September, 1–6.
Perry, C.W. and Gigliello, K. (1990) EPA perspective on current RCRA enforcement trends and their application to oil and gas production wastes. *Proceedings, First International Symposium on Oil and Gas Exploration and Production Waste Management Practices*, New Orleans, LA, 10–13 September, 307–18.
Smith, R.M. (1995) The Hunter Creek well: the successful drilling of an exploratory well in an environmentally sensitive area. *Proceedings, SPE/EPA Exploration and Production Environmental Conference*, Houston, TX, 27–29 March, 145–56.
Stacey, M.L. (1985) Marine pollution contingency planning – recent changes in the United Kingdom organization. *Proceedings, Oil Spill Conference*, Los Angeles, CA, 25–28 February, 89–92.
Stillwell, C.T. (1990) Area waste management plan for drilling and production operations. *Proceedings, First International Symposium on Oil and Gas Exploration and Production Waste Management Practices*, New Orleans, LA, 10–13 September, 93–108.

5 Decommissioning of offshore oil and gas installations

M.D. DAY and M.H. MARKS

5.1 Introduction

The offshore oil and gas industry had its beginnings in the Gulf of Mexico in 1947. The first offshore development used a multipiled steel jacket to support the topside production facilities, a design which has since been used extensively. Now there are more than 7000 drilling and production platforms located on the Continental Shelves of 53 countries [1]. Some of these structures have been installed in areas of deep water and treacherous climates, and consequently structure designs have adapted to withstand the environmental conditions of these areas. Some typical designs are shown in Figs 5.1–5.5. In the North Sea, which is an area that experiences some extreme environmental conditions, more than 200 structures have been installed, about 25% of which are in water depths greater than 75 m and can be exposed to maximum storm wave heights of 30 m. This combination of deep waters and extreme storm forces dictates large structures, some with component weights that exceed 50 000 tonnes [6]. One of the world's largest gravity base structures (GBS) is scheduled to be installed off the coast of Canada in 1996. It is designed to withstand impacts by icebergs and will weigh approximately 1.5 million tonnes including ballast [7]. Now, as oil and gas fields begin to deplete their reserves, the concern has turned to the removal and disposal of these structures at the end of their producing lives. Estimates indicate that the cost of some removals may exceed the cost of the original installation. The structures located on the Norwegian Continental Shelf contain only 1% of the world's offshore structures, but will account for nearly 20% of the worldwide removal costs [4]. Innovative removal and disposal techniques must be developed to limit costs and minimize the impact on the environment.

The Gulf of Mexico, the western and central costs of Africa, the Persian Gulf, the bulk of the Pacific region and the Mediterranean Sea are all examples of areas with more moderate environments. The majority of structures in these areas are in water depths from 3 to 300 m with maximum storm wave heights of 12 m. With a few exceptions, platforms in these areas will probably be totally removed at the end of their producing lives. The major implication with total removal is in choosing the method

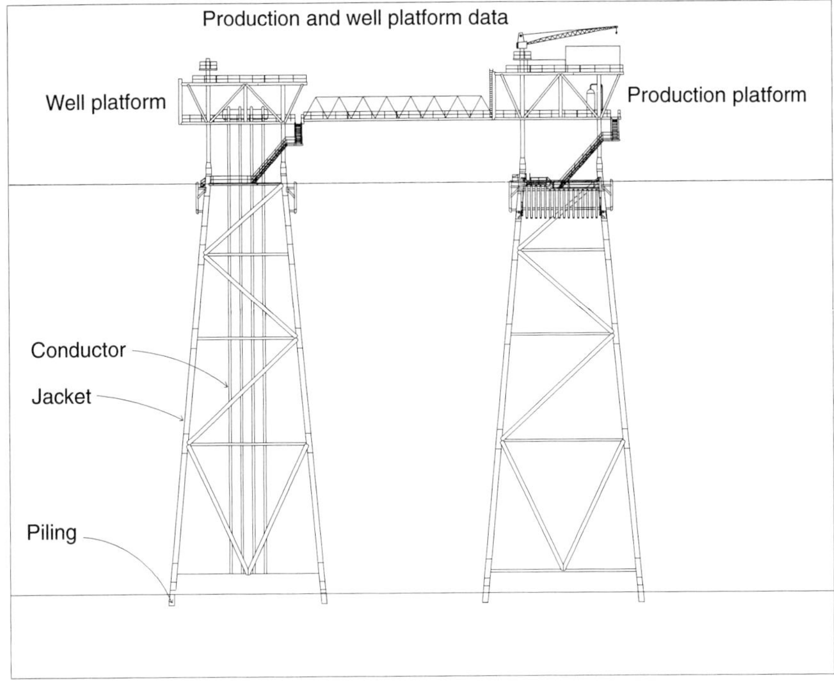

Figure 5.1 Steel-jacketed structure [2].

to dislodge the structure from the sea-bed and an issue in remote areas of the world is the availability of support equipment to perform the removals.

5.2 Legal framework of platform decommissioning

International law provides the basic foundation of the legal requirements for the removal and disposal of offshore structures. The removal of installations was addressed by the 1958 Geneva Convention on the Continental Shelf, which stated that any installations which are abandoned or disused must be entirely removed. However, several parties to the Convention were soon adopting some form of local standards to allow for partial or non-removal. The more widely accepted statement of international law is contained in the United Nations Convention on the Law of the Sea (UNCLOS), which allows for partial removal and has been widely accepted as it appears to represent customary international law in relation to abandonment [8]. The International Maritime Organization (IMO) guidelines were issued using UNCLOS as a basis. These guidelines state that if the structure exists in less than 75 m of water and weighs less than 4000 tonnes, it must be totally removed [8]. Structures installed after January 1988 will have a water depth criterion of 100 m, forcing the owner

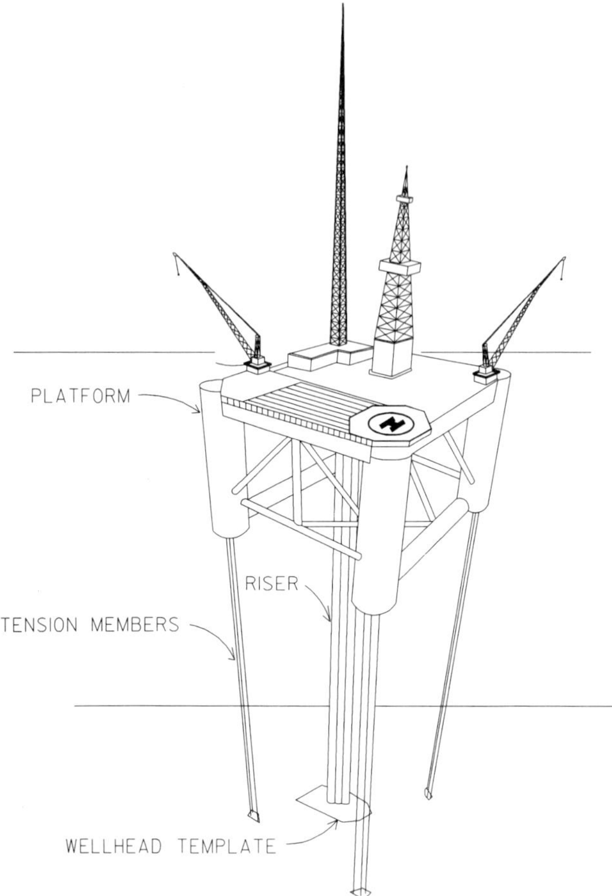

Figure 5.2 Tension leg platform [3].

to plan for the eventual abandonment in the initial design. If the removal is done partially, the installation must maintain a 55 m clear water column. There are exceptions in the guideline that allow for non-removal, e.g. if the structure can serve a new use after hydrocarbon production including enhancement of a living resource, if the structure can be left without causing undue interference with other uses of the sea or where removal is technically not feasible or an unacceptable risk to the environment or personnel [8]. If the installation is to remain in place, it must be adequately maintained to prevent structural failure.

Basic disposal stipulations can be traced to international dumping conventions. The Oslo Convention of 1972 for the Prevention of Marine Pollution by Dumping from Ships and Aircraft provides some guidelines.

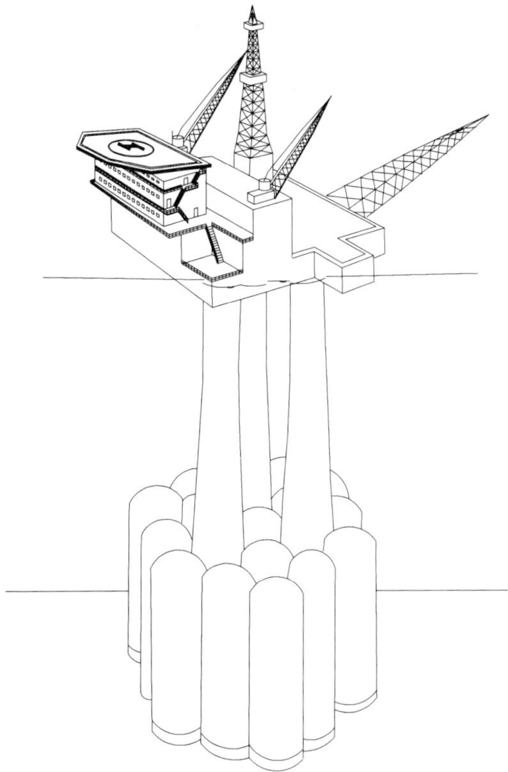

Figure 5.3 Concrete gravity base structure [3].

However, it is not clear if this Convention applies to dumping of platforms in place. The London Convention of 1972 on the Prevention of Marine Pollution by Dumping of Wastes and other Matter also supplies guidelines for deliberate disposal of platforms or other artificial structures at sea. UNCLOS deals with dumping, and states that 'dumping within the territorial sea and the exclusive zone or onto their continental shelf will not be carried out without the express prior approval of the coastal state . . .' [8].

The Convention for the Protection of the Marine Environment of the North East Atlantic (Paris, 1992) is relevant. It provides that 'no disused structures . . . be dumped and no disused offshore installation shall be left wholly or partly in place in the Maritime area without a permit issued by the appropriate competent authority of the contracting party on a case-by-case basis', and that 'dumping does not include the leaving wholly or partly in place of a disused installation . . . provided that such operation takes

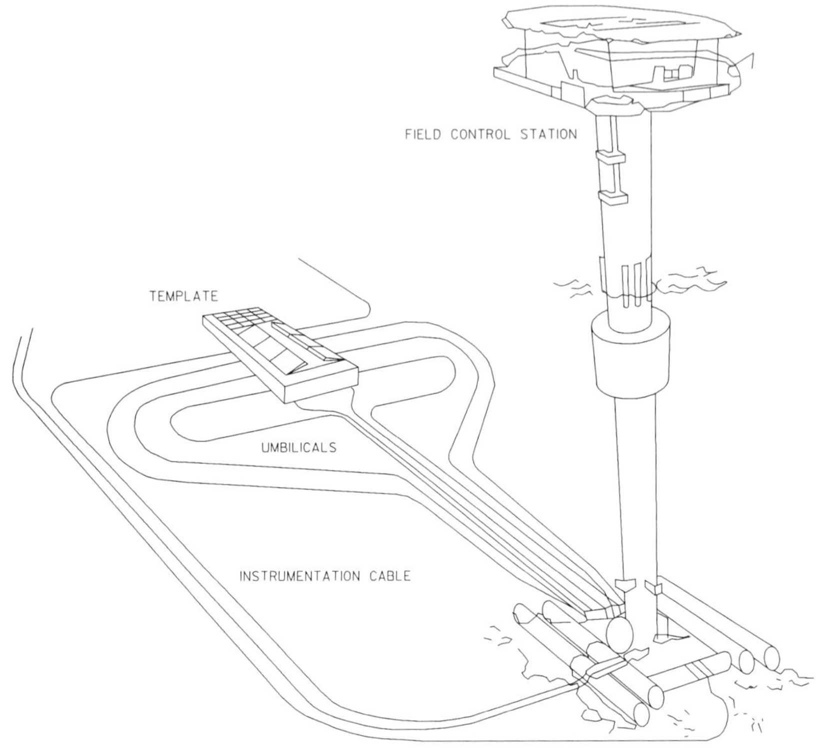

Figure 5.4 Floating production system [4].

place in accordance with any relevant Convention and with relevant international law' [8].

The body established by the 1991 Oslo Convention, the Oslo Commission, adopted guidelines on a trial basis to exercise overall supervision over the implementation of the Convention. These guidelines are complementary to the IMO guidelines and aim to minimize pollution to the sea by hazardous residues left in parts of installations disposed of at sea [8].

While all of the above are basic guidelines to removal and disposal, they do not account for all of the issues involved with the abandonment or disposal of offshore structures. Thus, local states are left to decipher the issues, and to generate legislation to cover loopholes in international law in accordance with their priorities. By 1992, 15 United Nations Environment Programme (UNEP) regional conventions had been held (Fig. 5.6). Here, local states have adopted varying degrees of guidelines for potential legal concerns such as determination of the party responsible for removal, responsibility and methods of payment, responsibility of owners in default

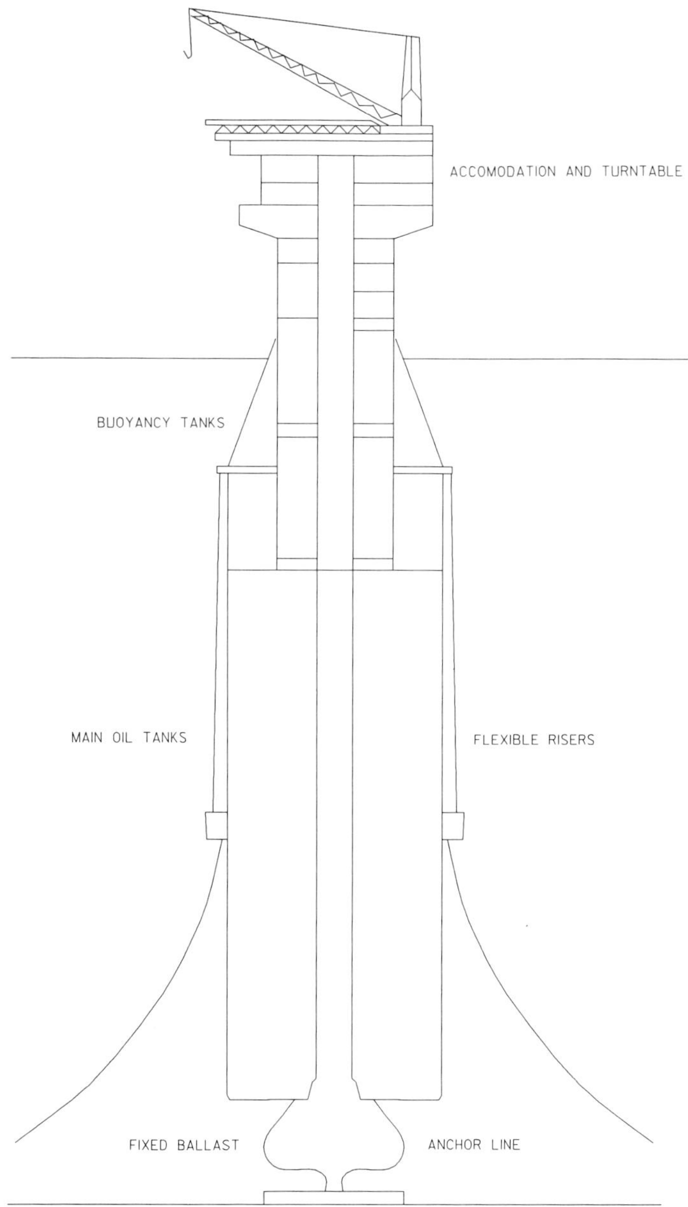

Figure 5.5 Spar buoy [5].

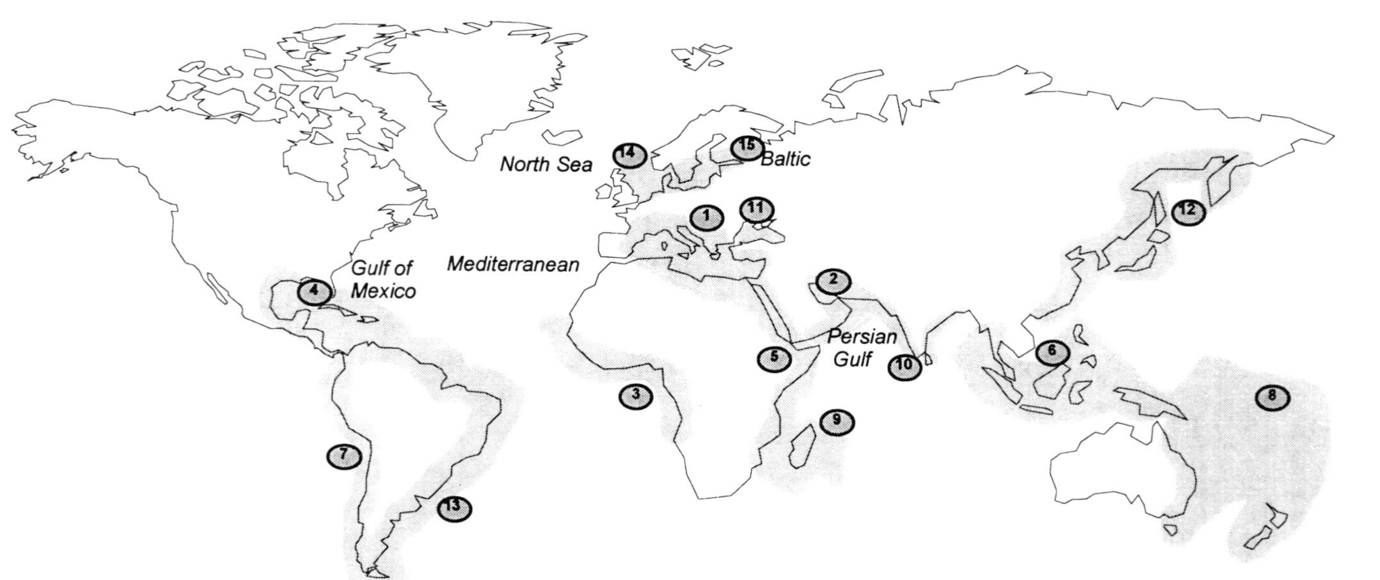

Figure 5.6 UNEP regional seas program and other conventions.

situations, owner designation upon non-use, maintenance responsibility and liability for items left in place and such site-specific issues as bottom debris removal and moratoriums for marine migrations.

The complexity of issues has stymied most countries from adopting specific guidelines and standards for platform removal, but most do require abandonment procedures to be submitted to designated regulatory agencies for approval on a case-by-case basis. Some countries, depending on their experience with removals, are fairly mature in their regulatory standards for abandonment, whereas others still have great strides to make in enacting requirements for removals within their coastal waters.

5.3 Planning

The most critical and time-consuming task of the abandonment process is the planning phase. This phase should be initiated years in advance when depletion plans for a field are recommended. The planning phase can be effectively organized with the aid of commercially available computer software. A software package which allows for input of schedules, tasks, resources and contingencies is recommended. This will be beneficial in establishing the critical path of the project and will help keep the project on schedule for the available construction weather window. A project management software package will enable the project engineer to maintain accurate cost accounting and to keep the project organized, on schedule and within budget.

5.4 Abandonment phases

The entire abandonment process can be broken down into seven discrete activities [9]:

1. *Well abandonment*: the permanent plugging and abandonment of non-productive well bores.
2. *Preabandonment surveys/data gathering*: information-gathering phase to gain knowledge about the existing platform and its condition. Governing ministries or standards organizations should be contacted to determine permit and environmental requirements.
3. *Engineering*: development of an abandonment plan based on information gathered during preabandonment surveys.
4. *Decommissioning*: the shutdown of all process equipment and facilities, removal of waste streams and associated activities to ready the platform for a safe and environmentally sound demolition.
5. *Structure removal*: removal of the deck or floating production facility

from the site, followed by removal of the jacket, bottom tether structures or gravity base.

6. *Disposal*: the disposal, recycle or reuse of platform components onshore or offshore.
7. *Site clearance*: final clean-up of sea-floor debris.

The following is a brief discussion of the sequence of processes involved with structure decommissioning.

5.4.1 *Well abandonment*

The exact timing of cessation of production can be difficult to predict. However, a close working relationship between the reservoir, downhole and salvage engineers should be developed to establish the timing of a well and platform abandonment project. Before abandonment can begin, the salvage engineer must confirm that all wells on the platform are abandoned. The wells should be permanently abandoned according to the recommended procedures of the governing body. Generally this means isolating productive zones of the well with cement, removing some or all of the production tubing and setting a surface cement plug in the well with the top of the plug approximately 30–50 m below the mudline. The inner casing string should be checked to ensure that adequate diameter and depths are available for the lowering of explosives or cutting tools. If the well plug and abandonment are not performed properly, removal of the conductor by explosive or mechanical means becomes unsafe and much more expensive.

To ensure no delays in structure removal, all well plug and abandonments should be completed several months prior to commencement of offshore decommissioning. After well plug and abandonment responsibility and schedules have been established, the next step is an information-gathering phase.

5.4.2 *Preabandonment surveys/data gathering*

Critical to a successful abandonment program is planning. Proper planning requires that as much as possible about the platform be known. Information must be gathered on the topside deck and support structure design, fabrication and installation as well as any structural modifications that may have occurred since installation. The preabandonment survey should assess the condition of the platform facilities and structure prior to beginning the abandonment. The survey should include the following:

(a) File surveys. All available documentation concerning the platform design, fabrication, installation, commissioning, start-up and continuing

operations should be investigated. The file survey will familiarize the project engineer with the other appurtenances to the platform facility such as living quarters, process equipment, piping, flare system and pipelines and any additions/deletions or structural repairs to the jacket or the topside since the original installation. The project engineer must remain aware that platform records may be incomplete or unreliable. After an extensive search of all available files, the engineer should be able to define the abandonment scope of work and the objectives of subsequent surveys.

(b) Geophysical survey. Depending on the results of the file survey, the engineer may choose to have additional data gathered by means of side-scan sonar. This survey will indicate the amount of debris on the sea-floor. In the case of deep-sea disposal, the sonar can determine if there are any obstructions at the dump site. Proximity of an available dump site or 'rigs to reef' site, water depths and obstructions along the tow route should be investigated as part of the geophysical survey.

(c) Environmental survey. This consists of an environmental audit of the offshore platform to identify waste streams or other government controlled materials. At this time items such as naturally occurring radioactive materials (NORM), asbestos, PCBs, sludges, slop oils and hazardous/toxic wastes should be identified and quantified. The problem of dealing with these waste streams should be addressed in the scope of work for handling during the decommissioning phase of the project. The project engineer should determine what permits or operating parameters are required by the host government or international standards.

(d) Structural survey. A structural engineer can use observation and non-destructive ultrasonic testing techniques to evaluate the structural integrity. Items inspected will include condition and accessibility of lifting eyes, obstructions on the deck which may require removal and interfaces between production modules/deck and deck/jacket which may require cutting for disassembly. Discrepancies between actual conditions and as-built information identified in the files should be noted during this phase. The platform legs should be checked for damage that may obstruct explosives or cutting tools from accessing the proper cutting depth. If obstruction from damage is anticipated or found, smaller diameter charges or cutting tools should be provided by the removal contractor as a contingency. Information concerning the underwater condition of the structure should be available from previous underwater inspections. If not available, consideration should be given for gathering this information by divers or remote-operated vehicles (ROVs).

5.4.3 Engineering

Upon completion of preabandonment surveys, a strategy for decommissioning and abandonment can be developed. The engineering phase takes all of the data previously gathered and pieces it together to form a logical, planned approach to a safe abandonment. Of major concern during the development of this strategy is the safety of the operations. As with all offshore operations, there exists a high potential for accidents involving bodily injury or loss of life and the accidental discharge of soil and flammable, corrosive or toxic material into the environment.

A risk analysis for all phases of the decommissioning should be performed. The results of this risk analysis are used to develop a decommissioning safety plan. Safety targets can be set and achieved provided the appropriate attention is devoted to the elements of the decommissioning plan. These procedural elements include the following items:

- Regularly scheduled safety meetings.
- Identification of safe work areas.
- Safety equipment and training for emergency situations.
- Working at high elevations and over water.
- Safe operations of cutting tools and explosives.
- Safe demolition to maintain structural integrity.
- Proper use of rescue and evacuation equipment.
- Diving and ROV operations.
- Testing for and monitoring of toxic/explosive gases.
- Pollution controls and containment.
- Methods for handling and disposal of oil wastes, corrosive, NORM or toxic materials.
- Weather monitoring/night watch procedures.

Addressing each of the above-mentioned elements will help in the development of a safe decommissioning and salvage plan. After all the safety and environmental aspects of the project have been considered, details of the salvage process need to be identified. The sequence of process equipment and structure decommissioning and the salvage and disposal methods need to be determined. Any required government permits should be submitted for approval.

A major determination for an effective and efficient abandonment program is proper selection of the salvage equipment. Equipment selection for lifting purposes is determined by maximum weights of components to be lifted. Oceangoing derrick barges currently available to the industry range from approximately 135 to 6500 tonnes (Fig. 5.7). Other lower capacity, less expensive lift spreads can be used if the lift weights can be broken down through equipment removal or by cutting the components into smaller lifts.

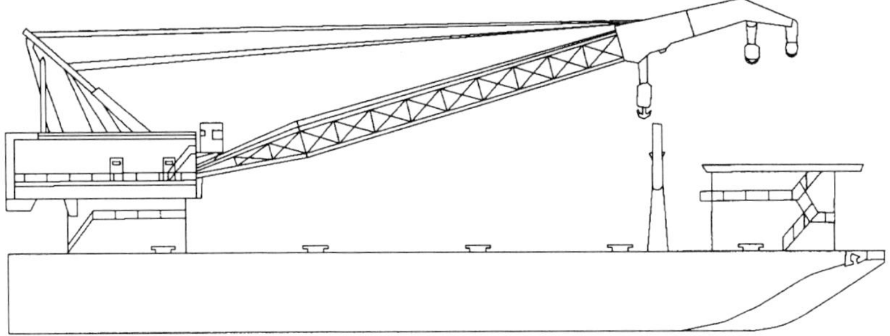

Figure 5.7 Derrick barge.

Cost comparisons must be made between the time savings afforded by heavier lift, more expensive equipment and time-consuming, lighter lift, less expensive equipment. In addition to costs, the project engineer must assess the safety and environmental risks associated with sectional removal. Sectional removal will require significant time at the site for dismemberment and removal of production piping and equipment prior to cutting the topside deck into pieces. Additional hazardous tasks involved with decommissioning, lifting and rigging operations need to be performed offshore in a sectional removal, thus the time during which personnel will be exposed to increased workplace hazards will be increased. More details pertaining to sectional removal will be addressed in section 5.4.5.

Once the sizing of equipment is complete, a qualified list of contractors can be generated based on equipment availability and the area of the world in which the salvage is to take place. Awarding of the job based on the list of qualified contractors can be carried out in many ways. Two often used methods are bidding out the job for award to the lowest bidder or by negotiating a contract with the contractor who is most capable of performing the work. The job scope could include all aspects of the abandonment from the well abandonment to the final site clearance. Another method might be to award each portion of the abandonment and salvage as individual components similar to the breakdown of the seven phases of abandonment.

5.4.4 Decommissioning

A primary objective during the decommissioning is to protect the marine environment and the ecosystem by properly collection, control, transport and disposal of various waste streams. Decommissioning is a dangerous phase of the abandonment operation and creates the possibility of environmental pollution. Decommissioning and removal or abandonment in place should be carried out by personnel who have specific knowledge and experience in safety, process flows, platform operations, marine transportation, structural systems and pipeline operations. All contractors involved with the decommissioning should be brought in early in the planning stage to further assure a smooth decommissioning project.

The sequence of decommissioning the process system, utilities, power supplies and life support systems is important. The platform's power, communications and life support systems should be maintained for as long as practicable to support the decommissioning effort.

Process systems throughout the platform will have to be flushed, purged and degassed in order to remove any trapped hydrocarbons. Safe lock-out, tag-out, hot work and vessel entry procedures must be in place to ensure safety. Procedures must outline all duties of the standby/rescue teams including the use of breathing apparatus, air purging and lighting and caution must be exercised in removing all amounts of gases, oils and solids which may still remain in valves, production headers, filter housings, vessels and pipework that could present hazards to the crew.

Platform decommissioning will result in large amounts of waste liquids and solids. Where possible, waste liquids can be dealt with most cost effectively by placing them in existing pipelines and sending them to existing operating facilities. If no ongoing operations are available, then the waste streams will have to be pumped into storage containers and transported onshore for disposal or recycling. The constituents of the waste stream will dictate the cost of disposal. Solid wastes such as discarded batteries, glycol filters and absorbent rags will also have to be handled onshore according to acceptable disposal practices. Many platforms will have chemical treatment additives as well as possible toxic/hazardous materials such as methanol, biocides, antifoams, oxygen scavengers, corrosion inhibitors, paints and solvents, some of which may cause damage to the marine environment if accidentally discharged. Therefore, the procedures for handling and containing should be followed. The presence of radioactive scale, NORM, PCBs, hydrogen sulfide, etc., should have been detected during the environmental survey and a disposal plan developed. Disposal will generally mean transporting this material in drums to disposal wells or approved landfills.

Prior to removal, a detailed plan on how each material will be disposed of should be developed. The plan should identify recyclable materials such as steel, rubber and aluminum and the recycling centers that will take delivery of these materials. For those items not to be recycled, the abandonment plan should include the environmental impact that disposal will have on the dump site.

After the process piping and vessels have been cleaned and it has been determined that there is no future utility for the pipelines, pipeline decommissioning should commence. Pipelines departing the platform will either board another platform or commingle with another pipeline via a sub-sea tie-in. A surface to surface decommissioning is the least costly to perform. This requires pigging the line to vacate any residual hydrocarbons followed by flushing with one line volume of detergent water followed by final rinsing with one line volume of sea water. Upon completion of the pipeline purging operation, pipeline ends should be cut, plugs inserted and the ends buried below the sea-bed. In the case of a sub-sea tie-in, details of the sub-sea tap will have to be obtained so that pipeline decommissioning plans can be developed. The flowline can be pigged, flushed and disconnected if the receiving platform can accept the fluids, otherwise the pipeline segment will have to be isolated from the adjoining trunkline and then decommissioned. This will generally involve a boat capable of mooring over the sub-sea tie-in, connecting flexible piping to the tie-in using divers or ROVs, then pumping pigs, detergent water and rinsing water toward the platform for handling.

Decommissioning involves a variety of waste streams, disposal handling methods and specialty contractors. This phase more than any other will determine the success of the abandonment and salvage.

5.4.5 Structure removal

The method of a structure removal will be determined by the structure design, availability of removal equipment, method of disposal and the legal requirements governing the jurisdiction in which the abandonment is to take place. The legal requirements will usually be based on the social, economic, environmental and safety concerns of the local governing bodies. All of these issues are interrelated and will have a direct effect on the overall cost of the removal operation. The economics of the removal are of prime importance to the party responsible for the removal, whether it is a contractor, local government or producer.

The majority of structures in moderate environments will be totally removed. Most regulatory bodies throughout the world require that the structure be removed anywhere from the mudline to 5 m below. The chief consideration when developing a removal procedure is to determine if the piles or well bores will be severed using explosive or non-explosive methods.

(a) Removals using explosives. Severing platform piles and well bores with explosives is relatively effective compared with using non-explosive methods, as multiple cuts can be made in a short period of time. This limits the amount of time that removal support equipment must be on the site and limits personnel exposure to unsafe working conditions. Generally, explosives are the least expensive and the method of choice for structure

removal. However, when explosives are used, more stringent regulations may become effective, including consultations with the local fishery or natural resource agencies. A project plan should allow lead time for consultations and permit approval from these agencies. Explosives emit high-energy shock waves that can be harmful to habitat fisheries immediately adjacent to a removal site and some endangered species, such as marine turtles or mammals, in close proximity to the detonations may be mortally affected by these shock waves. Local regulations should be researched to determine limits to the amount and size of charges allowed and to determine if moratorium periods exist during marine migration periods.

In some areas, a condition for approval requires that observers from the local regulatory agencies and/or resource groups be present at the removal site prior to detonations, to observe that permit requirements are being met and to ensure that no harm is done to endangered species that may be in the area. Other conditions that may be imposed to limit the effects of explosives on habitat fisheries are predetonation aerial surveys, daylight-only working hours and staggered detonations.

Numerous studies are ongoing to reduce the harmful effects on local fish populations during detonations. Focus or shaped charges concentrate the detonation energy to the target, requiring less explosive weight with the same cut efficiency. The disadvantage of focus charges is that they need to be properly set in the well bore or pile and corrosion scale or damage in the piles can inhibit the charge from applying its full energy to the target.

A technique to reduce the effects of explosives on habitat fisheries is to evacuate the platform piles of all water. This reduces the resistance of the shock wave from the charge to the target. Also, special shock-attenuating blankets can be placed at the mudline to limit the energy emitted from the sea floor. Another technique may be to deter fish from entering the blast area. Small, preset charges set off prior to the detonation of the severing charges, known as scare charges, have been used. However, there are risks that scare charges may actually draw some species of curious fish toward the blast site. The use of strobe lights similar to those used to keep fish away from dam intakes may be effective.

(b) Non-explosive removals. An option for the project engineer is to eliminate the use of explosives in the removal. Use of non-explosive removal techniques eliminates the impact due to shock waves. Consequently, costs and time associated with observers and additional permit conditions may be eliminated. However, salvages using non-explosive methods can be more costly since only one pile or well bore can in practice be severed at one time. Each non-explosive cut will typically take several hours to perform. The additional time and cost can be minimized depending on the scope of work and with proper project planning. The

project engineer should perform a precise cost estimate, evaluating the costs and risks between using explosive and non-explosive methods of severing. The following is a discussion of some non-explosive severing techniques.

High-pressure water/abrasive cutters. This system uses a high-pressure water jet operating at anywhere from 200 to 4000 bar to perform the cut. In some systems, sand, garnet or other type of abrasive is injected into the water stream to aid in the cutting process. The nozzle is lowered into the hole attached to an umbilical hose line or a hard pipe supply line. The nozzle is rotated 360° inside of the pile or well bore until the cut comes back on itself. One of the advantages to the system is its effective cutting ability. The casing strings do not have to be concentric in the well bore. The wall thickness of the platform piles is typically not a concern. The reaction of the water spray and the returns of the water give the operator an indication that the cut is actually being made. Some disadvantages are the tendency for system breakdowns due to the high working pressures, electrical and mechanical complexities, the delicate characteristics of the abrasive injection and wear and tear on the nozzle. Interrupting the cutting operation requires that the tool be placed in the exact location of the cut to avoid incomplete cuts. The effectiveness of these cuts is reduced at deeper cutting depths owing to the hydrostatic head that the water jet needs to overcome. As with all cutters, the tool must be centered in the pipe to maximize cutting efficiency. This can be difficult in heavily scaled pipes or in battered piles. Topside instrumentation can be used to monitor the position of the cutting tool during the cut. Camera technology has been used to inspect visually the status and effectiveness of a cut.

Mechanical cutters. Mechanical cutters use tungsten bit cutters that are extended from a housing tool with hydraulic rams. The tool is rotated continuously using friction to perform the cut. Disadvantages include frequent breakdowns of the tool due to frictional wear and tear, high labor intensity in handling heavy and bulky tools, the need for a work platform around the piling/well bore to be cut and poor cutting performance on non-concentric casing strings. Also, it can be difficult for the operator to determine if a cut is complete. Shifting of the well strings or platform piles downward can jam the tool into the kerf of the cut.

Diver cut. Internal or external pile or well bore cuts can be made with divers using underwater burning equipment. This type of cut can be made internally if there is access for the diver into a large-diameter casing or piling. If there is no internal access and the cut must be made below the mudline, a trench must be excavated to afford the diver access to the area to be severed. In some soils, keeping a trench open to the required 5 m depth may be impractical and may put the diver at undue risk from trench collapse. If the cut must be made below the mudline, the local regulatory agencies should be consulted as to the required depth of the cut. This may require obtaining a waiver to reduce the required cutting depth due to local soil characteristics and safety concerns for the diver personnel. Another concern to the diver's safety is oxygen entrapment in the soil near the cut or on the backside of the pipe being cut. Oxygen build-up can lead to an explosion if contacted with a flammable source such as a burning rod.

Cryogenics. Cryogenics is a little used technology that consists of freezing the platform pile in the area of a cut with CO_2. A relatively small explosive charge is then placed at the elevation to be cut and detonated. The brittle behavior of the frozen steel theoretically requires little energy to sever the pile. To use cryogenics, water must be completely evacuated from the pile, which can be a time-consuming operation. Also, the cutting efficiency is hindered by the freezing of the mud on the exterior of the pile to be severed.

Plasma arc cutting. Plasma arc cutting is achieved by an extremely high-velocity plasma gas jet formed by an arc and an inert gas flowing from a small-diameter orifice [10]. The arc energy is concentrated on a small area of metal, thus forcing the molten metal through the kerf and out of the backside of the pipe. Water can be used as a shielding agent to cool and constrict the arc [10]. The process requires a high arc voltage provided by specialized power sources. This method has not been used often, and is therefore not highly developed. For it to be effective, the tool must be set properly in the cut pipe. It is difficult to determine if a cut is being made unless camera technology is used.

Whether using explosives or non-explosive methods of severing, obstructions in the pile can hinder the proper placement of charges or cutting tools in the well bore or pile. Examples of obstructions include scale build-up, damaged piling, mud or pile stabbing guides. The removal of mud from the pile is generally accomplished with the use of a combination of a water jet and air lifting tools. When properly designed, these work well. This task is traditionally performed after the topside deck has been removed by the heavy lift contractor. A more cost-effective technique is the use of a submersible pump to excavate mud from the platform pile prior to removal. A small inexpensive work spread can be mobilized to the site prior to the arrival of the heavy lift equipment to perform this task. A window is cut into the jacket leg/pile and the submersible pump is then lowered down the jacket leg on a soft umbilical line.

(c) Alternative removal techniques. Most structures are removed with heavy lift equipment such as oceangoing derrick barges. In remote areas of the world, another concern in dislodging the platform from the sea-floor is

the availability of salvage support equipment. International Maritime Organization (IMO) guidelines permit the host government to allow a structure to remain in place provided that the structure is properly maintained to prevent failure. Maintenance costs over the life of the installation may eventually exceed the cost of the removal. When left in place, the platform may remain a hazard to navigation, exposed to collapse during storms or become a haven for refugees. These risks and liabilities may outweigh high removal costs to the host government and the operator, thus the decision to remove the platform may prevail.

Innovative methods of decommissioning, removal and disposal must be proposed to offset the lack of available salvage equipment and the high cost of equipment mobilization to remote areas. An alternative approach is cutting the platform into small, manageable components that lighter, more cost-effective equipment work spreads can handle. The equipment that may be used includes crawler cranes, A-frames and portable hydraulic cranes mounted on a cargo barge and these methods use readily available equipment that can be rigged up inexpensively.

Besides additional decommissioning hazards, other precautions must be taken during a sectional removal. Caution should be taken when cutting into a structural member as gases from scale or other sources may have built up over time inside of the member, and flame cutting into the member could result in an explosion. Each member should be drilled and checked for gases prior to any flame cutting operations. Sectional removal requires a detailed plan for lift sling connections and cut locations for each component to be removed. Lift slings should be properly attached so that a safe, level lift can be made, and a level, controlled lift will eliminate load shifting and allow for proper set-down on the transport barge without undo risk to personnel or equipment. Removal of a structure in sections may require multiple cuts underwater. The same concerns with load shifting and sling placement exist for underwater cuts as they do for above-water cuts. These cuts should be performed and/or supervised by skilled divers. Divers' activities can be reduced by using small shaped charges to sever members or by performing cuts with ROVs.

Other forms of less expensive salvage support equipment include barge-mounted 'stiff legs' and converted jack-up drilling rigs. Stiff legs have the capability to handle large lifts, but generally have limited hook height and are not easily maneuverable during the lifting and setting of components on transport barges. Stiff legs are generally built to work in protected waters and are affected by rough seas.

Converted jack-up drilling rigs are becoming more common in the abandonment industry. Companies are converting obsolete rigs to lift vessels to take advantage of the increased need to supply salvage support equipment. This type of equipment can work in heavy seas when in the jacked-up position, but in the floating condition maneuverability is limited.

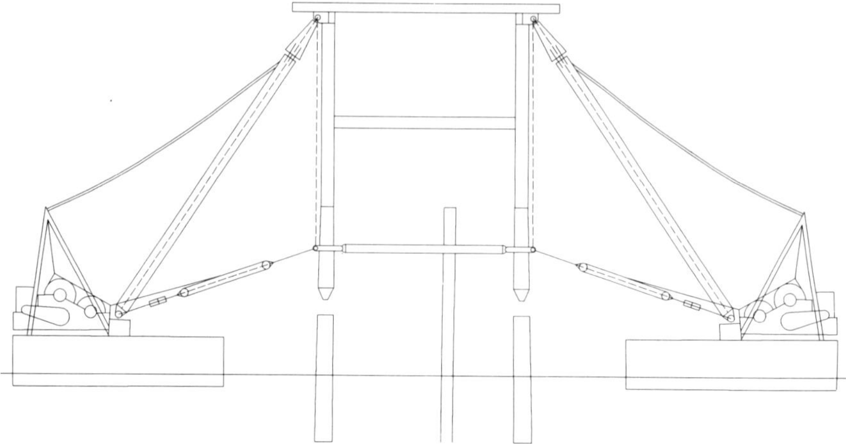

Figure 5.8 Versatruss method. Source: Versabar Inc.

Extreme caution must be taken when bringing transport barges near the jack-up rig to accept platform components. The legs of a jack-up rig cannot withstand any severe impact loading.

Another technique that can be used for the lifting of platform topsides is the Versatruss system (Fig. 5.8). The method uses a series of A-frames mounted on tandem cargo barges. The combination of the A-frames, tension slings and the topside deck create a catamaran and truss effect for lift stability. This lift method also uses available equipment and requires relatively low-cost preparation.

(d) Alternative structure uses. In some areas of the world, the host government is either wholly responsible for structure removal or, through participation by a national oil company, is partially responsible for the cost of structure removal. The political entity may not want to dedicate funds to a non-revenue generating project. These states may decide that leaving the structure in place is the only alternative. IMO guidelines give local states the discretion to allow offshore structures to remain in place if the removal is not economically feasible. In these situations, operators will need to review the contract terms for possible ongoing or future liabilities.

Alternative uses for the platform should be explored. The benefit of the alternative use should offset the costs to maintain the structure in place. Some alternative uses may be as follows:

- fish farm;
- marine laboratory;
- military radar support structure;
- weather station;

- oil loading station;
- spur for deep-water developments;
- aviation/navigation beacon;
- tourism/recreational;
- power generation, i.e. wind/wave.

Leaving the structure in place should not create a hazard to local fishing industries or to navigation in the area.

(e) Platform reuse. Reuse is another option. If a potential development can finance the removal of a structure, this relieves the non-revenue producing property from absorbing the salvage costs. Platform reuse can reduce the cycle time to get the new development in production, generating cash. However, an immediate reuse should be identified when decommissioning is undertaken. Storage of the platform onshore prior to identifying a reuse can result in costs that may offset the savings from reuse.

(f) Partial removals. In the North Sea, the abandonment issue is coming into focus as some of the area's fields are reaching the end of their productive lives. Some of the world's largest structures will need to be removed before the year 2005. The large component weights will result in removal costs that may exceed the cost of the original installation. These removal costs will largely be absorbed by the local governments because of tax breaks from removal costs available to the operator. Thus, local governments may need to regulate and monitor abandonment procedures to allow for a cost-effective removal strategy without compromising safety or the environment.

Any cost savings of abandonment in the North Sea will come from partial removal. Estimates show that total removal (Fig. 5.9) of structures now existing in the UK continental shelf will cost $6.6 billion and partial removal $4.5 billion [8]. These partial removal methods will consist of the following:

- partial removal of jacket component (Fig. 5.10);
- toppling in place (Fig. 5.11);
- total removal of topside and toppling in place of the jacket only (Fig. 5.12);
- emplacement (Fig. 5.13);
- transport to rigs to reef site;
- deep-water dumping.

The choice of removal method will depend on cost, proximity to disposal sites, availability of removal equipment, location of the removal relative to shipping lanes and fishing interests, and safety and environmental issues. In

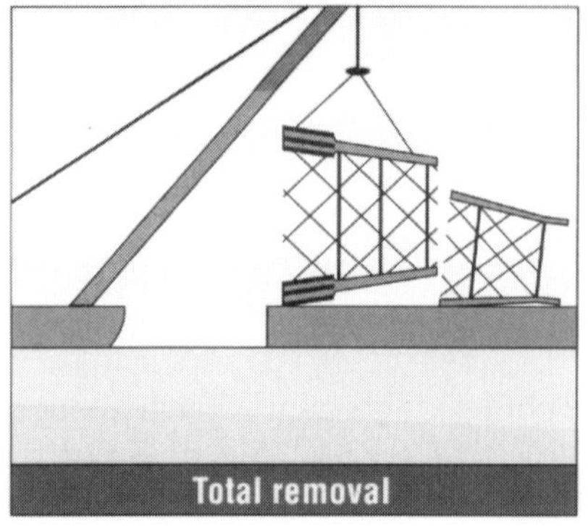

Figure 5.9 Total removal [11].

Figure 5.10 Partial removal [11].

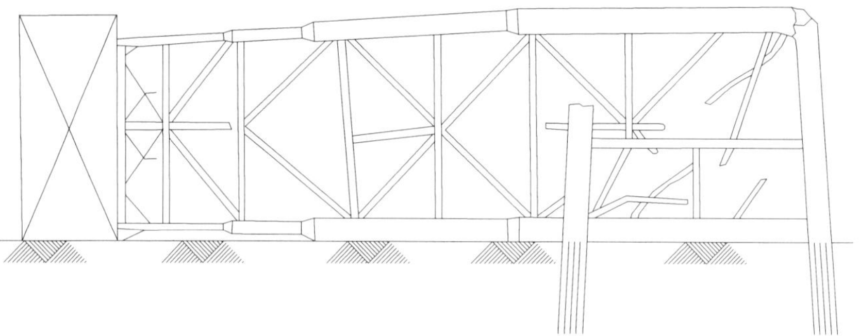

Figure 5.11 Hinge point in jacket leg.

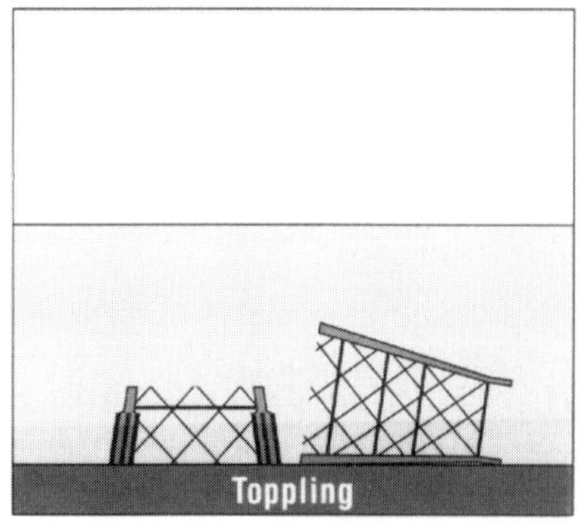

Figure 5.12 Toppling [12].

Figure 5.13 Emplacement [11].

addition, the disposal method will play a key role in the decision on the removal method. The next section summarizes the alternatives and key issues concerned with structure disposal.

5.4.6 Disposal

Once a platform or portions of a platform have been removed, the structure must be disposed of. Some disposal options include the following:

- transport inshore for disposal, storage or recycling;
- toppling in place;
- disposal at a remote rigs to reef site;
- emplacement;
- deep-water dumping.

The owner must be aware of the social and political climate in the area where abandonment and disposal are to occur. Public perception will play a key role in performing a successful disposal program. All environmental issues should be addressed by the operator up front, all stakeholder groups and regulatory agencies should be informed of the disposal plans and environmental effects of the plan and alternatives must be addressed. Miscommunication and misinformation to or from interested stakeholders could lead to the downfall of an otherwise well planned abandonment strategy.

Non-jacketed designs such as floating production systems, concrete structures, steel gravity structures and spar loading buoys will probably be refloated in whole or in part and towed away, and disposed of in deep-ocean disposal sites or brought inland for dismantling. Steel-jacketed structures will probably be disposed of in one or any combination of the ways mentioned above. Explanations of these methods are detailed below.

(a) Disposal inshore. Generally, topside deck facilities will be disposed of inshore because of the difficulty and expense in completely removing all of the hydrocarbons and their by-products at the installation site rather than shoreside. When disposal inshore is chosen, the structural component will be either totally or partially cut up for scrap. Portions may be disposed of in landfills or hazardous waste sites or recycled. The component may also be stored for future use or refurbished immediately if a reuse is identified. Once a structure has been removed for inland disposal, possession of the removed structure and their components is usually turned over to the removal contractor in exchange for a portion of the scrap value. The steel in offshore structures is of relatively good quality and is readily taken by steel mills for recycling. The handling and disposal of all other materials associated with the removal should be detailed in the pre-abandonment disposal plan.

The UK Offshore Operators Association performed a detailed assessment of the amount of waste materials projected from the disposal of offshore structures from the North Sea. Disposal amounts were calculated and the effect on the available landfill space was determined [13]. Another study, performed by planners for a removal in Norway, detailed costs and benefits of recycling an old structure. The study compared the emissions placed in the atmosphere by melting and breakdown to the cost of the energy and associated emissions generated if the same component was built new [4]. An environmental assessment could be made based on these studies. These types of analyses would be beneficial to the operator and regulatory bodies when the decision is made to bring offshore components inshore.

(b) Toppling in place. This method is generally performed after the topsides have been removed. The legs are severed selectively so that the jacket can be toppled with two legs acting as hinges (Fig. 5.11). The toppled structure must maintain 55 m of clear water column clearance as required by IMO guidelines. Another method is to cut the top section completely from the lower section, lift it off, place it on the bottom to the side and topple it with heavy-lift marine equipment (Fig. 5.12). In the Gulf of Mexico, toppling may only be performed in established reef sites. In the North Sea, a 500 m clear exclusion area must be maintained for the benefit of the fishing industry, much the same as when the platform is operational. The site should be clearly marked with buoys. In the Gulf of Mexico, the buoys are maintained by the state, whereas in the North Sea the responsibility remains with the operator to mark and maintain the site. In other parts of the world, marking is negotiable between the operator and the host government. The site should also be placed on navigation charts.

(c) Rigs to reef. When an offshore structure is removed, a habitat for fisheries and a source of recreational fishing is lost. It has been estimated by the Gulf of Mexico Fishery Management Council that oil and gas structures account for 23% of the hard bottom habitat in that area [2]. Prior to the emplacement of petroleum-related structures, suitable habitats in which new species could expand their range did not exist. Countries may establish a rigs to reef program to maintain the hard bottom habitats that these structures provide. When performing a cost comparison between dumping a platform at a reef site or disposal inshore, the size of the platform, location of the platform in relation to the placement site and the transport costs are the main factors.

A common method of transportation is to tow the structure while on the hook of the removal barge crane. Derrick barges are not constructed for this purpose, so extreme caution should be taken if this method is used. Weather and obstructions both below and above the water along the tow route should be anticipated. If the heavy-lift equipment has to accompany the structure to the placement site, this subjects the project to costly weather and operational delays. The need for the derrick barge at the disposal site can be avoided by setting up a winch and snatch block system to push the structure off the transport barge. These costs have to be weighed against the removal and transport of the platform components inshore. A rigs to reef program benefits the fish population and provides a popular source of recreational fishing while giving the project engineer an additional option to reduce platform removal costs.

(d) Emplacement. Emplacement (Fig. 5.13) is much the same procedure as toppling except that the top section is completely cut from the lower section, lifted off and placed next to the lower section.

(e) Deep-water dumping. This method is particularly reserved for huge floating systems located in the North Sea. Essentially, the structure is disconnected from its moorings and towed to the deep ocean waters where it is then flooded and sunk. Prior to any dumping operations, it is important to confirm that all components placed in the ocean waters are free of hydrocarbons in harmful quantities to avoid pollution of the open sea.

Partial removal may consist of any combination of the above-listed options. The method of structure and component disposal should be based on legal, environmental, safety, financial and timing issues. Identification of a disposal site and its proximity to the removal site must be considered to perform a cost analysis on the most effective disposal method.

An inherent concern with any disposal method is tying down the salvaged component on the transport barges, which can be particularly difficult and dangerous in rough weather. A well thought out plan has to be enacted to assure a safe and stable lift and placement on the transport barges. All components should be tied down with a system that provides the same integrity as when the platform was towed offshore for installation.

A marine surveyor should be available on-site to monitor the tie-down operations. The marine surveyor's responsibilities include confirming that the structure is secure for tow, certifying that the tow route is free of overhead, width or bottom obstructions and verifying proper ballast of the transport barge.

5.4.7 Site clearance

The final phase of the abandonment process involves restoration of the site to its original predevelopment conditions by clearing the sea floor of debris and obstructions after platform removal. If the abandonment was a partial removal, site clearance procedures may vary from a total removal. In the case of total removal, debris should be removed, leaving the site trawlable and safe for fishing or other maritime uses.

A site clearance plan may consist of two or three phases, depending on the information gathered during the preabandonment surveys and the water depth at the location. The first phase may occur before actual removal with divers making sector sweeps around the platform site during pipeline decommissioning. High-frequency sonar can be used to locate obstructions and direct divers to debris. Searches should be performed inside and outside the platform a distance of at least 100 m. Following this initial debris removal, site clearance can be discontinued until the structure removal has taken place.

Once the structure has been removed, the site is ready for a final clean-up if required. In shallow waters, a trawling vessel can be used to simulate typical trawling activities that may occur in the area after the platform removal.

Deeper water sites may not require trawling simulations to clear the area. Proper planning prior to the removal of debris can make a significant difference in controlling the costs. The geophysical survey performed with the side scan sonar during the preabandonment survey phase should identify the major debris, and this information will provide the basis for selecting the most effective equipment, personnel and timing. Equipment and personnel can range from a dive crew retrieving debris off a boat during pipeline abandonment, through a small derrick barge with divers to a boat capable of mooring over debris targets away from the platform. In deep waters, it is crucial to determine the amount and type of debris to size the equipment and work crews properly. Upon completion of the bottom clean-up, job completion summaries should be submitted to the proper governing body.

5.5 Conclusion

The offshore oil and gas industry will be faced with more than 7000 platform removals, each of which will include a multitude of tasks, involving interaction between operators, contractors, regulatory agencies, governing bodies and the public. Of importance to the operator will be the cost effectiveness of the removal. The operator will also share in the public and regulator's concern on the effect that the removal will have on the environment. The operator should focus on early interaction with regulatory agencies, detailed preremoval planning and engineering, efficient interface and timing of equipment and personnel movements, safety and disposal to assure a cost-effective removal with minimum impact on the environment. Finally, all stakeholders should continuously pursue advances in rulemaking and technology to ensure each abandonment program improves on the one that preceded it.

References

1. Offshore abandonment heats up: North Sea, Gulf of Mexico deepwater platforms are costly, difficult to remove. *Journal of Petroleum Technology*, August 1995, 643–5.
2. United Nations Economic and Social Commission for Asia and Pacific (1994), *Waste Recycling of Obsolete Oil and Gas Production Structures in Asia Pacific Waters (Overview)*, Volume 2, United Nations, New York.
3. United Nations Economic and Social Commission for Asia and Pacific (1994), *Waste Recycling of Obsolete Oil and Gas Production Structures in Asia Pacific Waters (Overview)*, Volume 1, United Nations, New York.
4. Habersholm, L. and Robberstad, L. (1994) Environmental impact from removal of offshore installations: the North East Frigg field case. *Second International Conference on Health, Safety and Environment in Oil and Gas Exploration and Production*, 25–27 January, Jakarta, Indonesia.
5. Deepsea disposal for *Brent Spar. Offshore Engineer*, March 1995, 12–9.
6. Dymond, P.F. (1995) The operators' perspective on the decommissioning of UKCS offshore installation. *Petroleum Review*, April, 176–8.
7. Hoff, G.C., Mobil Exploration and Producing Technology, personal communication, October 1995.
8. Roberts, M. (1994) Abandonment (field decommissioning): the legal requirements, *26th Annual OTC* 2–5 May, Houston, TX.
9. Shaw, K. (1994) Decommissioning and abandonment: the safety and environmental issues, *SPE Paper 27235*, 293–300.
10. American Welding Society (1987) *Welding Technology Welding Handbook*, 8th edn, Volume 1, American Welding Society, Miami, FL.
11. UKOOA (1994) *Decommissioning UK Offshore Installations, the Right Balance*, UKOOA, London.
12. Snell, R., Corr, R.B., Gorf, P.K. and Sharp, G.K. (1993) Options for abandonment; an operators perspective, IBC Technology Services, Ltd. *Decommissioning and Removal of Offshore Structures Conference*, 15–16 September, London.
13. UKOOA (1995) *An Assessment of the Environmental Impacts of Decommissioning Options for Oil and Gas Installations in the UK North Sea, Executive Summary*, Auris International, Aberdeen.

Further reading

Minerals Management Service (1992) *Notice to Lessees No. 92–02: Minimum Interim Requirements for Site Clearance (and Verification) of Abandoned Oil and Gas Structures in the Gulf of Mexico*, May.

Knott, D. (1995) North Sea operators tackling platform abandonment problems. *Oil and Gas Journal*, 20 March, 31–40.

Buckman, D. (1994) Abandonment – the North Sea's newest industry. *Petroleum Review*, September, 413–5.

Shaw, K. (1994) Decommissioning and abandonment: the safety and environmental issues. *Second International Conference on Health, Safety and Environment in Oil and Gas Exploration and Production*, 25–27 January, Jakarta, Indonesia.

Bartlett, T.W. (1994) Deconstruction of an offshore platform. *26th Annual OTC*, 2–5 May, Houston, TX.

6 Tanker design: recent developments from an environmental perspective

G. PEET

At four minutes past midnight on 24 March 1989, the *Exxon Valdez* went hard aground at Bligh Reef in Prince William Sound (Alaska). Oil began leaking from the vessel immediately, at a rate of tens of thousands of barrels per hour [1]. The *Exxon Valdez* was to become one of the icons of environmental disaster for years to come. Many publications, most with convincing illustrations, were telling the story to the world, e.g. some 40 pages of text and illustrations in *National Geographic* of January 1990:

> In the beginning, when the supertanker *Exxon Valdez* gutted herself on Bligh Reef and vomited 11 million gallons of crude oil into Alaska's exquisite Prince William Sound, it seemed truly like the ending of a world.
>
> Ashore it was war.

Vast quantities of oil on the shore do provide for excellent emotive publications and subsequently also for a strong drive for political action in response to these vast quantities of oil. This was also true in the case of the *Exxon Valdez*.

Whilst the human element was the major factor in causing the grounding of the *Exxon Valdez*, the political fall-out of this accident quickly focused on tanker design (i.e. on the double hull tanker design) as one of the major targets for political post-accident opportunism.

This chapter is focused on the post-*Exxon Valdez* tanker design discussions at the international level (i.e. the International Maritime Organization) as prompted by the developments in the USA, but not without first making an effort to put the importance of tanker accidents and tanker design in some perspective.

6.1 Tanker accidents

Many believe oil tanker accidents to be the major source of marine pollution. It is not; other substances also pollute the environment. As for oil, the situation is as follows.

Every year over 1500 million tonnes of oil are transported over the world's seas by ships as cargo (some 1480 million tonnes in 1989), as fuel oil (an estimated 110 000 tonnes in 1989) or for other uses [2]. A relatively small but in absolute numbers substantial amount of oil never reaches its destination as cargo or is effectively used in the ships' engines: it ends up in the marine environment either as a result of operational discharges or as a result of accidents with ships.

Various estimates have been made to assess the total amount of oil entering the world's seas and oceans. Table 6.1 summarizes some of these estimates, for the years 1973, 1981, 1990 and 1992. The estimate for 1981 provides an indication of the importance of shipping as a source of marine oil pollution. From an estimated total of 3.28 million tonnes of oil entering the marine environment, some 1.5 million tonnes (46%) were generated by shipping, of which some 410 000 tonnes (13%) were the result of (tanker) accidents.

The most recent (albeit not completely new) estimate was given in a report by GESAMP (the IMO/FAO/UNESCO/WMO/WHO/IAEA/UN/UNDP Joint Group of Experts on the Scientific Aspects of Marine Pollution) and was put at an annual introduction of 2.35 million tonnes of oil in the marine environment [3]. GESAMP did not develop new estimates; its report merely uses the most recent estimates available from previous studies (in Table 6.1 the figures given in either column 2 or 3).

The GESAMP report then continues to provide data (taken from other publications) regarding regional sea areas (see Table 6.2). When the estimates for these regional sea areas are added, a different picture emerges: at least 7.3 million tonnes of oil are entering the world's seas and oceans annually if those estimates are to be correct.

Table 6.1 Estimates of the quantities of oil annually entering the marine environment[a] (millions of tonnes)

Source	Year of estimate/publication			
	73/75	81/85	90/90	92/93
Natural sources	na[b]	0.25	na	0.25
Oil exploration	na	0.05	na	0.05
Shipping:				
Discharge of bilge and fuel oil plus operational losses from oil tankers	1.08	1.02	0.41	0.41
Tanker accidents	0.20	0.41	0.11	0.11
Accidents with other types of ships	0.10	–	0.01	0.01
Other (ports, shipyards, scrapping, etc.)	0.75	0.07	0.04	0.04
Atmospheric deposition	na	0.30	na	0.30
Land-based sources	na	1.18	na	1.18
Total amount of oil per year	–	3.28	–	2.35

[a]Sources: Refs 3 and 4.
[b]na = Not available.

Table 6.2 Estimates of the quantities of oil annually entering the marine environment in various regional sea areas (millions of tonnes)[a]

Area	Quantity
North Sea	260 000
Baltic Sea	21 000–60 000
Mediterranean Sea	500 000
North west Atlantic	–[b]
Wider Caribbean	950 000
West and Central Africa	–[b]
South Africa	–[b]
East African region	–[b]
Red Sea, Gulf of Aden	–[b]
Arabian/Persian Gulf	160 000[c]
Indian Ocean, Arabian Sea	5 000 000
Indian Sea, Bay of Bengal	400 000
South east Asia	–[b]
South east Pacific	–[b]
North east Pacific	–[b]
Arctic Ocean	–[b]
Antarctic Ocean	–[b]
Total	~7 300 000

[a]Source: Ref 3.
[b]No overall estimate.
[c]Pre-Iran–Iraq and Gulf wars.

More detailed estimates, providing information about the contribution of shipping, are also available from other sources, e.g. for the North Sea (Table 6.3). The fact that the North Sea States could not jointly agree on the amounts of oil 'produced' by illegal discharges and tanker accidents does not mean that there have been no estimates for, in particular, the amounts of oil discharged illegally. These estimates range from some 15 000 to as much as 60 000 tonnes per year for the whole North Sea. This, of course, adds considerably to the total input given in the above list.

From all these figures, it is clear that shipping is only one (albeit not unimportant) source of marine oil pollution and that (tanker) accidents are not the most important source of oil entering the sea from ships. It is important, however, to keep in mind that these figures are nothing but estimates and that, if these figures were to be prepared on the basis of information available now, the situation might be different. The recent accident with the oil tanker *Braer* in the Shetland Islands introduced some 85 000 tonnes of oil into the northeastern Atlantic (and North Sea). Had this accident happened in the North Sea, it would have more than doubled the total amount of oil in the lowest estimate, and it would have added more than half of the amount of oil to the highest estimate given in Table 6.3.

Table 6.3 Estimates of the quantities of oil annually entering the marine environment in the North Sea (millions of tonnes)[a]

Source	Quantity	Sub-total
Transportation:	0.001–0.002	
Legal discharges		0.001–0.002
Illegal discharges		No agreed estimate
Tanker accidents		No agreed estimate
Production platforms	0.029	
Atmospheric deposition	0.007–0.015	
Land-based sources	0.029–0.081	
Dumping operations:	0.004–0.022	
Sewage sludge		0.001–0.010
Industrial wastes		0.001–0.002
Dredge spoils		0.002–0.010
Natural seeps	0.001	
Total	0.071–0.150	

[a]Source: Ref 5.

Although, on a global scale, the amount of oil entering the marine environment as a result of accidents may be relatively small, on a regional or local scale the relative importance of such an accident may be substantially higher. In certain areas an accidental spill might even account for almost 100% of the total input of oil in that area. That is why the *Exxon Valdez* was an ecological disaster where it happened. That is how even a relatively small spill, such as the approximately 750 tonnes in the December 1988 accidental spill off Grays Harbour on the Washington State coast, could develop into a major environmental problem at the time it happened.

From these figures, it is also clear that there are no reliable data regarding the input of oil into the marine environment; the various estimates are too far apart. If it is possible to extract two completely different totals from just one publication (i.e. the 2.35 million tonnes estimate in the GESAMP report and the 7.3 million tonnes total that can be extracted from the same report), the only valid conclusion seems to be that we do not really know how much oil enters the marine environment, owing to a lack of reliable data.

The same is true with respect to the trends in the amounts of oil entering the marine environment. If one takes the figures from Table 6.1 it would seem that the total amount of oil introduced into the marine environment by shipping has dropped from an estimated 3.28 million tonnes in 1981 to 2.35 million tonnes in 1992; a substantial drop indeed and, if these figures are correct, the claim for success would be more than justified. The question is, of course, whether these figures are correct. It is likely that they are not.

An analysis of the method by which the figures for 1990 were determined shows that they are based on a large number of assumptions [6]. The estimate for operational discharges from oil tankers, for instance, is based on the amount of oil tankers can discharge legitimately under the MARPOL Convention [7] plus an estimate of the amount of oil that would be discharged unlawfully. The assumptions on which this assumption is based include the degree of compliance (estimated to range from 80% for oil tankers smaller than 20 000 DWT (deadweight tons) to 99% for oil tankers larger than 150 000 DWT) and on estimates of how much a non-complying ship would discharge unlawfully. Both estimates may be correct, but they may also be too optimistic.

It is important to note that it is not possible to conclude that less oil is being discharged into the marine environment if the figures used for that conclusion are based on the assumption that, in accordance with international regulations, less oil is being discharged.

6.2 Tanker design

If one looks closely at data with respect to the causes of major oil tanker accidents, one conclusion is evident: human error is a major cause. Human error may take the dimension of mistakes that provide the direct cause of an accident, but it also includes such (often implicit) decisions as allowing substandard maintenance of vessels, allowing substandard crews to run large and complex vessels, etc.

In a report written after the major tanker accident involving the grounding of the *Braer* on the rocks of Garths Ness in the Shetland Islands on 5 January 1993, important words are dedicated to the importance of the human element in shipping disasters [8]:

> In the last analysis it is individuals whose conduct leads directly or indirectly to pollution. It is generally accepted that human error is the cause of about four fifths of marine accidents. We are surprised that the figure is so low: we believe that human error, at some stage in a chain of events which could start with the design of a vessel, is the root cause of virtually all accidents. The only exception that we can see is the highly unusual case of unforeseeable forces overwhelming a vessel or her crew.

Whilst tanker design is definitely not a major cause of tanker accidents, the design could be an important element in determining the outflow of oil after an accident. Efforts to minimize operational pollution through design measures had already been introduced in the 1973 MARPOL Convention: requirements for segregated ballast tanks (SBT) were established for vessels of 70 000 DWT and larger to minimize accidental pollution. In 1978 the concept of protective location (PL) of SBT was introduced into the MARPOL Convention (Regulation 13E of Annex I at that time). It applied to vessels of 20 000 DWT and larger and was designed to provide protection for a percentage of the sides and the bottom of oil tankers. After 1978, additional accidents with subsequent pollution occurred, making a case for shielding the entire cargo block by protective spaces. These protective spaces would then have prevented or mitigated many of these accidents and subsequent pollution.

With the *Exxon Valdez* in mind, this concept was to be the driving force behind the post-*Exxon Valdez* efforts to develop improved tanker designs and to develop the necessary national and international regulations to ensure the future use of such designs in new tankers.

6.3 New tanker design standards: the USA takes the lead

As early as in October 1989, the International Maritime Organization (IMO) adopted a resolution calling for the development of an international convention on oil pollution preparedness and response. This resolution was a direct response to the *Exxon Valdez* accident. A draft text for such a convention was submitted by the USA to IMO's Marine Environment Protection Committee (MEPC) for discussion in March 1990, and the convention was adopted in November 1990 [9].

It was one of two major post-*Exxon Valdez* initiatives at the IMO by the USA. The second came in November 1990 when the USA delegation at IMO's MEPC presented proposals to amend the MARPOL Convention. At that time the 1990 Oil Pollution Act had already been developed and the proposal's intention was to make OPA 1990's double hull provisions part of the IMO instruments, i.e. part of the MARPOL Convention.

The USA-proposed amendments were designed 'to make double hull construction mandatory for new oil tankers and sought assistance of IMO in developing technically sound criteria for the construction of double hull tankers' [10]. In its proposal, the USA stated the following: 'Of immediate concern to the United States is the prevention of pollution from tankers that will be undergoing construction or major conversion under newly placed and future contracts. Given the current advanced age of the world tanker fleet and the large number of tankers now on order, it is reasonable to conclude that a significant number of tankers will be constructed in the near future'.

In the USA, a study had already been started by an *ad hoc* committee on Tank Vessel Design of the US National Academy of Sciences (NAS) Marine Board (in addition to 'traditional' experts, this committee also had one member from the environmental lobby, Sally Ann Lentz of the Oceanic Society and Friends of the Earth USA). This study [4] focused on how alternative tank vessel (tanker and barge) designs might influence the safety of personnel, property and the environment and at what cost. The study considered a wide range of engineering considerations (e.g. hull strength; tank proportions, arrangements and stability; salvage concerns; safety of life) and design alternatives (e.g. barriers; outflow management; penetration resistance). Some of its conclusions were as follows:

> Results of this study indicate that no single design is superior for all accident scenarios. Therefore, the Oil Pollution Act of 1990 which mandated double hulls for tankers travelling in US waters, should be viewed as only an interim step to reducing oil spills. More work remains to be done. [. . .]
>
> Existing design standards should be strengthened to ensure proper
>
> (1) corrosion protection [. . .];
> (2) dimensions of structural members; and
> (3) use of high-tensile steel. [. . .]
>
> Furthermore, naval architects traditionally have not designed tank vessels, at the detail level, to withstand collisions and groundings. Design based on the possibility of accidents, a practice common in many industries, should be considered for tank vessels. [. . .]
>
> Available information is inadequate for decision making. [. . .]
>
> [. . .] double hulls should save (in absence of other risk-reduction measures) an estimated 3000 to 5000 tons of oil spillage per year in US waters from collisions and groundings [. . .]
>
> Double hulls are particularly effective in low-energy (typically low-velocity) groundings and collisions. [. . .]
>
> [. . .] the committee does not favor hydrostatically balanced design options for new tank vessels. [. . .]

The committee could not agree with respect to the merits of the design alternative with intermediate oil-tight deck with double sides (IOTD w/ DS; also known as the mid-deck tanker).

> [. . .] some committee members consider the IOTD w/DS a practical and innovative application of available technology, which, if treated as an equivalent to the double hull and allowed to trade in commerce to the United States, would reduce pollution in several classes of accidents, including high energy groundings (which have caused some of the largest oil spills). Others of the committee hold the judgement that the gap between theoretical principles and practical application is wide, that the design and its application are unproven [. . .]
>
> Other design alternatives may be proposed in the course of future research. New proposals should be considered. [. . .]
>
> Double hulls need not increase incidence of fires or explosions, impair post-accident stability, or complicate salvage. [. . .]
>
> However, the risk cannot be ignored; planned maintenance and thorough inspection are critical. [. . .]
>
> Existing vessels will comprise the majority of the fleet serving the United States for many years.

The report of the committee was passed on to the International Maritime Organization.

6.4 New tanker designs: the international debate

The USA proposal prompted a vivid debate at the International Maritime Organization. It was not welcomed unanimously.

The International Chamber of Shipping, for example, argued that [11]:

- there was no single solution to preventing ship casualties and that of the many measures that might be taken, new construction standards would not make a major contribution;
- no particular design can be shown to be superior overall with respect to preventing ship casualties;
- there was no experience with the double hull construction for large tankers whilst there were concerns regarding the potential hazards of this design for oil tankers; and consequently that
- it would be wrong to amend the MARPOL Convention so as to stipulate a one-design requirement for all new oil tankers.

Similar concerns were expressed by many delegations.

Japan offered an alternative. The Japanese delegation reported the outcome of a Japanese study on the effectiveness of pollution prevention and safety aspects of three methods of design and construction of oil tankers, i.e. double hull construction, the underpressure method and the double-sided tankers with mid-height deck. According to this study, the mid-height deck tankers would be as effective as, or even superior to, double hull tankers in terms of reducing oil outflow after accidents.

Several delegations expressed their wish that, if double hulls were to be adopted as an IMO design standard for oil tankers, the mid-height deck design should also be accepted as an equivalent design.

The discussion then moved from the MEPC to the Maritime Safety Committee (MSC) and back to the MEPC, where many delegations stressed the importance that decisions be taken. A decision was taken by the MEPC at its 31st session: draft regulations with regard to design standards for oil tankers were approved for circulation to MARPOL member states with a view to adopting these at the next session. The double hull requirement in this draft regulation read as follows:

> The entire cargo tank length shall be protected by ballast tanks or spaces other than oil tanks as follows: [. . .] wing tanks or spaces shall extend for the full depth of the ship's side or from the deck [. . .] to the top of the double bottom [. . .]
>
> Double bottom tanks or spaces [. . .] may be dispensed with, provided that the design of the tanker is such that the cargo and vapour pressure exerted on the bottom shell plating forming a single boundary between the cargo and the sea does not exceed the external hydrostatic water pressure [. . .]
>
> Other methods of design and construction of oil tankers than those described [. . .] may also be accepted as alternative [. . .] provided that such alternative

provides the same level of protection against oil outflow in the event of collisions or strandings [. . .].

With respect to alternative designs (i.e. the mid-height deck design) a special Steering Committee was set up to carry out a comparative study of two oil tanker designs: the double hull and the mid-height deck designs (members of this Steering Committee included representatives from the industry, one of the members of the National Academy of Sciences committee on tanker design, a representative from the environmental organization Friends of the Earth International and several member state representatives). This Steering Committee should provide the information and recommendations as to whether or not the mid-height deck design would be accepted as an alternative to the double hull design at the same time as the new regulation for oil tanker design was adopted.

The Steering Committee carried out its work under extreme time pressure but managed to have several studies completed to serve as a basis for its conclusions [12]. Several different tanker types were compared: double hull and mid-height deck (low and high mid-decks) tankers of 40 000, 90 000, 150 000 and 280 000 DWT.

Analyses were made of oil outflow after collisions and groundings by Det Norske Veritas, Lloyd's Register of Shipping, Registro Italiano Navale and Nippon Kaiji Kyokai using probabilistic tools as well as simplified oil outflow calculations, all based in various assumptions with respect to, for example, penetration, extent of damage, tides, quantities of oil and dynamic effects. In addition, model tests were carried out in the USA (David Taylor Research Center) and Japan (Tsukuba Institute, Ship and Ocean Foundation) to assess oil outflow after accidents.

The most important conclusions from this comparison were as follows:

When the whole range of probable collisions and groundings are considered cumulatively, the oil outflow performance of mid-deck tankers is at least equivalent to that of double hull tankers, but the Committee recognized that within this overall conclusion each design gives better or worse oil outflow performance under certain conditions, in particular:

- in groundings which would result in the rupture of the bottom shell plating of double hull and mid-deck tankers but not the inner bottom of double hull tankers, which represent approximately 80% of the total grounding accidents resulting in hull penetration, no oil spill will occur in double hull tankers, but some oil outflow, normally small in relation to the ship's deadweight, would occur in mid-deck tankers;
- in groundings which would result in the rupture of the bottom shell plating of double hull and mid-deck tankers and the inner bottom of double hull tankers, the amount of oil outflow of mid-deck tankers, calculated on the assumptions using reasonable values of current and tide, is less than that of double hull tankers;

- in collisions which would not result in the rupture of the inner hull, no oil outflow will occur; mid-deck tankers have less probability of collisions resulting in the rupture of the inner hull because of the wider wing tank spaces in order to meet segregated ballast capacity requirements;
- the amount of oil outflow of double hull and mid-deck tankers after collisions which result in the rupture of the inner hull will depend on the actual tank arrangements.

> With respect to [. . .] fire and explosion, raking damage, operation of tankers and residual strength, double hull tankers are deemed comparable. [. . .] The Steering Committee, noting the present lack of knowledge to enable the evaluation of environmental performance of oil tanker designs with respect to fire and explosions and operational safety, recommends [. . .] studies in this field [. . .].
>
> The Steering Committee noted with concern the results of the analysis which showed that oil outflow of double hull tankers without longitudinal bulkheads inside cargo tanks is considerable [. . .].
>
> The Steering Committee [. . .] recommends that the MEPC consider whether the text [of the new regulation] should be modified to cover the other aspects [i.e. not just oil outflow but also fire and explosion, raking damage, operation of tankers, and residual strength].

In all fairness, it has to be said that the Steering Committee carried out an enormous amount of work in a very short period of time. One delegate at the 32nd session of the MEPC [13], where the findings of the Steering Committee were discussed, stated that 'the amount of work carried out [. . .] exceeded whatever anyone may have expected'. Still, the work of the Steering Committee had, in spite of its name which included the words 'tanker design', concentrated more on oil outflow than on design aspects.

The conclusions of this Steering Committee were not supported by unanimously. The Friends of the Earth International representative reserved its position with regard to the conclusion of equivalence of the two designs and expressed a preference for the double hull design. Arguments for this position were the conclusion that there would be fewer spills from double hull tankers in the case of groundings, and that the double hull design had more scope for further improvements (e.g. wider wing spaces to reduce the oil outflow risks in cases of collisions, the longitudinal bulkhead) than the mid-deck design.

The USA also, although on different grounds, distanced itself from the conclusion of equivalence, and proposed that the paragraph in the new regulation with respect to the mid-deck tanker be deleted. This proposal was not adopted.

No substantial changes to the draft regulations were made in spite of some of the recommendations by the Steering Committee. The recommendation with regard to longitudinal bulkheads was not discussed, and the implicit suggestion (made explicit during the debate by one or two delegations) that larger wing spaces for double hull tankers would increase the safety of these tankers in cases of collision also was not given any consequence.

The only real problem in the final discussion of the new double hull and mid-deck tanker design standards was brought up by the Republic of Korea at almost the last moment of the discussions, shortly before the formal adoption. The delegation of the Republic of Korea 'expressed its concern on the recent application by a company for an international patent on the mid-deck concept' and implicitly threatened (a threat not reflected in the official report) that they could not support the mid-deck tanker provisions in the new regulation because 'regulations of the Convention should not be utilized for commercial purposes'. Korea called for an assurance that the mid-deck tanker provisions, if adopted, could be implemented without financial consequences associated with an international patent. Many wondered why Korea had waited so long to bring up this issue, but whatever the reason for that, the intervention was effective. The Japanese delegation reacted that it had been informed 'by the inventor of the mid-deck design that, if the application for the patent of mid-deck design by them were ever accepted, they *would not* claim their right on the patent, i.e. any shipbuilder in any country will be free to build mid-deck tankers [. . .] without royalty and without permission by them'. The representative of Mitsubishi Heavy Industry in the Japanese delegation, the industry referred to by Japan as 'the inventor', confirmed this. He also stated that 'the intellectual right of inventors should normally be protected, even for inventions of this character'.

The equivalence of double hull and mid-deck tankers was in the end accepted by the MEPC (and by implication also by the IMO): the new regulation 13F of Annex I of the MARPOL Convention was adopted.

6.5 The present situation

The new oil tanker design standards (double hull and mid-deck) entered into force on 6 July 1993. No mid-deck tanker has been built or ordered. The major reason for that might well be the fear that, in spite of the international legal acceptance of the design, such ships will be denied entry into US ports (and waters).

The conclusions with regard to the equivalence of the two tanker designs are still being debated and will be as long as the US OPA '90 is not changed to accept also mid-deck tankers and other designs. Other issues of debate are whether the studies carried out by the Steering Committee are as valid as they seemed at the time. Some doubts have been cast on the validity of the model tests.

There is every reason for the debate on design standards to continue.

IMO's Steering Committee itself noted that more work needed to be done especially with regard to fire and explosion and operational safety. Its recommendations with regard to longitudinal bulkheads still needs a proper follow-up.

Many recommendations and conclusions have been made by the US National Academy of Sciences, e.g. existing design standards should be strengthened to ensure proper corrosion protection, proper dimensions of structural members and proper use of high-tensile steel; design based on the possibility of accidents should be considered for tank vessels. In addition, many suggestions have been made since in various publications.

New design ideas have also been brought forward. This was true already at the time the IMO Steering Committee on Oil Tanker Design was active, and it is still true. One of the interesting ideas in this respect is the so-called 'eco bulkhead' developed in The Netherlands (e.g. [14]).

The eco bulkhead is a modification of a swash bulkhead having a closed watertight plate structure running from the deck to the bottom except for relatively small holes near the bottom. In cases of accidents with side damage (at present one of the 'less-strong' points of the double hull design), this principle prevents (or limits) accidental oil outflow on the basis of the hydrostatic balance principle. The eco bulkhead prevents the outflow of oil from the undamaged part of a tank. To some extent it would seem that this principle provides a follow-up to the longitudinal bulkhead recommendations.

6.6 The need for further discussion on tanker design

The discussion on oil tanker design has not ended and should continue, as too many questions are still open. At the same time, the discussions regarding tanker design need to be put in perspective: the problem of (oil) pollution of the world's seas and oceans is not primarily a problem of oil tanker accidents. As far as ships are concerned, the most important problems are associated with operational discharges, not accidental discharges. In addition, the human factor, in terms of both quality and quantity (e.g. small crews), still needs to be addressed properly. Last but not least, it is not just the design of a ship that counts, its operation and maintenance are at least just as vital.

References

1. United States Coast Guard, Department of Transportation (1993) *Federal On Scene Coordinator's Report T/V Exxon Valdez Oil Spill*, US Coast Guard, Washington, DC, September.

2. IMO (1990) *Petroleum in the Marine Environment. Submitted by the United States of America (MEPC Document 30/INF, 13)*, IMO, London.
3. GESAMP (1993) *Impact of Oil and Related Chemicals and Wastes on the Marine Environment*, IMO, London.
4. National Research Council (1991) *Tanker Spills, Prevention by Design*, National Academy Press, Washington, DC.
5. Scientific and Technical Working Group, Second International Conference on the Protection of the North Sea (1987) *Quality Status of the North Sea, Status Report*, Scientific and Technical Working Group, London.
6. IMO (1990) *Petroleum in the Marine Environment*, IMO, London.
7. MARPOL (1973, 1978) *International Convention for the Prevention of Marine Pollution from Ships*, MARPOL, London.
8. *Safer Ships, Cleaner Seas (Report of Lord Donaldson's Inquiry into the Prevention of Pollution from Merchant Shipping)*, HMSO, London, 1994, p. 11.
9. See, e.g., IMO documents: *MEPC 29/22* (report of MEPC's 29th session, 12–16 March 1990) and *J/5893* (Status of Multilateral Conventions and Instruments in Respect of which the International Maritime Organization or its Secretary-General Performs Depositary or Other Functions as at 31 December 1994) for further details of this convention.
10. IMO documents: *MEPC 30/7/1* (submitted by the USA), *MEPC 30/INF.23* (submitted by the USA) and *MEPC 30/23* (report of MEPC's 30th session, 12–16 November 1990).
11. *IMO Document MEPC 30/7* (by the International Chamber of Shipping).
12. IMO (1992) *Report on IMO Comparative Study on Oil Tanker Design*, IMO, London, February.
13. *IMO Document MEPC 32/20* (Report of the 32nd session of the MEPC, 2–6 March 1992).
14. van der Laan, M. (1995) The eco bulkhead, outflow reduction by simple Archimedes. *HSB International*, February.

7 Pipeline technology

A.A. RYDER and S.C. RAPSON

7.1 Introduction

Pipelines have been in use for thousands of years for transporting liquids from A to B. Long before the Romans, ancient Chinese civilizations were making use of timber to construct primitive interlinked conduits/pipelines for the transportation of irrigation water. However, it is only in the last century that pipeline design, construction and operation have made major advances in order to provide a safe and reliable method of transporting vast quantities of hydrocarbons over long distances.

The assertion that pipelines are a safe and reliable method of transportation is often made but unfortunately there has been no rigorous and quantitative analysis to compare transportation methods, so we cannot directly compare pipelines with road, rail and shipping. However, the much publicized disasters of the *Exxon Valdez*, the *Torrey Canyon* and the *Braer* illustrate the scale of environmental problems associated with sea-going tanker accidents. The volumes of product transported in a single road tanker are much smaller than a potential release from a pipeline or tanker and so the consequences of accidents from road transport tend to be much less publicized, but disasters involving out-of-control road vehicles have been reported. For example, on 11 July 1978 at San Carlos in Spain 200 people died when a road tanker ruptured. The incident occurred when the tanker, designed to carry ammonia, was overloaded with propylene. The pressure caused it to burst, spreading 22 tonnes of propylene over a campsite. There were several ignition sources on the campsite and an enormous fireball resulted (Health and Safety Commission, 1991). Similarly with rail transport accidents do happen; one well publicized example was in Canada. The incident happened in Mississauga, a city on the north shore of Lake Ontario, west of Toronto, on 10 November 1979, where 24 cars of a 106-car freight train (2 km long) were derailed at a level crossing. Twenty-two of the cars contained chemical products including propane, toluene, caustic soda, chlorine and styrene. Explosions and fire ruptured propane tankers and a chlorine tanker, which leaked its contents. Tankers were thrown up to 700 m in all directions. The incident resulted in the evacuation of 217 000 people, 75% of the population of Mississauga (Burrows, 1990).

The pipeline industry has, in the main, been free of major disasters. There have, of course, been pipeline failures leading to pollution and in some cases to fatalities. In recent years, the most publicized incidents have been in Russia, where problems associated with pipeline monitoring have become acute because of the age of the pipelines, their length and the fact that they are subject to severe climatic regimes. In addition, Russian pipelines are characterized by a lack of pigging facilities, and most of them are not piggable anyway owing to varying pipe diameters and other problems. Typical pipeline failures result from pipes floating up in bogs, or in frozen ground they occur because of metal fatigue resulting from the development of ravines and crevasses in the ground (Gritzenko and Kharionovsky, 1994). Perhaps the best known incident was when a Trans-Siberian train ignited a gas cloud from a pipeline. The massive explosion that resulted left a reported 706 people hospitalized and 462 dead or missing. The incident, on 3 June 1989, happened when a pipeline from the gas fields of Western Siberia to Ufa in the Urals developed a leak. Instead of investigating the leak, engineers increased the pumping rate to keep up the pressure. The escaping LPG formed pockets in two low-lying areas. Two trains were in the area when the incident happened. The turbulence they caused mixed the LPG with the air to form a flammable cloud and one of the trains sparked off the cloud. Two massive explosions were followed by a wall of fire and the trains were blown apart; trees were flattened 4 km away and windows 13 km away were broken (Det Norske Veritas, 1993).

Another well publicized case was in Russia's Arctic Komi Republic. The 43 km long Vozey–Usinsk pipeline, built in 1975, began leaking in 1988. By 1994, leaks averaged 17 a month, and a record 23 leaks were recorded in August of that year. Dams were built to contain the oil which soaked into the marshy ground. In August 1994 multiple leaks resulted in several spills totalling an estimated 30 000 tonnes. On 16 September further spills occurred and on 28 September one of the dams broke. Other dams collapsed soon after. Ironically, the Russian media learned of the disaster from US sources (*Russian Petroleum Investor*, Dec. 1994/Jan. 1995). Estimates of the total amount lost varied from the official low of 14 000 tonnes up to 270 000 tonnes claimed by American oil workers in the area (*Moscow Tribune*, 9 November 1994).

Fortunately, problems with pipelines in Western Europe are much less severe. The oil industry group CONCAWE reports annually on oil industry pipeline failures, and in their most recent report (4/95) they recorded only 13 incidents in 1993 and 11 incidents in 1994. The average number of incidents per year is 12.8 for the period 1989–94, which is an improvement when compared with an average of 13.8 for the period 1971–94. Volumes spilled in the period 1989–94 were in all but two of 64 instances less than 1000 m^3 and in nearly 80% of cases were less than 100 m^3.

In this chapter, we consider the environmental pressures on pipeline owners and operators in the 1990s and examine the ways in which the industry is responding to these pressures during design, construction and operation. It focuses on European and mainly UK experience, but in many cases this has international implications. For convenience, the remainder of this chapter is divided into four parts: Environmental Pressures looks briefly at the reasons behind increasing environmental awareness in the pipeline industry. The sections on Onshore Pipelines, Offshore Pipelines and Pipeline Landfalls all consider the measures taken by pipeline operators to initiate, implement and monitor environmentally sound working practices. These last three sections draw heavily on examples to illustrate the ways in which the industry is responding to the pressures.

7.2 Environmental pressures

Increasing academic and public concern over the state of our environment has led to an explosion of legislation in recent years designed to control the impact of new developments. Pipelines in and around the UK have always been subject to control in order to ensure their safe operation; however, increasingly there are demands, reflected in the legislation, for much closer attention to issues which affect not only their safety but also their visual and environmental impact.

The authorizations, consents and specifications for building and operating pipelines in the UK are contained in a number of Acts of Parliament; the principle one for onshore pipelines is The Pipe-lines Act 1962 and for offshore pipelines it is The Petroleum and Submarine Pipe-lines Act 1975. Various clauses in these acts state that steps must be taken to avoid or reduce danger to wildlife and human activity. In addition, a pipeline must be designed, constructed and operated in a manner that ensures that it is protected from damage. These requirements ensured that pipelines were constructed and operated in a safe manner and that the environment was given due consideration. However, it was not until June 1985 when the member states of the European Community adopted Directive 85/337/EEC on the assessment of the effects of certain public and private projects on the environment that environmental assessment became formalized, widely recognized and methodically implemented.

The Directive made assessment of the environmental effects of a proposed project compulsory for those projects likely to have a significant effect on the environment. The result of this Directive has been the widespread adoption of what is known as ‘environmental assessment’, in which information about the environmental effects of a project is gathered and evaluated. Where significant effects are identified, measures for reducing those effects are also included. The developer normally commissions environmental specialists, who produce an environmental statement which the developer includes as part of an application to the planning authority.

EC Directives are not legislation, but each country has an obligation to make its own laws implementing the directives. The UK has its own regulations on these matters, and cross-country pipelines are dealt with under the Electricity and Pipe-line Works (Assessment of Environmental Effects) Regulations 1990. These regulations do not actually make environmental assessment of new pipeline projects compulsory; however, any developer applying to the government without such an assessment is likely to be asked to provide one, and failure to do so automatically causes the application to be refused. It is therefore compulsory in all but name and, so far, all applications to build such pipelines have been accompanied by an environmental statement. The Regulations state that such an environmental statement must comprise:

- a description of the proposed pipeline;
- the data necessary to identify and assess the main effects of the pipeline on the environment;
- a description of the likely significant effects; and
- a description of measures envisaged to avoid or remedy those effects.

Aspects of the environment that should be considered include human beings, plants, animals, soil, water, air, climate, landscape, material assets and the cultural heritage.

Offshore oil- and gas-field development, including any associated developments, are required to be subject to environmental assessment through special additions attached to a licence. Environmental statements have to comply with the requirements of the Directive (Cobb, 1993).

7.3 Onshore pipelines

Onshore pipelines are generally one of three types: those built within the oil and gas fields for the collection of oil, known as infield lines; those built to cover longer distances between the point of production and consumption, known as cross-country pipelines; and smaller diameter pipelines used for distribution and supply. These smaller diameter, and usually low-pressure, pipelines are normally associated with the distribution of natural gas.

With the exception of the low-pressure distribution and supply pipelines, the material used in construction is normally high tensile steel. Individual joints of pipe are welded together to form a continuous tube. Valves and tee pieces may be installed along the length of the pipeline and pig traps may also be installed at intervals along the pipeline as well as at the ends of the pipeline. A valve is used to restrict or prevent flow, a tee piece for diverting flow and a pig trap for the launch and recovery of pipeline 'pigs' – devices put into the pipeline for construction or operational reasons. Pigs are often spherical or cylindrical in shape and have all manner of uses, e.g. cleaning, separating batches of product and the gathering of information. A pipeline should therefore be thought of not just as the pipe but also the associated apparatus which makes up the system, including valves, tees and pig traps as well as pumps or compressors and other miscellaneous ironmongery.

7.3.1 Design

(a) Preliminary design. The real opportunity to minimize the environmental impact of a pipeline is at the early design stage; it is then that the route of a pipeline is being chosen and it is then that environmentally sensitive areas can be avoided. By avoiding sensitive ecological areas such as ancient woodlands, species-rich grasslands, heaths and archaeological sites, many impacts can be avoided completely. It is often much easier and less costly to avoid a site than it is to develop special construction and reinstatement practices.

Today it is usual to undertake an environmental impact assessment during the preliminary design stage of an onshore pipeline project. Typically, a 2 km wide corridor, avoiding centres of population, and focusing on topographically defined corridors, is established and is screened for features that would have a direct bearing on route considerations. At this stage, environmental impact assessment has a dual purpose: the identification of environments on which a pipeline would have a significant impact and also the identification of environments which would have a significant impact on the pipeline. The opportunity is also taken to identify other linear developments, including pipelines, as there are obvious advantages in parallel and adjacent routing. In particular, it minimizes the cumulative effect on land use and offers definite advantages in woodland areas which may already have an easement cut through them.

The issues addressed at the route concept stage are:

- existing linear developments and established corridors including motorways, trunk roads, railways, canals, overhead electricity cables and pipelines;
- historic buildings;
- archaeological sites;
- areas subject to subsidence;
- geographical features;
- estuaries and rivers;
- geology and mineral resources;

- aquifers and water resources;
- conservation areas and landscape;
- areas of woodland.

With correct design, construction, materials and restoration techniques, there are few onshore environments through which it is not possible to lay a pipeline. However, each of the aforementioned features presents varying degrees of construction difficulty and may necessitate greater expenditure on construction materials and/or restoration. The objective at this stage in the routing operation is to achieve the most cost-effective route by attempting to minimize the length, while ensuring a minimal risk to the public and no significant adverse environmental impact.

The assessment is usually conducted as a desk-top exercise, complemented by an aerial video of the whole route. The video film provides a rapid means for updating maps and provides a ready visual reference for the project team. By screening a 2 km wide corridor for fundamental features it allows the routing engineers a degree of flexibility, and there is no need to undertake a reassessment so long as the route remains within the appraised corridor.

Consultation with organizations who have a responsibility for environmental protection at national, regional and local level will assist in the identification of areas that require protection and should form an important part of the assessment process. Consultation of this nature should lead to the identification of all major sites of environmental interest. It is then possible to undertake ecological and archaeological surveys of a preferred route. Consultation, field survey and the study of maps, aerial photographs and aerial videos should enable a composite set of constraint maps to be produced. This might best be achieved through the use of a geographical information system in which the physical extent of a constraint can be mapped and a database record can be attached to provide details on the source of the information, the nature of the constraint and its implications for pipeline construction.

(b) Detailed design. Once the principal features affecting the pipeline route have been identified, the role of the environmental assessment is to identify in detail the possible impacts of the proposal. The corridor principle still applies. However, its width can be reduced, say to 500 m. Within this zone, significant environmental features are identified.

(i) Consultation. Once a preliminary route has been established, it is usual for representatives from the pipeline company to visit all relevant statutory authorities along the proposed route to discuss the possible implications. After these preliminary discussions, numerous meetings are then organized to focus on a range of regional and local issues associated

with pipeline construction. The authorities are usually extremely helpful in providing detailed information about their districts. Consultation with many other statutory and non-statutory bodies responsible for nature conservation, archaeology, landscape and recreation is useful and should be maintained throughout the project.

(ii) Examination of the existing environment. For many of the environmental issues associated with pipeline development, a vital part of the assessment is the acquisition of good baseline data. A description, both qualitative and quantitative, of all aspects of the environment is required to provide a basis for design and assessment and a record of the existing situation. The following phased studies are typically associated with a pipeline development:

- nature and distribution of land cover;
- nature and distribution of land forms;
- a geological investigation of the pipeline route;
- a landscape assessment;
- an ecological assessment in three or more phases;
- archaeological assessment in five or more phases;
- an agricultural assessment;
- the distribution of soils;
- a study of hydrological implications.

The different phases of work reflect an increasing amount of detail and effort. For example, an archaeological assessment may involve five phases. Phase 1 would be screening a 2 km wide corridor for known sites of national importance and phase 2 screening a 500 m wide corridor for all other known monuments and sites. Phase 3 would be detailed field survey work along a 40 m wide corridor. Phase 4 would involve the excavation of sites which are threatened by construction. Phase 5 would comprise a watching brief throughout construction and the subsequent publication of results.

(iii) Impact appraisal and prediction. The environmental data generated by baseline survey work, when considered in conjunction with detailed project studies, are used to identify the probable environmental implications of the development. Examples of studies that might be undertaken are as follows:

- atmospheric emissions during construction and operation;
- noise implication of construction and operation;
- blasting and vibration;
- agricultural implications of construction;
- socio-economic implications of pipeline construction;

- strategic-economic appraisal of the project;
- safety.

The techniques used for impact prediction are established and well documented. Some aspects are relatively easily modelled, giving quantitative outputs to reasonable degrees of accuracy, such as the propagation of noise and dispersal of contaminants in the atmosphere. Others require a more qualitative approach and rely more on the judgement of experts than on comparison with accepted criteria.

Some issues have fairly well defined criteria, established by standards against which to assess impacts. These, again, are associated with the physical, chemical and hydrological impacts connected with construction and operation. The impacts associated with land-take and ecology do not have easily definable criteria against which to assess impact. The predictions of impact in these situations tend to be presented as qualitative descriptions and the demonstrations that impacts have been minimized by design and other mitigative measures.

Criteria for assessing environmental risk are not well established, although presentation of risk helps to put certain impacts into perspective. The overall perception of environmental risk is influenced by the concept of risk acceptability, which has been applied when considering the effects of major accidents on the people living adjacent to the route.

(iv) Identification of mitigative measures. With a large pipeline project it is usually inevitable that there will be some adverse environmental impacts as a result of disturbance of the land surface. These effects can be minimized by considering details of routing, construction techniques and site-specific reinstatement and aftercare programmes. For example, with one recently constructed pipeline in the UK it was possible, when crossing most moorland sites, to reduce the normal working width of 20 m to 12 m and topsoil stripping operations were restricted to the width of the pipe track only. Instead of stripping topsoil across the whole width, a sand, bog mat or subsoil road was constructed directly on the vegetation. Turves were lifted from the pipe trench area, stored to one side and put back. The road was then lifted and the working area scavenged for debris.

(v) Proposals for future monitoring. The environmental statement broadly identifies the potential environmental impacts of the development. Monitoring programmes need to be established to:

- obtain, where appropriate, baseline data for the environment prior to the construction, commissioning and operation of the pipeline;
- monitor any significant alteration to the biological, chemical and physical characteristics of the local environment;

- monitor emissions and discharges at all stages of the development to ensure they meet the national, local and company management standards;
- monitor any alteration to the inter-relationships of different aspects of the environment;
- determine whether any environmental changes which may occur are the result of the development or result from natural variation.

The intention is to determine, where appropriate, both the natural fluctuations of environmental parameters and the extent of other anthropogenically induced changes before, during and after construction of the pipeline and throughout its operational life.

(vi) Preparation of the environmental statement. Environmental assessment is the process of environmental input to project planning and prediction of likely impacts. The products of the process are often a series of technical reports which are summarized, in a more user-friendly form, in an environmental statement written by an environmental specialist. The report may be submitted to the statutory authorities in draft form. Then, after further consultation, the final document can be made available to the public and other interested parties.

(vii) Contract documentation. The key to effective environmental management is to turn the products of the environmental assessment, i.e. the series of technical reports and the environmental statement, into action. For environmental assessment to have some impact upon the reality of construction, the results must be built into the technical specifications and included where necessary in contract documents including alignment sheets.

There is no one way to achieve this and, as with all contractual matters, a balance has to be sought between providing the contractor with too much and too little information. It has been argued that if too much environmental information is given to potential contractors they will react adversely and charge a premium on the basis that perceived environmental sensitivity presents a risk. Conversely, if inadequate information is provided then there is a risk of claims for additional work which was not adequately specified and there is the possibility that adequate environmental controls will not be implemented. One solution is the production, by the contractor, of a method statement. In the contract documents the basic requirements needed to comply with the environmental statement are set out. The contractor is then invited to produce a method statement which adds detail to the information provided by the design engineers in the contract.

7.3.2 Construction

From the above discussion it has become clear that during the planning and design phase of a pipeline considerable effort is expended in the identification of potential environmental impacts, the identification of suitable mitigation measures, the inclusion of mitigation measures into the design, and where appropriate their stipulation in contract documentation. For those mitigation measures to be implemented effectively during construction they must be *known*, *understood* and *implemented* by all relevant personnel. In practice these three basic requirements, the cornerstone of effective environmental management, are perhaps the most difficult to meet.

(a) Raising awareness and understanding. Raising awareness is perhaps the first step in achieving satisfactory environmental performance. Management must appreciate the significance of environmental issues and be committed to achieving a high standard of environmental performance. This commitment in the management of a pipeline project will be strongly influenced by the level of importance given to environmental issues by the senior management of the company. A company with a strong commitment to environmental protection and a visible environmental policy is more likely to achieve the commitment of its project management team.

A project's workforce will need to become familiar with those environmental issues that are specific to that project. This can be achieved in a number of ways, for example:

- A full-time environmental scientist may be appointed to the management team from the outset of the project. He or she would have the responsibility for briefing the project, construction and engineering managers on environmental issues.
- Monthly health, safety and environment meetings may be held, allowing issues of concern to be discussed by the management team.
- Informal workshops may take place. For example, on one recent project an archaeological dig took place along the pipeline prior to construction and many members of the project team took part under the supervision of trained archaeologists. In the evenings there were presentations about archaeology and what had been found along the pipeline route during the preconstruction surveys and what was likely to be found during construction.
- Health, safety and environment workshops may be held once construction contractors have been selected. Members of the client team and the construction contractors participate to ensure that all senior management on the project appreciate the importance of environmental issues on that project and understand the mitigating

measures which have been designed and incorporated into the contract documents.

- All personnel should go through a programme of induction training before they are allowed to work on-site. This may take the form of a talk from the site safety and/or environmental officer.
- Tool-box talks may be held on an as-required basis with different construction crews. Typically these are held on weekly basis, or before entering a special section. The talks are given by the supervisor or foreman although, if a special environmental crossing is about to be encountered, an environmental officer would explain what is important about a site and how to protect it.
- Signs should be erected along the spread indicating the beginning and end points of areas where special precautions have to be taken.

(b) Site supervision. The number of inspection staff required is always contentious. Financial constraints will always mean that there is pressure to reduce the number of such staff. Quality assurance philosophy maintains that well written procedures and the use of appropriately trained staff help to reduce the number of inspection staff required. Experience suggests that the higher the level of supervision, the better is the end product. Of course, the question must be what level of supervision is necessary.

It is essential that environment, like safety, is perceived as a line responsibility and not the sole responsibility of the environmental officer. All supervisors and inspectors can help ensure that environmental requirements are implemented. However, the effectiveness of this is dependent upon the supervisor appreciating and implementing a project's environmental controls. In sensitive areas a greater input will be needed from an environmental officer. He or she will probably have been involved in designing mitigation measures and will therefore know how flexible those measures are. When problems arise the environmental officer, with a knowledge of the site, can advise on how to overcome the problem.

(c) Reporting. The construction of most pipelines is managed by a project team, where the promoting company appoints a project manager. Reporting to that manager will be various management disciplines such as construction, engineering, health, safety and environment (HSE) and so forth. Most organizations have a corporate HSE group and it may be useful to maintain a link between a project's HSE group and the corporate HSE group. This provides a mechanism whereby a project manager can be circumvented if need be.

(d) Contractor plans. Contractors should be encouraged to prepare their own environmental management plans. These plans will allow the

contractor to implement procedures that are tailored to their organization and way of working. Plans may be required that cover:

- archaeology – what to do in the event of an archaeological find;
- waste management – including waste minimization, reuse, recycling and disposal;
- pollution prevention – including avoidance, containment, clean-up and reporting arrangements.

(e) Construction methods. The standard method for the construction of welded steel cross-country pipelines across normal agricultural land is based upon the spread technique. A 'spread' for a pipeline consists of all the people and equipment necessary to conduct the construction operation from surveying the route to restoration. The work is carried out on a continually moving assembly line basis, with each sequential activity maintaining a consistent rate of progress. On a long pipeline there may be a number of spreads, with work being undertaken by different contractors on different spreads. Progress may be as much as 1 km per day. In the UK construction is usually confined to the period March to October when weather conditions are most favourable.

Each spread contractor will need a number of different crews. They will undertake the following tasks:

- Fencing the working width – all construction activities take place within the fenced area of land referred to as the working width. This is normally within the region of 20–40 m wide.
- Pegging – the centre line of the pipeline will be marked out on the ground with pegs.
- Removal of trees, hedgerows and walls and fluming of ditches – where necessary trees, hedgerows and walls will be removed. Established trees will be protected by fencing where possible, and there will be minimal removal of hedges with wildlife interest. Stone from walls will be kept for later replacement. Ditches are piped or 'flumed' to allow vehicles to run over them.
- Topsoil stripping – topsoil is then removed from the working width and stored to one side. Part of the working width is designed as the running track for vehicles.
- Transport and distribution of pipes – the pipes are transported to the working width from the pipe storage depots. They are laid out on wooden sleepers alongside where the pipe trench will be.
- Pipe bending and welding – where the pipeline route turns sharply factory-bent pipes are used. Where the change of direction is only slight the pipe can be bent in the field. After the pipes have been welded together they are subject to radiographic inspection, with

detected faults being repaired or cut out. The welds are then wrapped or coated in a similar manner to the rest of the pipe.

- Trenching – a trench is excavated to allow the pipeline to be buried. Its exact width and depth will depend on the pipe. In rocky areas a bed of sand will be laid for the pipeline to rest on. The subsoil that is excavated is stored on the opposite side of the working width to the topsoil to avoid mixing of the two.
- Lower and lay – the pipeline is then lowered into the trench using side-boom tractors and extreme care is taken to avoid damaging the pipe coating.
- Backfilling – the subsoil and subsequently the topsoil are replaced in the trench and compacted to their original state.
- Reinstatement – in normal agricultural land the working width is reinstated to the requirements of the landowner or occupier.

In environmentally sensitive areas such as Sites of Special Scientific Interest (SSSIs) and other conservation areas, special construction methods are needed. Each of the areas of concern will need to be the subject of a separate study prior to construction to determine the best crossing method. Possible suitable methods at such sites include reducing the working width, use of temporary roads, stripping the pipe trench only and not the whole of the working width, turfing, fluming and boring beneath.

(f) Monitoring. Most environmental monitoring will take the form of checks to ensure that the contractor is complying with contractual requirements, for example, waste management and use of designated disposal sites. Some special forms of environmental monitoring may be required at particular locations; for example, at river crossings it may be necessary to monitor dissolved oxygen and suspended solids. When working in close proximity to residential areas it will be important to monitor noise levels.

(g) Audits. Any management system should be subjected to audits to allow shortcomings to be identified and, importantly, to allow improvements to be made. For example, corporate HSE may audit the project's HSE group; the project's HSE group may audit the construction contractor or specialist environmental contractors.

(h) Case study: the North Western Ethylene Pipeline, UK. Such special construction methods are well illustrated and were rigorously tested during the construction of the Shell North Western Ethylene Pipeline in 1991–1992, which is the longest pipeline to be built in the UK (Fig. 7.1). The 10 in (25 cm) diameter pipeline was built because ethylene, which is made

from natural gas from the North Sea, needed to be transported from Grangemouth (near Edinburgh) to Shell's petrochemicals plant at Stanlow in Cheshire, where it is used in the manufacture of plastics and solvents. It was the first pipeline to be subject to the Electricity and Pipe-line Works (Assessment of Environmental Effects) Regulations, 1989, which emerged as a result of EC Directive 85/337/EEC. Now, 5 years later, the pipeline, 411 km in length, 10 in in diameter and containing 17 100 tons of steel, lies buried 1 m underground, its path invisible to all but the informed eye.

Because it is such a long pipeline, and because it had to follow a line which was already littered with other pipelines, railways and roads, it was impossible during the design stage to establish a route which did not affect any important areas. In particular, it had to cross two Roman walls, the Antonine Wall and Hadrian's Wall, both of which are Scheduled Ancient Monuments and are protected by law. In addition, it had to cross four Sites of Special Scientific Interest (SSSIs) which are also protected by law. Only with careful negotiation, and after a public inquiry, was Shell allowed to cross these and other important features (Rapson, 1994).

Shell took care to ensure that all construction was undertaken in an environmentally sound manner. For example, trees near the working width had their roots protected from vehicles by fencing, fuel containers were kept in trays to avoid spillage and sediment in water had to be allowed to settle out before it could be discharged into watercourses. Four environmentalists and four archaeologists monitored day-to-day construction.

Special construction methods were agreed for all the sensitive environmental and archaeological sites. Carstairs Kames, near Lanark in Scotland, important geomorphological features surviving from the last ice age, designated as an SSSI by Scottish Natural Heritage, had to be crossed. A low point was chosen for the crossing, and where the pipeline had to run parallel to the edge of the kames, the width of the working area was reduced to as little as 4 m.

Lazonby Fell, an area of heathland near Penrith, is another SSSI which required special attention. Before construction began, the heather was cut to promote new growth in the following year. A strip 12 m wide was fenced off, a temporary road was laid and turves were removed only from the area of the pipe trench. After the turves had been replaced, heather cuttings collected from nearby were spread over the area to help new growth. By the summer of 1992 many new heather plants were growing in the thinly vegetated areas, showing how successful reinstatement had been.

Similar methods were used at Crosby Ravensworth Fell, near Shap in Cumbria, also an SSSI. It is a large area of upland grassland and it was not considered practicable or necessary to turf the whole area; instead, turfing was confined to those floristically rich areas identified by botanists before construction began. In remaining areas the vegetation layer was scraped off using an excavator bucket and stored separately from the topsoil and

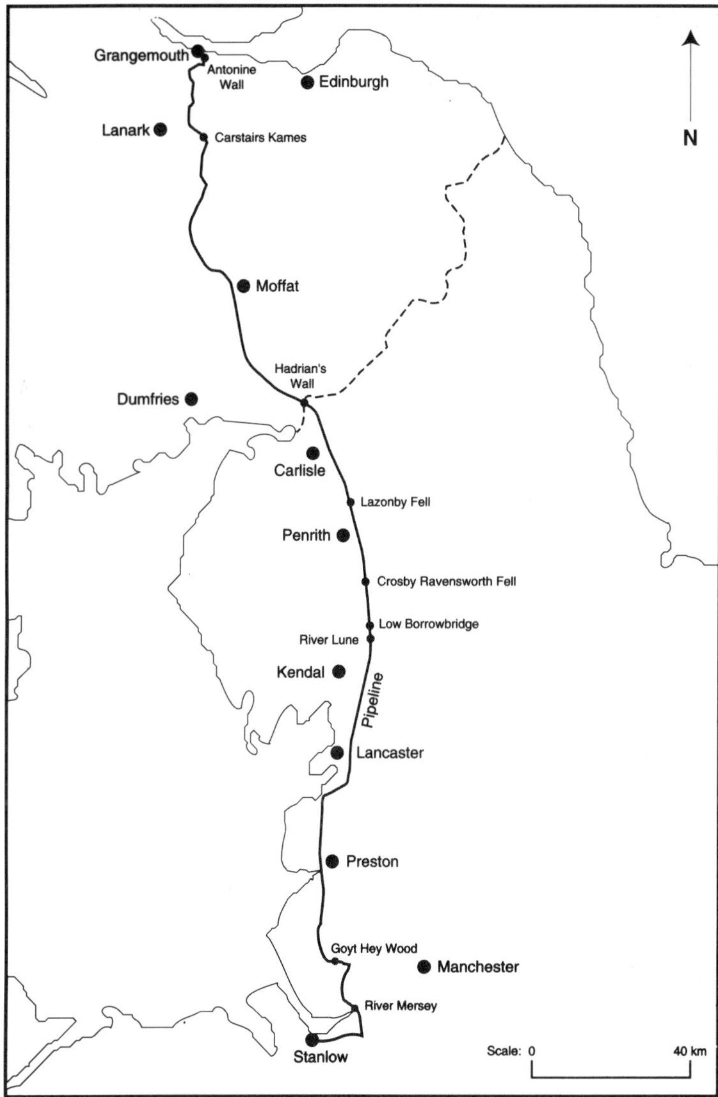

Figure 7.1 Map showing the route of the North Western Ethylene Pipeline and the sites mentioned in the text.

subsoil beneath so that later it could be replaced on the surface, thus encouraging existing plants to grow and to maintain the plant rhizomes and the seedbank. Subsequent monitoring has shown that the turved areas recovered extremely well within a very short time. The remaining areas fared more poorly; the vegetation is recovering more slowly and it has been necessary to seed some parts with a special upland grassland seed-mix. However, this was expected, as wet upland areas do take a long time to recover and in general reinstatement is considered satisfactory.

A narrow strip of woodland called Goyt Hey Wood, near St Helens, had to be crossed. A point was chosen where it was not necessary to fell any mature trees, the working width was reduced to only 4.5 m and special small excavators were used. The soil containing the bulbs and seeds of the important ground flora was carefully stored and replaced. It was encouraging to see bluebells growing on the working width the following spring.

Shell had to cross several rivers along the pipeline route, including the River Lune in the Tebay Gorge. Here excavators working in the river caused much sediment to be disturbed, but by working very quickly and by stopping for periods to allow clear water through, the effect was reduced. However, the River Mersey was too large for this method and the horizontal directional drilling technique was used.

Archaeological sites needed a different approach. Some required excavation before construction began. For example, near Grangemouth, where archaeologists knew the exact location of the Antonine Wall, an excavation was carried out. Evidence of the wall and of a fort was found. Other sites could not be excavated beforehand because no one knew they were there. This was the case at Low Borrowbridge, where construction stopped whilst a Roman cemetery was uncovered.

Shell have a commitment to looking after the land along the pipeline route for the life of the pipeline, which is at least 25 years. For 5 years they are monitoring the success of reinstatement of its 40 most sensitive sites. They will also have to check on the growth of the hedges and trees planted to replace those felled.

7.3.3 Operation

Pipelines are generally believed to be the safest means of transporting large quantities of hazardous fluids and gases over long distances. From an environmental viewpoint pipelines are also the preferred mode of transport: there is a reduced likelihood of accidents and spillage of products, and the environmental impact of operating pipelines is less than for rail or road transport. However, as a follow-on effect from increasing public awareness of environmental issues and tightening legislation throughout the world, pipeline operators are under continual pressure to make pipelines even safer. This becomes increasingly important as pressures on land increase and pipelines become squeezed into narrower and narrower corridors.

Pipelines can fail through material defect, corrosion, natural causes (e.g. earthquakes) and third-party interference (CONCAWE, 1995). Through rigorous adherence to design code standards and by continual monitoring of pipelines in recent years, failure from material defect, corrosion and natural causes are now much less of an issue with regard to recently built pipelines. To ensure that pipelines have a minimal impact during operation there are two areas that require action:

- avoidance of spills resulting from pipeline failure and adoption of plans to deal with a leak should one occur;
- preparation and implementation of a restoration plan.

To achieve these aims, a number of actions are required, many of which are simple components of a good management system, e.g. pipeline integrity monitoring and maintenance, prevention of third-party interference, emergency planning, record keeping, monitoring and audits and reviews.

(a) Testing, commissioning and operation. After pipelaying, a pipeline must be cleaned and checked. Pipeline pigs are used to clean and check the pipe in the initial stages, then hydrostatic testing takes place. Where possible water for testing is drawn from a nearby river after agreement with the relevant authority, but where this is not possible tankers will be required. Pressure is generated by a diesel-driven reciprocating pump, so there is some noise but, given that normal standards of noise control are in place, e.g. exhaust silencer and standard enclosures, noise levels would be expected to be no greater than from other normal pipeline construction activities. The water is then discharged at a controlled rate to a site agreed with the appropriate authority. After dewatering, pumps may be used to dry the pipe. These pumps may have to operate over a period of several days and so strict noise targets may have to be imposed. As an alternative to or in addition to vacuum drying, the tested sections may be swabbed to remove residual water by passing through specially designed pigs propelled by compressed air/gas.

During normal operation there will be no significant impacts on the environment resulting from an onshore pipeline, although there may be some noise from pump units. Careful planning at the design stage should ensure that such noise is not sufficient to cause nuisance to nearby residents.

(b) Pipeline integrity monitoring. Pipelines built today have sophisticated loss monitoring detection systems such as the supervisory control and data acquisition (SCADA) system, which can detect when a leak occurs through a drop in pressure. These systems allow very early detection of a leak and allow the operators to shut down the pipeline, identify the location of the leak and isolate it by shutting off block valves on either side. Remote

operation of the compressors and block valves from a central control unit means that a shut-down can take place within minutes rather than the hours that would be required to handle the operation manually. Sophisticated telemetry allows continual checking of the system to ensure that any failures are quickly identified and rectified.

Also built into today's pipelines are facilities for the prevention and detection of corrosion and the detection of other defects. Pipelines are protected from corrosion first by the application of a protective coating or wrapping in the factory and the application of a similar coating or wrapping of joints in the field, and second by cathodic protection. Cathodic protection stops corrosion by the prevention of current flow from the pipe (the cause of corrosion is the removal of metal ions by the flow of current). The metal is made electronegative with respect to its environment to such a degree that no current can leave at any point. The current, which under natural conditions would leave the metal, is opposed by the flow of the current in the opposite direction and this opposing current is either equal to or greater than the total of all the currents naturally leaving the structure. The power-impressed systems usually used on onshore pipelines comprise a DC power supply, with the negative connected to the pipeline and the positive connected to an earth electrode. The latter is normally referred to as the groundbed.

Another measure which helps maintain the integrity of a pipeline is regular checking of the state of the inside of the pipe using a pig. This remotely operated device, usually spherical or cylindrical, is sent down a pipeline to perform a variety of functions. Some pigs will simply clean the pipeline, whilst more sophisticated ones will record information about such things as wall thickness, corrosion or the location and size of dents or other deformities in the pipeline.

(c) Prevention of third-party interference. Third-party interference is widely recognized as the single most probable cause of pipeline failure. It can arise from four major sources: landowners and tenants, utility companies, contractors and local authorities. With regard to interference from landowners and tenants, research on recent UK pipelines has shown that, despite the pipeline operators expending considerable time and money informing this group of people, a third of those questioned did not tell staff or contractors about precautions to take when working near pipelines, and they were not clear about the type of work that should be notified to the pipeline operators or the safe working distance from a pipeline. Most interviewed judged the pipeline route from marker posts and did not have accurate maps showing the route. In addition, although most did have an emergency contact telephone number to hand, one third were unaware of the full range of services and advice that is provided free of charge by the operators (Sljivic, 1995).

The research also found that many cross-country pipeline operators are not included in the routine contacts made by utility companies and their contractors before beginning an excavation. It is possible that the current trend towards deregulation of the utility companies could make this situation worse. In addition, local authority planners, responsible for identifying planning applications lying close to pipelines, often hold poor information on the pipeline routes.

Education about the risks associated with pipelines and the continual supply of information about the pipelines to third parties in order to reduce ignorance and misunderstanding are clearly an essential part of a pipeline operator's job and will help to reduce risk of pipeline failure. In the 1990s and beyond there is, however, much more that can be done by the implementation of information technology within the industry. In particular, many pipeline operators are now implementing, or considering implementing, geographical information systems, one-call systems and improved surveillance techniques.

Geographical information systems (GIS) are useful in a number of respects. First, they allow root cause analysis to be carried out on excavation work, authorized and unauthorized, near a pipeline. Regular analysis of the cause and nature of infringements will clearly help pipeline operators target more effectively those parties most likely to offend. Affordable PC-based systems specifically tailored to the job which can manage data relating to the day-to-day operation and inspection of pipelines are now readily available, and analysis of third-party activity is quick and efficient. In addition, thematic maps can be produced showing the location and type of offenders, excavation hot spots and notifications of works in roads, in the vicinity of rivers, etc. An additional benefit of GIS is that systems can output customized maps; these are particularly good for use by third parties. In the USA the Office of Pipeline Safety is in the process of implementing a national pipeline mapping programme using GIS.

One-call systems, which allow anyone wanting to carry out an excavation to telephone a central number to register their intentions, have become increasingly popular in recent years. Two types of system are in operation: those which cover a defined geographical area and include all or most utilities, and those which are utility specific and provide data only on the location of their particular underground pipe. The former requires a great deal of investment and the co-operation of all utility companies in order to be successful. The latter is much quicker and cheaper to set up but is of limited use. In The Netherlands there already exists a legal requirement to subscribe to a country-wide, all utility scheme, and in the USA similar legislation is proposed. In the UK there is no such government-led incentive, but companies are moving towards such schemes as a means of fulfilling their safety obligations.

Surveillance techniques to detect third-party interference have traditionally involved regular helicopter or aeroplane flights along pipelines. These enable an observer to spot any violations of the easement from the air and, if a helicopter is used, to land in order to stop the work if necessary. It is recognized, however, that such flights, even if done on a regular basis, only identify infringements which occur within a very short time span, and it is much more likely that a major infringement would be missed. In addition, such flights do not allow an observer to examine the pipeline in detail and vital clues may be missed. This latter failing can be overcome by the use of a real-time video record made at the time of the flight. The technique is currently being adopted in the USA and is being considered for use in the UK.

At least one pipeline operator in the UK has decided to increase the effectiveness of its ground survey techniques. To this end it commissioned a risk analysis to determine which parts of its pipeline were likely to cause most risk to people. Heavily built-up areas were deemed to pose most threat whilst remote upland areas pose least threat. It then decided to concentrate its ground survey crew on the areas most at risk, developing a strategy which involved frequent monitoring of the highest risk areas, less frequent survey of medium risk areas and infrequent monitoring of low risk areas. In order to satisfy itself that this regime was being implemented, the operator armed the ground survey crew with bar code recorders, and when they walk the pipeline route they have to read the bar codes located on the marker posts. The bar code data are downloaded to a GIS system and analysis of the information allows the operator to identify areas where the surveillance programme is behind schedule and to reallocate resources appropriately.

(d) Emergency planning in the event of a spill. All pipeline operators have plans which can be acted upon in an emergency. These clearly state the line of responsibility in such an event and detail what will happen. Emergency response vehicles containing necessary equipment are held at convenient locations by the operators. Regular training is given to staff involved. In addition, the emergency services will be familiarized with the plans.

(e) Record keeping. It has already been shown how good keeping of records on a GIS system can help prevent third-party interference by analysis of trends and how it can help record and monitor ground surveillance. However, in its simplest form a GIS is no more than a store of information, and as such it allows huge amounts of information to be easily accessed and readily updated. In order for pipeline managers to access information, all they have to do is look at a VDU screen with a map of the pipeline route, point the cursor to a location of interest on the pipeline

route and ask for the information wanted. Basic information could include:

- name, address and telephone number of the landowner;
- engineering data, e.g. depth of burial, pipe wall thickness;
- crop compensation data since pipe installation;
- aerial photographs, video images or other photographs of the site.

Any information that a pipeline manager wants can be added to the GIS, making it a central store for everything relating to a pipeline.

The potential use of this technology is enormous, and the data retrieval option described above is the least demanding of the capabilities offered by such systems. GIS can also be used to give the answers to 'what if' questions. For example, if the pipeline were to leak at a particular location, the GIS could tell:

- the best access route to that section of pipeline;
- who to contact, with name and telephone number;
- which settlements fall within the area affected by the release.

Furthermore, GIS can interface with simulation models, and then present the results of a simulation run in an easily understood form; for example, in the case of a gas cloud, how big it is and where it will travel under certain weather conditions.

However, the wizardry of GIS cannot compensate for poor information on a pipeline – GIS records are only as accurate and comprehensive as the data that are put in. Complete, up-to-date records are essential whether or not GIS is used. Such records will be diverse and many-fold; two examples are environmental records and waste management records.

Information about the state of the environment is essential. Data should be kept on the location of archaeological sites, recreational areas, water resources including aquifer protection zones, areas of conservation importance including Sites of Special Scientific Interest, landfill sites, landscape features and so forth. Information on the location of these sites and the reason for their sensitivity is useful when planning maintenance work or responding to emergencies. If this information is not available it may be a good investment to undertake an environmental review of the pipeline system. Such a review would focus upon the location of environmentally sensitive sites, the company's relationship with third parties and the availability of emergency response equipment.

It is vital that records are kept on the subject of waste management. In the UK, the 'Duty of Care' requires that the originators of waste keep records of what was disposed of, who transported it and what the final destination was. The exact requirements are specified in the Duty of Care published by the Department of the Environment.

(f) Monitoring. It is important that the success of reinstatement is measured and areas which are unsatisfactory are improved. In agricultural land this is often a question of repairing damage to soil structure and/or drainage. Environmentally sensitive areas have sometimes been overlooked in the past, probably owing to their low economic value from an agricultural perspective. Such areas may include moorland, heathland, unimproved grasslands, species-rich wetlands and deciduous woodlands. Perhaps hedgerows should also be added to this category because far too often new hedges are planted at the end of construction and then not maintained; many die and are not replaced and unfortunately it is the hedgerow that is most often seen by the public. The type of monitoring that is required will depend on the nature of the site and the purpose of the monitoring. In some cases a simple 'look see' and brief report will suffice. In other cases a detailed ecological survey will be needed, using, for example, quadrats across a permanent transect.

The only other monitoring that is likely to be required will be noise monitoring in the vicinty of pumps/compressors. This will be especially important if the pump house or compressor station is located near to a residential area. If there are other emission sources it may be necessary to undertake monitoring, although these are likely to be associated with activities other than the pipeline.

(g) Audits and reviews. Audits should be undertaken to assess compliance with the company's environmental policy or legislative requirements. Some pipeline operators have been doing this, such as British Pipelines Agency and Shell Chemicals UK. British Pipelines Agency have been undertaking audits to help them set priorities for remedial maintenance work (Barr, 1993). In some case, a more general review may be appropriate. A review will not be testing procedures, it will be collecting information. An audit will be verifying actual practice against a yardstick such as a company's environmental policy.

7.3.4 Decommissioning

To date, few onshore oil and gas pipelines have been decommissioned. Generally they are cleansed and simply left *in situ*. It is important to ensure that all product is removed from the line in order to prevent pollution of soil and groundwater. The removal of the pipeline would cause greater environmental impact than leaving it in place.

7.4 Offshore pipelines

Offshore pipeline have three functions (Haldane *et al.*, 1992). First, intrafield lines carry product from sub-sea installations to either another subsea installation or a production platform. Pipelines between two neighbouring platforms within the same field are also usually classified in this manner. Second, interfield pipelines carry product from one production facility to another or to a connection to another pipeline, and their function is normally to transport the oil or gas to the next link in the system, another pipeline or perhaps a tanker. Third, trunk lines link the pipeline transportation system to the shore terminal.

Most UK Continental Shelf pipelines are constructed of carbon–manganese steel or low-alloy steel. They are cathodically protected, most commonly by the use of zinc- or aluminium-based sacrificial anodes. They are externally coated to protect against erosion and many have a concrete coating which provides additional protection but is in the main designed to add weight to the pipeline to prevent buoyancy. When building pipelines in the UK sector of the North Sea it is a mandatory requirement to comply with the Submarine Pipelines Guidance Notes issued by the Department of Trade and Industry in addition to other regulatory documents.

7.4.1 *Design*

As with cross-country pipelines, environmental assessment is a process that begins at the preliminary design stage and continues throughout detailed design. Consultations with relevant statutory and non-statutory bodies are essential throughout. Examination of the existing environment is required together with impact appraisal, impact prediction and identification of mitigative measures where necessary. In addition, proposals for future monitoring of the environment will be needed together with an environmental management programme which will ensure that contract documentation takes account of the findings of the assessment. The final result, the environmental statement, will be required as part of the application for Pipeline Construction Authorization.

There will, however, be essential differences in the nature of the existing environment and, in consequence, the resultant impacts, the proposed mitigation measures and the requirements for future monitoring will differ. These are discussed in some detail below.

(a) Preliminary design. During the preliminary design stage, engineers and environmental scientists are concerned with finding a broad corridor for the pipeline route. This is normally determined by:

- sea-bed topography – a sea-bed which is too rough could lead to spanning of the pipe;
- potential landfall sites – these will limit the location of the end-points of a pipeline route;
- Flora and fauna of the area – known sensitive sites should be avoided if at all possible at an early stage;

- any military activity in the area – including military exercises and munitions dumps.

(b) Detailed design

(i) Consultation. Once again, the importance of consultations with both statutory and non-statutory bodies through the detailed design stage of a pipeline project cannot be overemphasized.

(ii) Examination of the existing environment. As with onshore pipelines, at this stage it is essential to gather good baseline data. However, as the offshore environment is very different from the onshore environment, there is clearly a need for a different set of criteria. For an offshore pipeline these will generally include:

- physical conditions – bathymetry, sea-bed geology and sediments, sediment transport, water levels, water currents, water temperature, winds and waves;
- biological environment – nearshore benthic communities, offshore benthic communities, nearshore and offshore fish, plankton, seabirds and shorebirds, marine mammals;
- human activities – commercial fishing, shipping and navigation, Ministry of Defence areas, cables and oil and gas exploration, minerals and dredging, marine archaeology, conservation designations, recreation, waste disposal and planning policies.

(iii) Impact appraisal and prediction. The baseline surveys and more detailed project work generate the data required to appraise and predict the likely impacts of a subsea pipeline and, in the same way as with onshore pipelines, some techniques used for predictions are necessarily qualitative and some quantitative. The main studies at this stage are likely to concentrate on:

- physical conditions – the effect of anchor mounds;
- biological conditions – the effects of physical intervention, sediment disturbance and noise on benthic communities, fish, plankton, bird and mammal communities and the reef effect of the pipeline (a well documented phenomenon whereby fish are attracted to structures providing shelter; in turn some fishermen trawl the length of the pipeline to benefit from the extra fish);
- human activities – the effects of exclusion of vessels from an area during construction, on fishing, on cables, on munitions dumps, minerals and dredging, marine archaeology and waste disposal.

(iv) Identification of mitigative measures. As with onshore pipelines, mitigative measures are often not needed if the pipeline route has been

carefully selected in the initial phases of design. However, it is likely that there will still be a need for some special measures to be taken. These might include such things as removal or partial removal of anchor mounds, avoidance of environmentally sensitive areas or adaptation of working methods within them to minimize disturbance, carefully planned crossing of obstacles such as cables and avoidance of munitions dumps and known archaeological sites.

(v) Proposals for future monitoring, preparation of the environmental statement and contract documentation. There will also be a need for construction and post-construction monitoring, a requirement for the production of an environmental statement to accompany the construction authorization application and subsequently a need for contract documentation incorporating the environmental requirements of a project.

7.4.2 Construction

Construction methods for an offshore pipeline are clearly very different from those required for an onshore pipeline. A brief summary of the main methods employed is given below together with two case studies.

(a) Good site practice. Raising awareness, site supervision, good reporting procedures, preparation of contractor plans and regular monitoring and audits are all essential elements of a good health, safety and environment programme and should be well established before the start of construction.

(b) Construction methods. There are three standard methods of laying submarine pipelines:

- reel barge method – this is only used for laying small diameter pipelines in shallow waters;
- bottom pull method – this is used in inshore waters; the pipe is fully prepared on land and is pulled into the sea by barge; welding and concrete coating can take place on land, and any damage to the pipe is more likely to result from friction along the seabed than from bending the pipe;
- lay-barge method – this is the most common way of laying pipe. The barge acts as the pipeline factory, where pipelines are welded and X-rayed and the joints coated. Lowering the pipe into the sea is difficult, and the pipe may need to be supported by a ‘stinger’ so that the bend does not exceed a maximum permitted curvature. Such barges need anchors at each corner.

There are several different methods for laying pipelines on the sea-bed; these include partial trenching, complete trenching, trenching and back-filling and rock armouring. In UK shallow waters pipelines are trenched and buried where possible, regardless of their dimensions, because of potential damage from currents and waves. In soft, sandy sediments the trench tends to backfill itself with time. The most common method of trenching is 'jetting', where a 'trencher', a saddle-shaped construction, is placed on top of the pipe which has already been lowered on to the sea-bed. The trencher is equipped with water jetting nozzles or a rotating cutterhead (depending on the sediment), and the apparatus is towed along the pipeline. The spoils are sucked into a pipe, discharged and carried away by any prevailing currents, and the pipeline then automatically lowers into the newly cut trench. Where trenching is not possible because of a hard sea-bed, the pipeline may be laid on the sea-bed and rock armoured. In deeper waters (>60 m) it is usually unnecessary to trench or bury larger pipelines for engineering reasons. However, even where the pipeline has been left proud, it may sink over time.

In recent years there has been much dispute over the value of trenching and burial. Trenching and burial do have the advantage of protecting a pipeline from some of the most frequent physical impacts such as those from fishing gear, strong currents and, more rarely, dropped objects, and clearly there will be no impact on fishing gear if the pipeline is buried. Trenching and burial may also minimize problems associated with scouring and spanning, and thus it might be possible to offset some of the additional costs of burial against costs incurred in span correction. In addition, burial may make future abandonment a more viable option. However, when Shell and Esso began plans for the 36 in Flags gas line, they initiated a series of studies to test the notion that large pipelines in deep water did not need to be trenched. In summary, the studies found that impacts from fishing gear were unlikely to result in serious damage to pipelines, that buried pipelines were not protected from anchors from large ships, and that concrete coating does a more reliable job of weighting the pipe to provide stability than does trenching. It is now generally accepted that large diameter, proud pipelines in deep water are not problematical.

Special construction methods in environmentally sensitive areas can involve minimization of disturbance of sediments, minimization of rock blasting, reduction of noise levels emitted from plant and machinery, careful timing of operations in order to avoid, for example, bird breeding periods and even drilling under particularly sensitive nearshore habitats.

(c) Case study: Scotland to Northern Ireland Natural Gas Pipeline (SNIPS). This 42 km long pipeline (Fig. 7.2), built in 1995, was laid using a pipeline largely anchored to the sea-bed. Where feasible the pipeline was

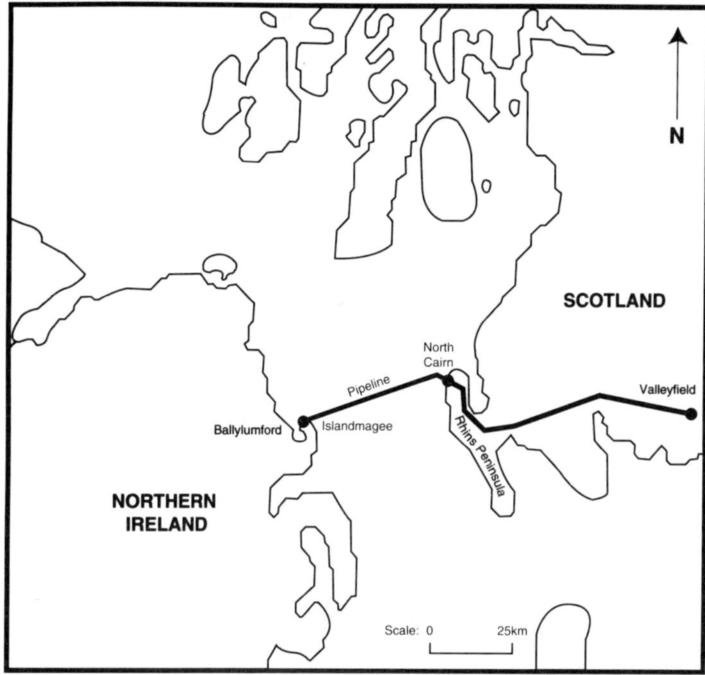

Figure 7.2 Map showing the route of the Scotland to Northern Ireland Natural Gas Pipeline (SNIPS).

trenched, although in certain areas the nature of the sea-bed did not permit this and it was necessary to place rock over the pipeline.

At the preliminary design stage, 16 possible crossings of the North Channel from the Rhins Peninsula in South West Scotland to Islandmagee on the east coast of Northern Ireland were considered (Premier Transco Limited, 1994). The cliffs along much of the coastline precluded wide areas of the coast as landfall sites but it was possible to identify two potential areas on each side of the North Channel.

Coastal surveys were undertaken at each of the sites to assess the sea-bed and coastal conditions. Offshore surveys were undertaken to assess the physical conditions along the potential routes across the North Sea. The main factors influencing the selection of the corridor were:

- the operational risks associated with laying a pipeline across Beaufort's Dyke, owing to the steep slopes on the faces of the dyke and the sediment conditions;
- an independent assessment of the landfall options which identified areas within the North Cairn, Browns Bay, Ferris Bay and Port Muck survey areas as the most suitable for the pipeline landfalls;

- the desire for the selected route to minimize the amount of disruption to the sea-bed that would be required;
- the need for the selected route to avoid Danger Area D411 and the munitions dump area identified by the Ministry of Defence.

During the detailed design stage of the SNIPs pipeline, a number of possible environmental impacts were identified by environmental scientists. Many of the impacts were ones that would be associated with any offshore pipeline, such as the creation of anchor mounds, temporary exclusion of fishing, changes in the habitats of benthic flora and fauna, stress to plankton organisms, change in behaviour of fish species and localized avoidance of the area by some seabirds. Others, discussed below, were specific to the area concerned.

With regard to the physical environment, one of the main concerns was the presence of a dredge spoil dump to the north of Larne. It was believed that the dumping may have led to contamination of sediment within the pipeline corridor. Dredging close to Larne could disturb the sediments and cause pollution and if contaminated sediments had to be removed from the sea-bed there could be problems obtaining a licence for disposal.

Also of concern were three environmentally sensitive areas close to the pipeline route: the bird nesting and roosting areas on the Isle of Muck and Skernaghan Point the diverse benthic habitats of Castle Robin, all on the Irish coast. At the Isle of Muck, a bird reserve, and Skernaghan Point there was concern about the effects of construction noise. At Castle Robin there was concern about the effects on the diverse benthic communities of nearshore blasting through hard rock. Options for minimizing the effects on birds included avoidance of sensitive areas where possible, careful timing of construction to avoid the breeding season in spring and early summer, careful choice of plant in order to minimize noise and the fitting of noise attenuators to plant. Mitigation measures identified for Castle Robin included possible avoidance of blasting if a suitable route through softer sediments could be identified, minimizing the use of explosives and rock-ripping and using controlled rock splitting where possible, and avoidance of the most sensitive benthic communities where possible.

It was apparent that before construction could begin there was a need for further studies of the pipeline route close to the sensitive areas described above. These studies led to the inclusion of special specifications in contractors documentation which ensured that the correct procedures were carried out during construction.

(d) Case study: the Gas Interconnector Pipeline (GIP). The Gas Interconnector Pipeline has recently been built by Bord Gais Eireann between Moffat, South West Scotland, and Ballough, 20 km north of Dublin, Eire, in order to supply gas to Eire from the North Sea (Fig. 7.3). As with the

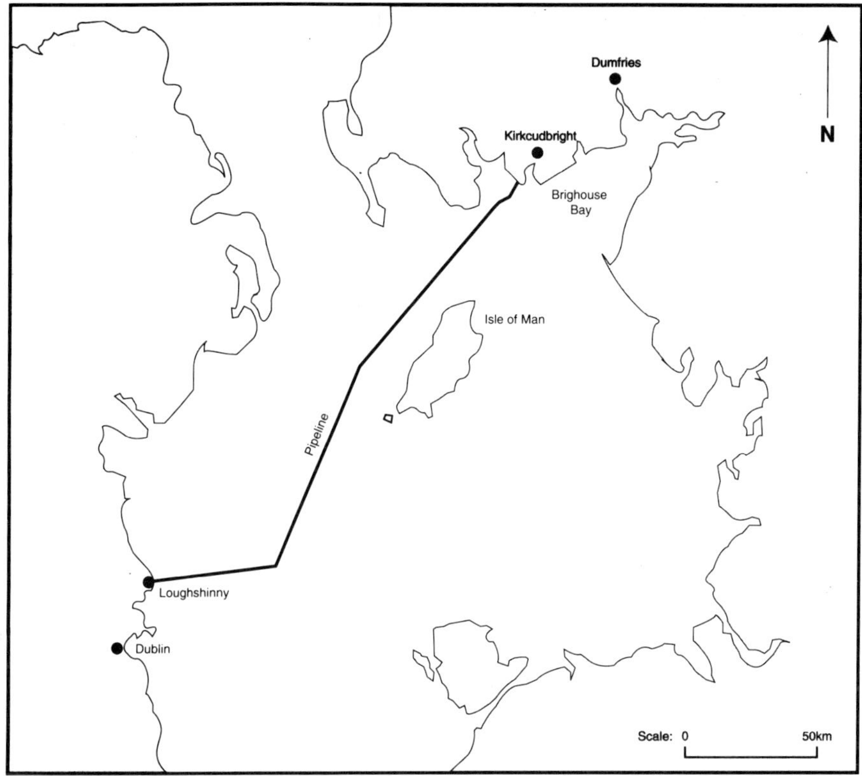

Figure 7.3 Map showing the route of the Gas Interconnector Pipeline.

SNIPs project, the detailed design stage of this pipeline was concerned primarily with the identification of landfalls on the Irish and Scottish coasts and the identification of broad corridors across the North Sea suitable for the pipeline. In particular, the landfalls were the subject of extensive study; the exact methods employed for the Scottish landfall at Brighouse Bay are discussed later and will therefore not be further elaborated upon here.

During consultations with affected bodies, Bord Gais Eireann found that various groups of fishermen were concerned about the proposal not to bury some of the pipeline. In the end, the project managers had to agree to bury the whole length of the pipe; however, as much of the pipeline was expected to sink in soft clays it did not involve trenching the whole length.

During detailed design, the environment assessment identified many of the same probable impacts as on the SNIPs line. However, there were some which were specific to the GIP. One aspect needing consideration was the temporary loss of access to a 15 m wide strip along the nearshore fishing grounds at Kirkcudbright. To enable the pipeline to be laid, fixed gear such as creel pots had to be moved and mobile gear such as trawling nets would have to be restricted in their operations. Unlike fishermen working farther out to sea, the shellfish fleet operating out of Kirkcudbright have limited alternative areas in which to fish. In order to minimize problems associated with the shellfish fishing, the fishermen were fully involved in the decision making process, they were given early notice of the timing of the work, and they were contracted to lay the approach channel buoys. However, during the actual pipelaying operation there was no choice but to exclude the fishermen from the pipelaying zone.

7.4.3 Operation

During testing, commissioning and normal operation there will be little effect on the marine environment. Severe effects will be felt only in the event of a spill. In order to prevent such a spill, pipeline operators commission regular inspections of their subsea pipelines, and subsequently carry out repair and maintenance. This will clearly be much more difficult in a subsea environment than on land and has resulted in the development of sophisticated subsea equipment and machinery.

(a) Testing, commissioning and normal operation. The possible effects on the physical and biological environment and on human activities of offshore pipeline commissioning are related primarily to the discharge of test water. The composition of the test water for a pipeline will need to be the subject of study, and dispersion modelling will be required in order to determine its effects on the area concerned. The use of test chemicals such as biocides and corrosion inhibitors is subject to prior statutory or regulatory authorization. However, in general it is likely that any effects, relating principally to the toxicity of the test waters, will be minor and short term.

During operation the main concern is the effect of the pipelines and associated debris on fishermen. The Scottish Fishermen's Federation claims that debris on the sea floor is much more damaging to fishing gear than the damage caused by pipelines and their associated rock dumps, yet some fishermen regularly claim that they lose or damage their gear on submarine pipelines. However, it is difficult to prove that this is the true origin of the damage. According to de Groot (1982), fishing gear often hits rocks and ship wrecks which cause the same sort of damage and effect as a pipeline would.

(b) Emergency planning in the event of a spill. Severe and long-term damage to the offshore environment, and in turn to human activities, can occur in the event of a pipeline spill. As a result, operators of subsea

pipelines are required to prepare emergency response plans. In the North Sea there have only been two significant spills from pipelines (Haldane *et al.*, 1992). One was on 7 April 1980 from the Thistle–Dunlin pipeline. The rupture, which was believed to have been caused by a vessel dragging an anchor over the line, was identified after a drop in pressure in the pipeline, and it was thought that about 1000 tonnes of oil was lost over a period of 25 min. The other was from Occidental's Claymore pipeline on 26 November 1986. In this instance the leak was from a valve spool and it was estimated that between 1000 and 2000 tonnes of oil was spilled. The slick moved towards the Norwegian coast, and when after 8 days of extremely rough weather it had still not broken up, Norwegian pollution control vessels were mobilized to monitor and to attempt recovery of the oil as it approached the coast. On 6 December the wind changed direction taking the slick offshore and by 7 December the slick had dispersed. In view of the nature of the Norwegian coastline and in particular the large numbers of fish farms, environmental teams were mobilized to survey the area, but only minor traces of oil which may have resulted from the spill were found.

(c) Pipeline integrity monitoring. Subsea pipelines are regularly monitored to check for corrosion (to which they are particularly subject because of the salt water, despite cathodic protection measures), third-party interference (interaction with vessel anchors and fishing gear) and scouring and spanning (removal of the sea bed from beneath the pipe due to currents). A range of monitoring techniques are available:

- visual;
- electrical potential difference;
- magnetic particle inspection;
- ultrasonics.

These are carried out using a variety of undersea vehicles or by pigging. Undersea vehicles can be manned or they can be unmanned, remotely operated vehicles (ROVs). ROVs usually have on board a video camera, a trench profiler, a pipe tracker and a cathodic protection probing system. Pigs may be used for cleaning and for measuring pipe diameter, roundness and wall thickness.

(d) Pipeline maintenance and repair. Pipelines in a marine environment will require more maintenance and repair than land-based pipelines. In particular they will need protection against scouring and spanning. This can be done by a number of methods (Haldane *et al.*, 1992):

- mechanical supports can be installed using diverless installation systems;
- grout bag supports can be installed by divers or by ROVs;

- rock infill is particularly suitable where the sea-bed is hard and where long distances are involved;
- jetting can be used but is diver intensive and is limited to relatively shallow waters;
- trenching of shoulders is useful for short spans.

Other techniques tried include anti-scour mattresses and artificial seaweed. In some instances it may be necessary to anchor a pipeline to the sea-bed using concrete, piles or clamps to prevent it moving.

Pipelines may also need additional protection from third-party interference. This can be done by trenching the pipeline, although special sections such as tie-ins can be covered with protective covers. In addition, cathodic protection anodes may need periodic replacement owing to excessive use, loss or damage.

Repair of a subsea pipeline may require a section to be cut out and replaced, which can be a very difficult or a relatively simple operation depending on the conditions. In shallow water divers may be used, but in deeper water this will not be possible and remote-controlled repair systems must be used.

(e) Record keeping, monitoring and audits and reviews. Again, GIS is rapidly becoming the way forward in record keeping. Monitoring and auditing are becoming more important as companies have to become more and more accountable for their actions and, where information is not available, perhaps because it was not collected at the pipeline design and construction stage, reviews sometimes need to be carried out.

7.4.4 Decommissioning

Decommissioning of offshore pipelines is at present an issue of the future. There may be some pipelines where removal is the favoured option because of the possible interference with fishing gear. However, this may not be a viable option where the pipeline is buried, as its removal may cause more disturbance to the sea-bed, and thus to fishermen, than if it was left in place. It may become necessary to accurately map broken sections of decommissioned pipelines in order to make the information available to fishermen and other users of the sea (Haldane *et al.*, 1992).

7.5 Pipeline landfalls

The point where an offshore pipeline comes ashore is known as a landfall. This interface of the land and the sea is the single critical element in a pipeline route which crosses the boundary between the two; often the energy levels impacting on the pipe from the marine environment, and hence the potential to damage the integrity of the pipeline, are greatest at this point and therefore the decision of where the landfall should be located requires considerable forward planning. Because of the critical nature of the landfall it is considered here, in detail, as a separate issue.

7.5.1 Design

The planning process required essentially follows the same steps as for an onshore pipeline through the preliminary and detailed design phases, and an application for permission to construct a landfall is generally included within the same environmental statement as the application for the onshore pipeline associated with it. However, because both environmental and engineering constraints are often severe, it is particularly important that both are considered in great detail and that neither is considered in isolation. Environmental protection measures during landfall construction, while also following the same basic principles as for onshore pipelines, may require specialized techniques not used elsewhere. For these reasons, the landfall is often subject to separate study and separate technical reports can be produced.

(a) Preliminary design. At the preliminary design stage it will, as with the remainder of the pipeline, be necessary to carry out consultations and undertake surveys to identify a location for the landfall. As the route of the cross-country pipeline and the subsea pipeline will depend on the location of the landfall, it is clear that getting the siting of the landfall right as soon as possible is of fundamental importance. The nature of the coastline bears a direct relationship to the ease of construction of the pipeline, and therefore a study of its physical characteristics will prove invaluable in helping identify a suitable location. However, in addition to such a study, a number of other parameters play a controlling role in the suitability of a particular stretch of coast for the construction of a landfall:

- the form and nature of the sea-bed close to the coast;
- marine energy levels;
- technical constraints.

In order to determine the suitability of a particular location for a landfall, a list of features that are considered desirable can be compiled together with a list of features that would be considered undesirable. These are shown in Table 7.1. In simplistic terms, it is easier to construct a pipeline across a narrow, sandy beach than a coastline in which rocky outcrops predominate. In addition, sandy beaches are generally far easier to reinstate than rocky shores. Sandy beaches need little extra protection for the pipe whereas a rocky landfall needs the importation of sand for

Table 7.1 Desirable and undesirable features of a landfall

Desirable features	Undesirable features
Stable beach – long-term integrity of the pipeline is preserved due to the sediment transport being minimal	*Population centres* – it is preferable to avoid population centres due to the effect of construction on the quality of life of residents
Water depth – a water depth of 15 m is preferable within 2–3 km of the shore; this reduces the amount and scale of excavation/ dredging	*Rocky coastline* – span problems can occur; blasting may be required and restoration becomes difficult
Direct routing (linearity) of the shore approach – this would minimize length and ensure a less complicated construction technique; there would be less disturbance to the intertidal zone	*Exposed area of coastline* – exposed coasts may lead to the exposure of the pipeline by marine processes
Ease of reinstatement – the ability to achieve 'full' restoration of the landfall is of importance	*Long shallow approach* – extensive dredging required and hence the impact on marine life is greater; scale of construction operations would be larger
Trenchable sea-bed to deep water – a trenchable sea-bed avoids 'free spans' which may lead to stress failure of the pipe	*Steep slopes* – pipeline installation and long-term stable reinstatement difficult
Sandy beach – sandy beach provides a soft bedding for the pipe, it is easily excavated and can be readily reinstated	*Non-cohesive sediments* – these are unstable and susceptible to bearing strength failures and sediment mass movement
Good land access – minimize the upgrading of the road that is necessary to allow plant access to the beach	*High-velocity nearshore currents* – these can interfere with pipe-laying activities and may entail additional protective measures
	Coastline designated as having landscape value and possibly experiencing recreational pressure – disruption must be minimized
	Nature conservation areas – the potential disruption of species and loss of habitat are to be avoided

bedding the pipe. Marine energy levels are often higher on rocky coastlines.

However, environmental constraints cannot be categorized according to the coastline landform alone. All constraints need to be identified and by careful planning individual constraints must be minimized or avoided. In general terms, landfalls should avoid population centres, specific wildlife sites and areas of outstanding scenic beauty. The planning of a prospective landfall must also assess the surrounding land in terms of access for heavy construction plant and for any infrastructure that will be necessary for the operation of the pipeline, such as a receiving terminal, a compressor station or a pressure reduction station. Other particular problems encountered in some types of coastline include the low load-bearing capabilities of some intertidal muds and salt marshes and the longer term stability and reinstatement problems associated with cliffs. The visual element is equally important in this regard, where the results of pipeline construction may be visible for a number of years. If hard structures are required to protect the pipeline and provide stability, then the construction may be visible for as long as the life of the pipeline. An ideal landfall would be a stable, sandy, sheltered, low-angle beach with no statuary designations relating to flora, fauna or scenic value.

It can be seen that, with all the constraints discussed above, a long stretch of coastline may have to be investigated before a suitable location for a landfall can be found.

(b) Detailed design. Once a suitable landfall site has been found, detailed field studies of the existing environment will be required, a precise route chosen and impact appraisal and prediction carried out in the same manner as for onshore pipelines. However, proposals for mitigative measures to overcome predicted impacts will often have to be innovative and very site specific.

7.5.2 Construction

(a) Construction methods. The pull method is the basic installation technique used on many types of coasts, including beaches (assorted grain size), salt marshes, intertidal flats and raised beaches. A cut will need to be excavated through a raised beach or cliff, provided the composition of the material is unlithified and can be easily excavated. Two contrasting techniques can be used: pull onshore and pull offshore.

The basic pull onshore technique involves the pulling ashore of the pipeline from a laybarge anchored offshore. The length of pull can be up to 5 km, depending on the diameter of the pipe. A winch is erected on land and anchored by piles or wires grouted into bedrock. The trench is excavated by land-based equipment to the low-water mark and by a dredger seaward of this point. The pipe is then pulled by the winches through the trench. Variations on this theme occur depending on site-specific parameters. For instance, the trench across the intertidal area is often 'pre-proved'. This involves excavating the trench to ensure that no hard rock will prevent land-based excavators digging the sediments from underneath the pipe. Rollers are then laid out over the backfilled trench to reduce friction during the pull.

In essence, the pull offshore is the reverse of the pull onshore. The main difference is that a string fabrication area is required for welding individual pipe lengths into pipe strings. These can be several hundred metres long.

The landtake and vehicle movements into site are therefore correspondingly greater with a pull offshore than a pull onshore. In isolated areas, where landfalls often occur, the road system is usually not ideal for the many extra vehicle movements that a pull offshore would entail.

The choice of construction method will be determined by the nature of the onshore and nearshore environment. As each landfall is unique, whichever is chosen will have to be adapted to suit the individual needs of the site. The examples discussed below serve to illustrate some of the ways that coastal environment environmental constraints have been managed in recent years.

(b) Case study: the Gas Interconnector Project (GIP) landfall at Brighouse Bay. The critical points on the GIP pipeline which determined its route were the landfalls on the Irish and Scottish coasts. In particular, the landfall on the Scottish coast is in an area of significant environmental sensitivity and is classified as an area of Regional Scenic Significance and as a Site of Special Scientific Interest (SSSI); it is that landfall that is considered here (Fig. 7.4).

The steps outlined below indicate the planning stages followed to determine the optimum landfall location:

- Determination of the area of interest, in this case the coastline of Dumfries and Galloway (obeying the straight-line principle between Moffat and Ballough).
- Identification of all marine and terrestrial constraints in the vicinity of the landfall that may impact on the construction of the landfall. These were plotted on a constraints map.
- Hand-in-hand with the above approach, constraints had to be identified on both the subsea and landfall routes that may preclude a particular landfall option. This was considered by tabulating matrices having identified the parameters crucial to the construction of the pipeline.

Subsea constraints ranged from water depth and the nature and topography of the sea-bed to the presence of fishing grounds and military bombing ranges. Landline constraints included such considerations as landform relief, protected areas and the number of road and river crossings. An additional consideration in the early planning stages was the requirement for a compressor station as close to the Scottish landfall as possible. This was needed to generate sufficient pressure to permit the transmission of gas to Eire. Owing to the attractiveness of the Dumfries and Galloway coastline, the siting of the compressor station was a crucial issue. Zones of Visual Influence (ZVIs) were identified to help identify the optimum location and hidden line perspectives were generated to ensure that any visual intrusion was minimized.

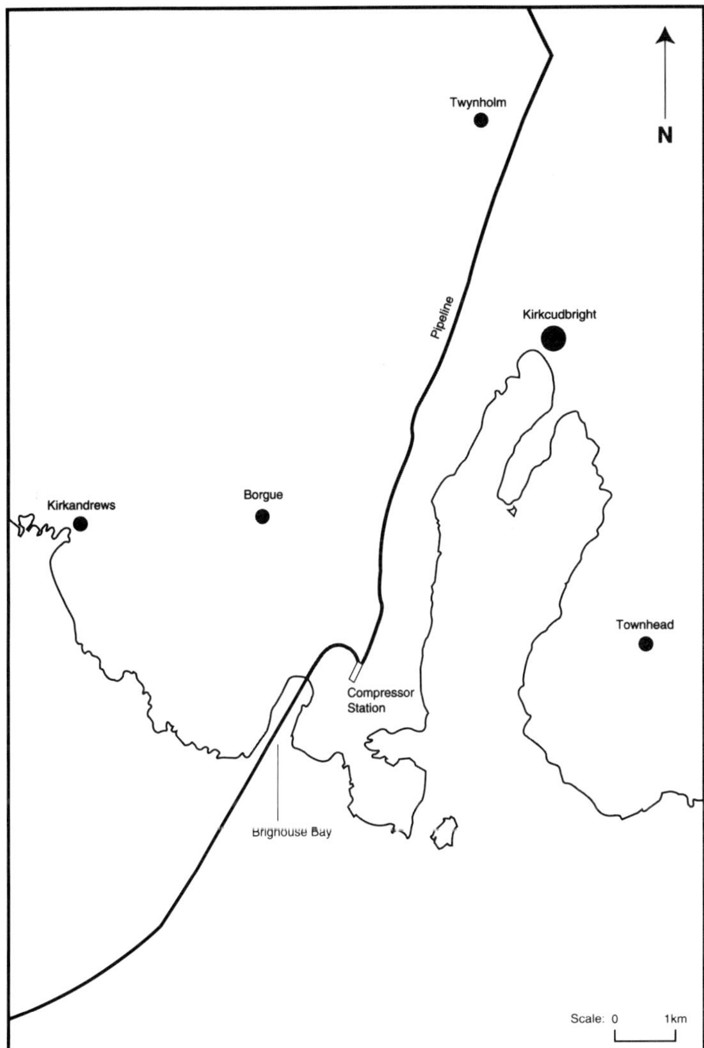

Figure 7.4 Map showing the Gas Interconnector Pipeline landfall at Brighouse Bay.

The exercise outlined above indicated that 13 potential sites were worthy of further consideration. In order to evaluate these sites, a matrix was completed which identified the following parameters grouped under six headings:

- marine environment: tidal streams, maximum fetch, site exposure, wave activity, tidal range, predominant wind direction, military activity;

- physical constraints: beach composition, beach dynamics, beach and nearshore profile, shore topography, sea access, presence of bedrock, water depth, land access;
- biological constraints: fragile habitats (land and marine);
- environmental constraints: recreational pressure, land designation, archaeology;
- availability of land for compressor;
- construction: technical notes, resultant impact, restoration problems, relative extent of landfall construction.

The completion of a matrix identifying all the parameters considered to have an impact on the landfall location enabled a more considered judgement to be made as to the optimum landfall location. One of the most important criteria was the ability of a site to be fully reinstated. Having considered all the options, Brighouse Bay, near Kirkcudbright, was selected as being the optimum landfall location on the Solway coast.

Because of the environmental sensitivity of the Brighouse Bay landfall, it was more important that the constraints and specifications particular to the landfall location were made known to the contractor prior to their appointment. The constraints arose from a number of sources, including:

- the environmental statement;
- specific technical reports commissioned during the environmental assessment;
- stipulations attached to the Pipeline Construction Authorization;
- planning permission requirements;
- requirements detailed by statutory and non-statutory bodies;
- general requirements as a result of far-ranging dialogue.

From the wealth of information arising from the project, the contractor for the Brighouse Bay landfall had to ascertain the environmental and engineering controls imposed on construction and devise measures by which the environment and particularly sensitive areas would be protected. These were written into method statements by the contractor prior to construction and had to be submitted to Bord Gais Eireann, the planning authorities and other statutory bodies for approval.

Brighouse Bay was identified as the optimum landfall location in the south west of Scotland primarily because of the sandy nature of the bay and hence the ability of the landfall to be fully reinstated. However, because it experiences heavy recreational pressure and is designated as important for its landscape, wildlife and geological value, special construction methods were essential.

Scheduling construction activities for the winter months overcame, to a large extent, the problems associated with recreational pressure. However, Brighouse Bay is part of the extensive Borgue Coast SSSI and is particularly sensitive on botanical grounds. In particular, the presence of perennial blue flax, the pyramidal orchid and lesser meadow rue provides considerable botanical interest. A detailed botanical survey identified a zone where the distribution of the above species was sparse and therefore the line of the pipe was centred on this area. In fact, not one flax plant was identified within the 26 m working width. Therefore, although at first sight the bay appeared an unlikely choice for a landfall, the fact that full reinstatement could be achieved and the botanical interest was not being compromised determined that Brighouse Bay was the optimum choice.

It was decided by environmental specialists that the best way to ensure rapid reinstatement of the sensitive dune area crossed by the pipeline at Brighouse Bay was to turf it. From the nine specific habitats that were identified by an ecologist, turfs 1 m square and 20 cm deep were cut, lifted on to pallets and transported some 500 m to a laydown area for the duration of the construction period. The location of the habitats and turves from each habitat were clearly identified and the turves from the different habitats were stored separately. In total, almost 3000 m^2 of turf were lifted and stored for reinstatement. In addition, hawthorn bushes up to 2 m high were transplanted, using a large excavator bucket to dig out the complete root system. The method proved to be very successful, and the following year it was only necessary to supplement the turves with seed collected from the site the previous summer.

(c) Case study: the Gas Interconnector Project (GIP) landfall at Loughshinny. Loughshinny is the location of the Gas Interconnector Pipeline landfall on the Eire coast just to the north of Dublin (Fig. 7.5). It provides an example of landfall construction through a boulder clay cliff some 15 m high. Slumping on the face indicated that there was potential for erosion, although historical records showed that the current position of the cliff face was within 1 m of a survey conducted over 150 years ago.

The main concern at Loughshinny was to ensure that the methods utilized to stabilize the cliff face were in keeping with the surrounding area so as not to create a visually intrusive monument. The neighbouring headland and Martello Tower are a favoured area for walkers. To this end, a gabion base at the toe of the cliff was constructed as the main support. Layers of terram folded back on itself provided stability at the face. The cliff was seeded to provide protection against erosion and to blend in with the surrounding cliffs.

(d) Case study: the Theddlethorpe landfall, Lincolnshire. In 1992, Conoco UK installed a 26 in diameter natural gas pipe from the Murdoch Platform to the Theddlethorpe Gas Terminal (Fig. 7.6). This was the fourth landfall to be brought ashore over a short stretch of coastline.

The flat, sandy beach at Theddlethorpe is backed by an ancient dune

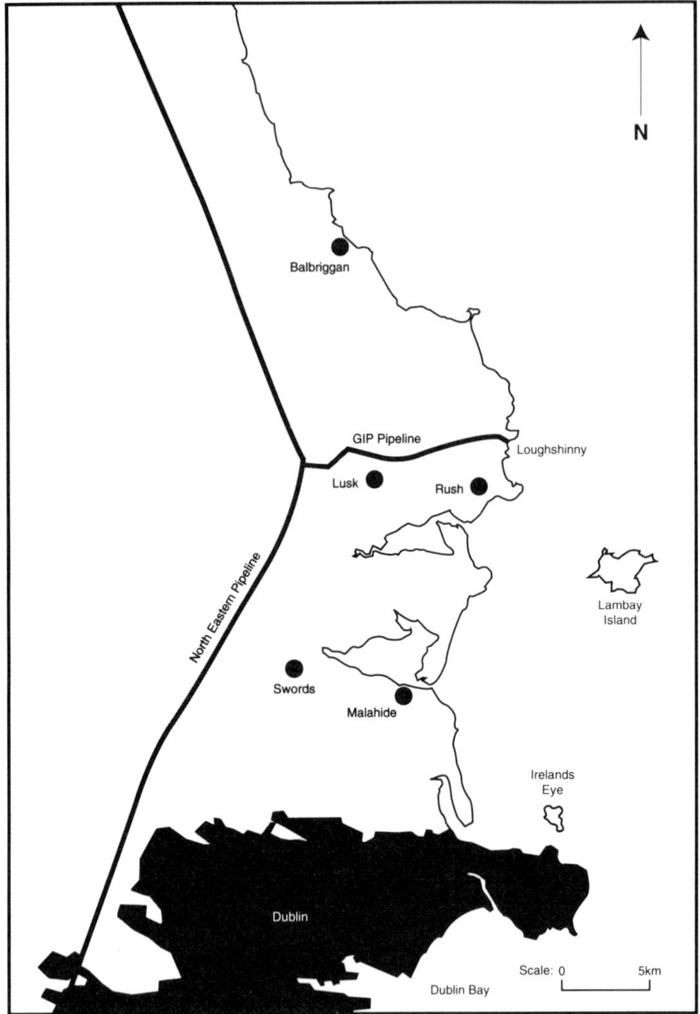

Figure 7.5 Map showing the Gas Interconnector Pipeline landfall at Loughshinny.

system which is designated as a National Nature Reserve. The previous three landfalls had utilized an open cut through the dune system. However, for this landfall Conoco proposed the construction of a concrete-lined tunnel, using conventional pipe-jacking techniques, in order to leave the ancient dune system intact. The pipe-jacking operation was successful with the majority of the ancient dunes remaining undisturbed, although the final 75 m of the operation had to be open-cut owing to a survey problem resulting in the concrete casing being off-line. The dunes which were

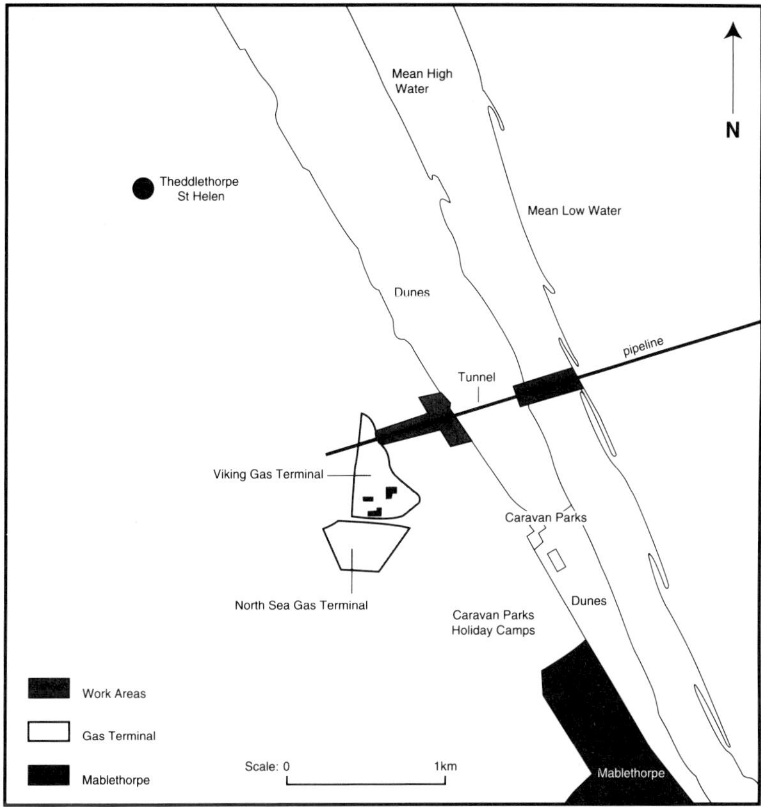

Figure 7.6 Map showing the Theddlethorpe pipeline landfall.

affected, at the edge of the ancient dune system, were covered by sea buckthorn, which grows vigorously, and were considered to be the least sensitive part of the system by English Nature. In order to aid dune stability, a marram replanting exercise was undertaken over the embryonic dunes adjacent to the beach.

(e) Case study: the Walney Island landfall, Cumbria. British Gas has recently constructed a 3.5 km long pipeline across Walney Channel, a tidal estuary comprising saltmarsh and intertidal flats near Barrow-in-Furness (Fig. 7.7). The pipe, which was winched from a string fabrication yard on the mainland across to Walney Island, forms part of a larger project to bring gas from the North Morecambe Bay gas field to a new gas terminal at Westfield Point, to the south-east of Barrow-in-Furness.

Virtually the whole length of the pipeline falls within the South Walney and Piel Channel Flats SSSI, and therefore it was not surprising that special

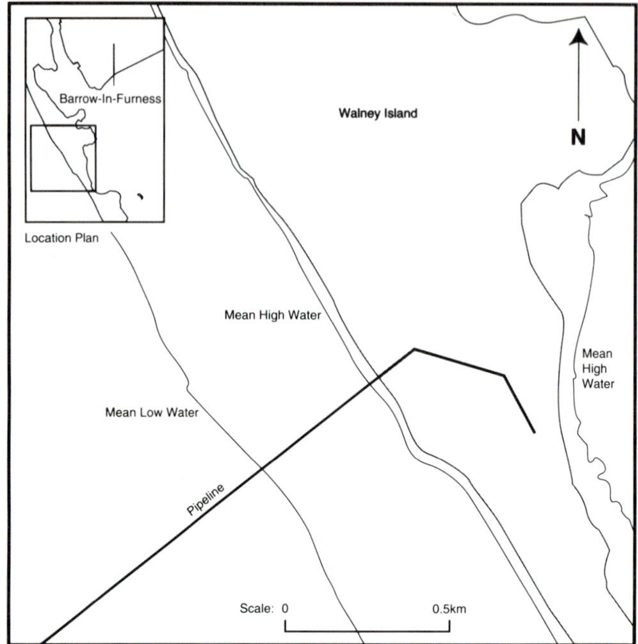

Figure 7.7 Map showing the Walney Island pipeline landfall.

construction techniques were needed to overcome a number of botanical issues. The most sensitive of these was the fact that the intertidal area contained the only recorded site in north-western England of the nationally scarce *Zostera angustifolia* (narrow eelgrass). It was decided that it was not possible to store the sediment containing the *Zostera* for the duration of construction in a manner which would allow the diurnal tidal inundation on which it thrives and at the same time prevent erosion of the material. Instead, in the hope that the plant would survive, and in the knowledge that this layer would contain the rhizomes and seed of the *Zostera* as well as a significant invertebrate population, shallow (10 cm) cuts of sediment were transferred from the most densely populated areas on the pipeline route to a ready excavated site away from the working area which contained relatively few plants but had similar tidal conditions. In this way, although the plants could not be replaced in their original location, disturbance was minimized.

References

Barr, J. (1993) On-line inspection: BPA's experience. *Pipes and Pipelines International*, **38**(1), 14–19.

Burrows, D.K. (1990) The great Mississauga evacuation, in *States of Emergency: Technological Failures and Social Destabilisation* (ed. P. Lagadec), Butterworth-Heinemann, London, pp. 66–73.
Cobb, G. (1993) Administration of UK pipelines' legislation. *Pipes and Pipelines International*, **38**, 3.
CONCAWE (1995) *CONCAWE Report*, 4/95.
de Groot, S.J. (1982) The impact of laying and maintenance of offshore pipelines on the marine environment and the North Sea fisheries. *Ocean Management*, **8**, 1–27.
Det Norske Veritas (1993) Upgrading Russia's 140,000 km of gas pipeline. *DNV Forum*, No. 3/93.
Gritzenko, A.I. and Kharionovsky, V.V. (1994) Problems of pipeline integrity in Russia. *Pipes and Pipelines International*, **39**, 6.
Haldane, D., Paul, M.A., Reuben, R.L. and Side, J.C. (1992) Submarine pipelines and the North Sea environment, in *North Sea Oil and the Environment* (ed. W.J. Cairns), Elsevier Applied Science, London, pp. 481–522.
Health and Safety Commission (1991) *Major Hazard Aspects of the Transport of Dangerous Substances*, Advisory Committee on Dangerous Substances, H.M. Stationery Office, London.
Premier Transco Ltd (1994) *Scotland to Northern Ireland Natural Gas Pipeline Project: Offshore Pipeline Environmental Appraisal Report*, Premier Transco Ltd, Larne, Co. Antrim.
Rapson, S. (1994) Environmental assessment of cross-country pipelines. *Geography Review*, **7**(5), 21–4.
Sljivic, S.C. (1995) Pipeline safety management and the prevention of third-party interference. *Pipes and Pipelines International*, **40**, 6.

Further reading

Berry, K.G., Bianchini, M., Muller, B. *et al.* (1995) *Performance of Oil Industry Cross-Country Pipelines in Western Europe*, Report 4/95, November, CONCAWE, Brussels.
British Standards Institution (1992) *Specification for Environmental Management Systems BS 7750:1992*, BSI, London.
Department of Energy (1989) *Electricity and Pipe-line Works (Assessment of Environmental Effects) Regulations 1989*, Statutory Instrument 1989 No. 167, H.M. Stationery Office, London.
Department of Energy (1989) *Electricity and Pipe-line Works (Assessment of Environmental Effects) Regulations 1989*, Statutory Instrument 1990 No. 442, H.M. Stationery Office, London.
Department of the Environment (1988) *Environmental Assessment Implementation of EC Directive Draft Circular*, H.M. Stationery Office, London.
Department of the Environment (1991) *Waste Management: the Duty of Care. A Code of Practice*, H.M. Stationery Office, London.
Department of Trade and Industry (1992) *Guidelines for the Environmental Assessment of Cross-Country Pipelines*, H.M. Stationery Office, London.
Eberlein, J. (1986) Implications for project design and management. Presented at EEC Environmental Assessment Directive: Towards Implementation, January, London.
European Community (1985) *Council Directive on the Assessment of the Effects of Certain Public and Private Projects on the Environment*, 85/337/EEC, *Official Journal*, EC, Brussels.
Fairclough, A. (1986) Objectives and Requirements of the Environmental Assessment Directives. Presented at EEC Environmental Assessment Directive: Towards Implementation, January, London.
Lednor, M.C. (1993) Environmental input to pipeline landfalls. *Pipelines, Risk Assessment and the Environment*, 9–11 June, Manchester, (*Pipe and Pipelines International*).
MCS International/ICT (1992) *Gas Interconnector Project: Subsea Pipeline*. Bord Gais Eireann, Dublin.

Pipe nightmares. *Russian Petroleum Investor*, December 1994/January 1995, 39.
Richardson, S.C. (1995) *Pipeline Safety Management Programme: Once Call Feasibility Study*, RSK Environment, Helsby, unpublished report.
RSK Environment Ltd (1992) *Gas Interconnector Project: Irish Onshore Pipeline*. Bord Gais Eireann, Dublin.
RKS Environment Ltd (1992) *Gas Interconnector Project: South West Scotland Onshore Pipeline*. Bord Gais Eireann, Dublin.
Ryder, A. and Smyth, M.A. (1992) *UK Environmental Legislation Affecting Pipelines*, RSK Environment Ltd, Helsby, unpublished report.
Ryder, A.A. (1993) Implementing environmental controls in the planning and operation of new cross-country pipeline. *Pipelines, Risk Assessment and the Environment*, 9–11 June, Manchester (*Pipeline and Pipelines International*).
Shell Chemicals UK Ltd (1989) *The North Western Ethylene Pipeline, Environmental Statement*, Shell Chemicals, Chester.
Smyth, M.A. (1992) *Impact of Offshore Pipelines on Trawling*, RSK Environment Ltd, Helsby, unpublished report.
The Pipeline Inspectorate Petroleum Engineering Division (1984) *Submarine Pipeline Guidance Notes*, Department of Energy, H.M. Stationery Office, London.

8 Environmental management and technology in oil refineries

H. AMIRY, H. SUTHERLAND and E. MARTIN

8.1 Function of an oil refinery

The purpose of oil refineries is to produce marketable products from crude oil or other hydrocarbon feedstocks. Crude oil, the basic feedstock, is a mixture of a large number of different hydrocarbons and other organic compounds, with widely ranging molecular structures from gases to substances with very high boiling points. Crude oils can vary greatly in their physical and chemical characteristics, depending on their origin.

A refinery produces a wide variety of products including:

- fuels such as LPGs, gasolines, aviation fuels, gasoils/diesels, fuel oils and marine fuels;
- chemical feedstock – naphtha, gasoils and gases;
- lubricating oils, greases and waxes;
- bitumen and asphalts;
- petroleum coke;
- sulphur.

A refinery consists of a number of plants using a variety of chemical and physical processes. Each plant has its own specific function, the output of some plants often forming the inputs for others, as well as providing the components of products. These plants are supported by a number of utilities supplying steam, power, water, hydrogen, etc.

Refinery process plants fall into the following major categories:

- physical separation process, such as distillation, extraction and various separation techniques;
- chemical conversion processes, such as reforming, catalytic cracking, isomerization, alkylation, etherification, hydrocracking and thermal cracking/visbreaking;
- purification processes, such as hydrodesulphurization, desalting, gas sweetening, sour water treatment, sulphur removal and recovery and effluent treatment;
- utilities and general facilities – steam and power supply, fuel system, flare system, water, air, hydrogen and nitrogen supply, hydrocarbon

slops treatment, blending, storage and loading facilities for road, rail and sea;

- environmental controls – aqueous effluent treatment, combustion and other air emission controls, waste disposal, odour and noise controls.

All of the above activities may have an impact on the environment.

8.2 Overview

In order to maintain our quality of life and sustainable development, it is vital that the environmental impact of all industrial practices is reduced as far as is reasonably practicable. Thus, as refiners, environmental management has become a major part of our culture both inside and outside the refinery, resulting in significant progress in performance with respect to neighbours and the general public.

Nevertheless, over the past 10–15 years a steady stream of legislation world-wide has been introduced aimed at further cleaning up industrial operations and reducing their environmental impact. The growing public awareness of the adverse consequences of poor environmental management and disposal practices is a major factor. Most recently the main driving forces for environmental control throughout industry are the current perceived global environmental risks such as the depletion of the ozone layer, acid rain, the quality of the rivers and lakes and the poor air quality in some large cities.

There are approximately 100 refineries in Western Europe. These vary enormously in the processes used, type of oil processed, size, throughput, location and age. Because of this variation, there is no such thing as a best environmental practice applicable to all refineries. Therefore, many of the techniques mentioned in this chapter may not be practised or be appropriate in many refineries.

One thing that many refineries have in common is that they were established before the time when environmental concerns became a major issue for the public, government or industry, with some having been commissioned over 80 years ago. Even then, measures were taken to protect the environment such as the spin-offs from the essential precaution to ensure closed containment of petroleum during processing, although refinery design did not include many elements which today would be considered essential for environmental protection.

The process by which environmental standards are determined utilize the expertise of a number of professions, including legislators, scientists, engineers and planners. As well as the setting of emission standards, decisions have to be made on the most appropriate technology and techniques to be applied and the appropriate monitoring programmes to be implemented. Whilst complete elimination of all pollution is not an attainable goal, steps can be taken to minimize environmental impact through pollution reduction at source, recycling, emissions control and responsible waste disposal. The preferred option is reduction at source, which encompasses both good operating practices and good housekeeping, as well as technological change.

The refining processes have environmental impact on their neighbours and on the air, water and land, and it is important that refiners at least meet the standards set and implement continuous improvements to minimize their impact if they are to retain community acceptance. Some improvements to environmental protection can be readily integrated into an existing plant, whereas others are very difficult, if not impossible, to implement in an existing refinery. In general, the most obvious and lowest cost steps have already been taken and successive steps can only be achieved at progressively increasing cost. This chapter looks at the technologies and techniques available to enable refineries to meet environmental standards through pollution prevention.

8.3 Control of atmospheric emissions

Emissions to the atmosphere are the most obvious form of pollution and have been the first target for control in many countries. Consequently, there are comprehensive regulations in most parts of the world to cover this topic. The quantity and type of atmospheric emisssions by refineries vary greatly, depending on crude oil capacity and type, the particular refining processes used, air pollution control measures in effect and the general level of maintenance and housekeeping.

The principle emission sources from a refinery include:

- sulphur removal units;
- catalytic cracking units;
- coke plants;
- storage and loading operations;
- combustion processes;
- emissions from regeneration of catalyst;
- fugitive emission sources;
- flares.

In industrial plants with multiple emission sources, there are two main philosophies of control; either standards can be set for each individual source or alternatively a 'bubble' concept can be applied. In the latter case the refinery is treated as a single source with an overall mass limit for each pollutant over a given time period. With this system, process units, feedstocks and emission controls can be managed most efficiently to reduce the overall emission levels.

Air emissions from refineries include the following:

- sulphur dioxide (SO_2) and other sulphur compounds (SO_x);
- oxides of nitrogen (NO_x);
- particulates (including smoke);
- carbon dioxide and other carbon compounds (CO_x);
- VOCs (volatile organic compounds);
- malodorous materials.

The main techniques for minimizing and controlling air emissions are discussed below.

8.3.1 Minimizing combustion-related emissions

Between 4% and 9% of refinery crude oil feedstock is used as fuel at the refinery. This generates flue gas that is discharged to the air. In Europe, many air quality regulations initially specified minimum discharge stack heights to ensure good dispersion of the flue gas, but did nothing to control total emissions. Now, however, regulations passed in many countries demand control of specific pollutants, such as SO_x, NO_x, particulates and even CO_x in both mass emissions and concentration permitted in the flue gas.

In a refinery, up to 80% and in some cases even 90% of all SO_2 emissions and also a major part of the NO_x and particulate emissions of the refinery are directly related to the type of fuels used and their respective share in the total fuel consumption of the refinery. Thus the efficient use of energy is ideally a first step to control air pollution, and energy conservation can be a very cost-effective form of emission control.

SO_x and NO_x tend to be the largest atmospheric emissions from refineries. However, the amount of SO_2 emitted by refineries has decreased in the last 15 years (Fig. 8.1). To put this in context, in 1992 refineries in Western Europe contributed only 3.5% of the total annual SO_2 emitted to atmosphere.

(a) SO_x control. Sulphur oxide emissions typically emanate from fuels consumed, the catalytic cracking process and sulphur removal and recovery operations. To control SO_x emissions, refiners can adopt one of the following measures:

- process low-sulphur crude oils;
- burn low-sulphur fuels; in refineries there may be the flexibility to fire more gas in place of oil and sulphur emission limits can often be met by appropriately mixed gas/oil firing;

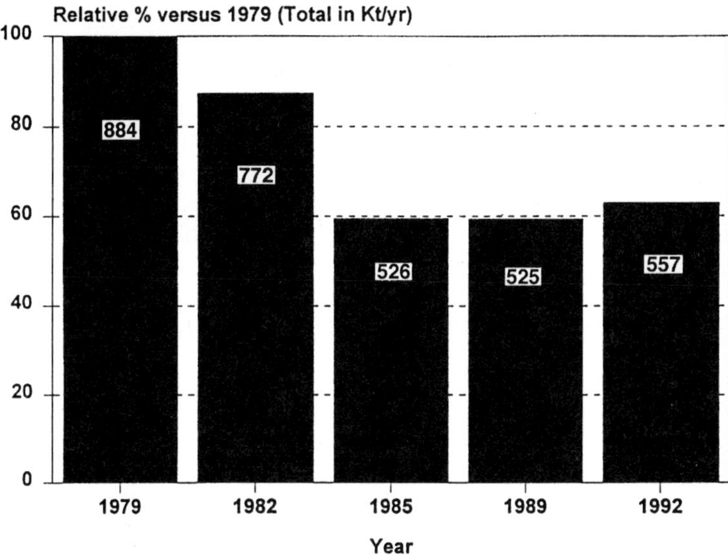

Figure 8.1 SO_2 emissions from refineries in Western Europe [1].

- pretreat fuels and feedstock to remove sulphur;
- install exhaust gas clean-up (flue gas desulphurization) systems; methods available for SO_2 removal from combustion plant are generally very expensive and involve downstream flue gas clean-up, by either wet processes (non-regenerable or regenerable) or dry processes;
- modify catalyst formulations;
- install enhanced sulphur recovery units (i.e. tail gas treatment).

It has been observed that coals containing alkaline ash will burn giving reduced SO_2 and SO_3 emissions due to the formation of metallic sulphates. Recent studies have shown that the addition of alkali metal compounds to fuel oils has the same effect. However, in addition to the increased cost, the disadvantage is that high levels of alkali metals will increase particulate emissions and deposits in the heater which will eventually need to be disposed of.

(b) NO_x control. Nitrogen oxides, from refineries, are generated in the combustion processes and originate from the oxidation of atmospheric nitrogen (thermal NO_x) and the indigenous nitrogen (fuel NO_x) in the fuel. Apart from removing nitrogen compounds prior to combustion by hydrotreating processes, there is little that can be done to avoid emissions due to the fuel quality. The formation of thermal NO_x is promoted by the

intensity (temperatures) of combustion, the availability of oxygen and the residence time in the combustion zone.

Unfortunately, what is good practice for efficient combustion and minimum particulate emissions is not always good for nitrogen oxides prevention. For instance, forced draught burners increase burnout temperatures and reduce residence times, which can reduce particulate emissions, but overall produce more NO_x. In the combustion of sweet natural gas, the formation of thermal NO_x can double with air preheat. Results on test rigs show the effects of combustion air preheat, fuel-bound nitrogen and excess oxygen levels.

The general rules to reduce NO_x are the opposite to its formation:

- reduce nitrogen in feed – not always a practical consideration;
- reduce oxygen – with less oxygen available, less NO_x is formed (Fig. 8.2); this is generally a very good policy since it is also energy saving, but it does run the risk of increasing particulate emissions;
- reduce the residence time – this is a design feature and burner manufacturers have for at least the last decade been developing low NO_x burners (Fig. 8.3), which often used staged combustion or flue gas recycling to limit the production of NO_x at the burner by reducing peak flare temperatures;
- reduce the combustion temperature – for the same excess air this will reduce the peak temperature, and thus reduce NO_x, but this is not a very convenient route for operations. Reducing the air preheat

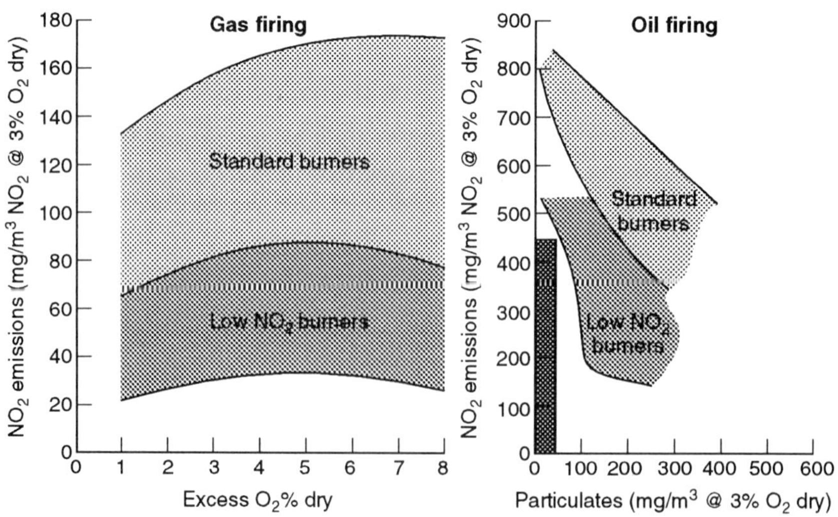

Figure 8.2 Performance of low NO_x burners and conventional burners firing natural gas and heavy fuel oil on a burner test rig.

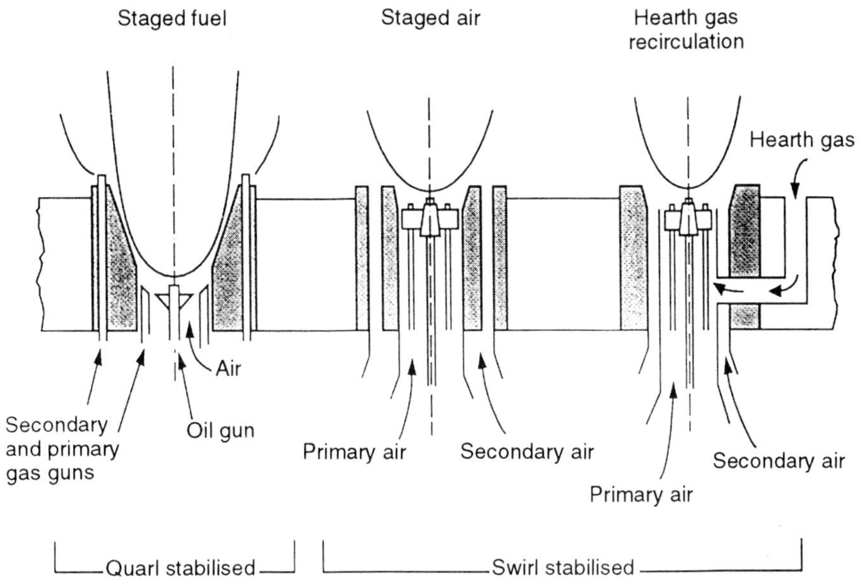

Figure 8.3 Schematic diagrams of the three types of low NO_x burners.

temperature has a dramatic effect on NO_x emissions. However, this is counter-productive with energy conservation and particulate emissions, and should only be considered as a last resort since heater efficiencies can drop by up to 12%.

In addition to the above NO_x reduction technologies, other preventive technologies such as catalytic and non-catalytic flue gas treatment are available. However, these are expensive to implement and generally economically viable only on plants larger than those usually found in refineries.

(c) CO_x emissions. There is at present no treatment method for reducing CO_x emissions, and the main abatement instrument is increasing the efficiency with which energy is used on the plant. It should be borne in mind that control techniques for other emission pollutants and product upgrading may be significant energy users and therefore cause CO_x emissions to increase. In additions, demands for improved fuels to reduce emissions from the use of oil products has meant that more energy has to be used in the refinery. The effects of this on total European refinery energy consumption for two different scenarios are shown in Fig. 8.4.

CO emissions can be controlled by good management of fuel combustion and are generally not a problem in refineries.

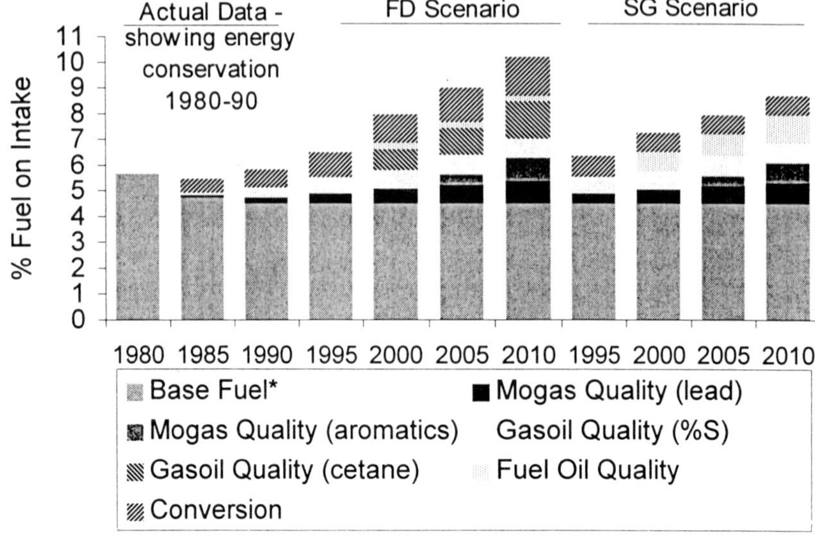

FD - Fuel Oil Decline Scenario
SG - Sustainable Growth Scenario

Figure 8.4 Effect of environmentally driven product changes on refinery energy consumption [2].

(d) Particulates control. Particulate emissions from refineries come from three sources:

- fuel combustion;
- process units;
- flares – a special case (section 8.3.2).

(i) Fuel combustion. Ash and unburned hydrocarbons agglomerate to form particulates (typically 1–100 μm in diameter) when gas or oil are combusted. The level of particulate emissions is related to the intensity of combustion and the fuel quality. Low-quality fuel oil has higher levels of 'unreactive carbon' such as asphaltenes and sulphur, giving slower burnout, which increase particulate emissions.

Particulate emissions can be reduced by suitable changes to the burner or to fuel technology, and primary low-cost techniques such as those described below should be the first avenue of approach to emission reduction.

Combination firing With some burners, combination firing of oil and gas is known to reduce overall particulate emissions compared with burning in separate burners.

Improve atomization Finer atomization of fuel oil will increase the carbon burnout rates by making a larger oil surface available in the flame and so lowering particle emissions. Modern F-jet atomizers can give a ten-fold decrease in particulates formation. Increasing atomizing steam will also improve atomization and burnout, therefore reducing particulate emissions, although each incremental increase in steam is less effective than the previous one. Alternatively, increasing oil preheat gives a thinner oil which is easier to atomize. However, once the viscosity reaches 10–15 cSt (10–15 mm^2 s) at the burner tip there is little further benefit in additional heating.

Lowering particulates by better atomization has a drawback since finer atomization leads to higher flame temperatures owing to more intense combustion, which tends to increase the formation of thermal NO_x.

Improved air–fuel mixing Increasing the system energy will improve combustion and reduce particulates. The draught loss across a burner is an indication of the system mixing energy. Higher draught is achieved through improved burner design and increased combustion air pressure or increased oil pressure. However, as with atomization, the better combustion also increases thermal NO_x production.

Increase excess air The more excess oxygen is available, the greater is the rate of carbon burnout. However, there is a very high cost penalty in terms of energy usage and this really should only be considered as a last resort. Also, at high excess air rates flame chilling can occur, which tends to increase particulates.

Use of selected combustion improver additives Recent studies have shown that it is possible to introduce selected metal-containing additives to the fuel and thereby catalyse the combustion process. Although the additives add overall to the inorganic ash produced, they can substantially improve carbon burnout and therefore reduce total particulate emissions.

Water–oil emulsions There is good evidence that water–oil emulsions containing 25–30% of water improve combustion. The theory is that the water droplets explode as they enter the combustion zone, giving a substantial increase in fuel oil atomization. Currently, water–oil emulsions are being used on boilers firing fuel oil but are not widely used in refineries. Trials have shown that it is possible to achieve energy savings whilst maintaining the same level of particulates, or for the same oxygen level decrease particulates. The other benefit is that NO_x levels also decrease, probably owing to the quenching impact of water vaporization.

Increase air pre-heat Increasing the air temperature increases the volume of air and achieves better mixing and burning. However, it is an expensive option and also increases the NO_x formation.

(ii) Process units. Particulates can be a major emission from some refinery process units. Typical sources are catalytic cracking units and cokers.

In catalytic crackers, the particulate emissions are mainly catalyst fines with associated carbons and as such are mainly inorganic (e.g. zeolites). The use of cyclones and electrostatic precipitators and careful catalyst selection all help to minimize particulates emissions from this source.

In cokers, where petroleum coke and some lighter oils are produced from heavy oil, the particulate emissions are mainly organic (e.g. coke). The coke is formed in drums and then cut out using high-pressure water. The coke is maintained in a damp condition to minimize escape of these fine particles.

8.3.2 Minimizing flare-related emissions

The main purpose of the flare is to collect and dispose safely of gases produced during upset conditions. Owing to the intermittent nature of flare gas flows, it may be uneconomical and not energy efficient to recover all of the gases. Flare operation is a special case of combustion emissions. Flares in refineries contribute to SO_x, NO_x and particulate emissions. Flaring represents a direct loss to the business and can be the biggest single hydrocarbon loss from refineries in monetary terms.

Traditionally, flares in refineries have been high-level open pipes. Emission of light is an inevitable consequence of high-level flare operation. The impact can be reduced by the use of ground flares (enclosed combustion incinerator) for a standard flaring load, with sequential release to a high-level flare for emergency conditions. Flickering of the flare flame should be avoided by using properly designed water seal pots at the base of the flare, which also act as a safety feature in preventing any flashback from the flare to the process units.

All the problems described above can be reduced by minimizing hydrocarbons entering the flare at source and avoiding unnecessary flaring. The management of flares not only makes good business sense, it also makes good environmental sense. Leakage from relief or safety valves contributes to a base load. Management of flares should concentrate on reducing the base load to the flare by regular checks on relief valves to determine those passing to flare. There are various techniques available to identify leaking valves, including inspection and ultrasonic detection. Frequent emergency flaring also needs investigation as to its cause and process solutions should be put in place to reduce it. Also, steps should be

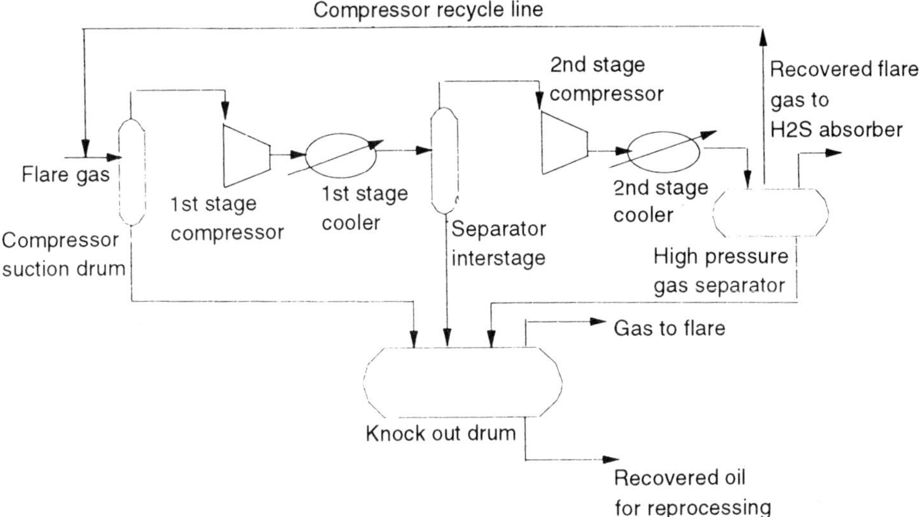

Figure 8.5 Simplified scheme of refinery flare gas removal unit.

taken to improve combustion by injecting steam to give good mixing with air. Steam injection rates can be automated using luminosity, infrared or flow detectors.

As an end-of-pipe solution, flare gas recovery systems can be used (Fig. 8.5). These typically compress flare gas back to the fuel gas system, assuming there is sufficient capacity to receive this. Flare gas can contain valuable products such as LPG, and modern recovery systems include hydrocarbon condensate recovery as well as compression and sweetening facilities.

8.3.3 Minimizing fugitive emissions

Fugitive emissions are volatile organic compounds (VOCs) that escape mainly from the process and offsite areas such as tankage and oily water sewer/effluent treatment systems. If not well controlled, fugitive emissions can represent a significant loss to the refinery. Not only are fugitive emissions a financial loss, but they also have an environmental impact owing to the part they play in photochemical smog and ground-level ozone production. As such, they can attract regulatory authority attention; for example, The Netherlands have proposed a 50% reduction in emission levels by the year 2000, a challenge for the oil industry.

There are many discrete emission sources in a typical refinery emitting hydrocarbons to the atmosphere. A detailed study carried out by the Radian Corporation for the US Environmental Protection Agency measured fugitive emissions from all detected sources in five US refineries and calculated the total emission levels. Interestingly, they found that 70% of all emissions from a process unit originated from around 1% of the leaks sources. Typical sources are control valve stems, flanges, compressor/pump seals, tanks and loading facilities.

Valves represent the largest source of process fugitive emissions in a typical refinery. Furthermore, a very high proportion of the losses is contributed by a small number of large leakers. Simple tightening of valve seals can often eliminate leakage once detected. For known troublesome valves or those which cannot be repaired on-stream, graphite gland packings can significantly reduce both fugitive emissions and maintenance costs. A number of special low-emission packings are now commercially available.

Evaporative losses from refinery tankage also represent a significant proportion of the total loss. Volatile products are generally stored in external floating roof tanks, and this design of tank can experience significant losses and offer the opportunity for reductions to be realized. Evaporative losses from external floating roof tanks (EFRT) are dependent on the properties of the stored product (composition, volatility) and external conditions (wind speed, ambient temperature and solar temperature gain on the roof). Refineries with high average ambient temperatures, solar radiation and also those with high average wind speeds are more susceptible to these losses. The effects of ambient conditions can be minimized by modification to fittings and roof seals.

Reductions in VOC emissions from tanks can be achieved through the use of techniques such as vapour recovery on fixed roof tanks and the use of secondary tank seals and stillwell covers for hydrocarbon control on floating roof tanks. Other areas where hydrocarbons can gather on the surface of water, where they can be lost as fugitive emissions, are API (gravity) separators and interceptors. Floating covers can be fitted to the API separators to reduce VOC emissions. Atmospheric vents on oily sewers are another major source of VOC emissions and these can be controlled by closing the sewers or by installing carbon scrubbers on atmospheric vents. There are two methods that can be used to determine the quantity of VOCs emitted. The differential absorption (DIAL) method relies on the absorption of laser beams by airborne hydrocarbons to measure atmospheric concentrations. The American Petroleum Institute (API) method estimates emissions using average emission factors for different sources and hydrocarbon service. CONCAWE have carried out a trial of the DIAL method at an oil terminal and compared the results (Fig. 8.6) with calculations using the API method. The results showed good agreement between the two methods provided that the DIAL results were averaged over a long enough period of time (30 h).

The fugitive emissions which have attracted the most attention are the

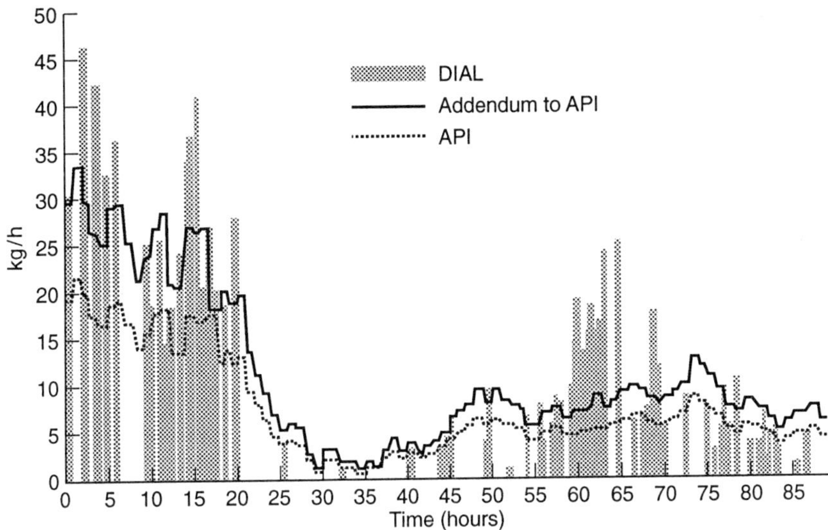

Figure 8.6 Emissions: DIAL measurements versus API calculations [3].

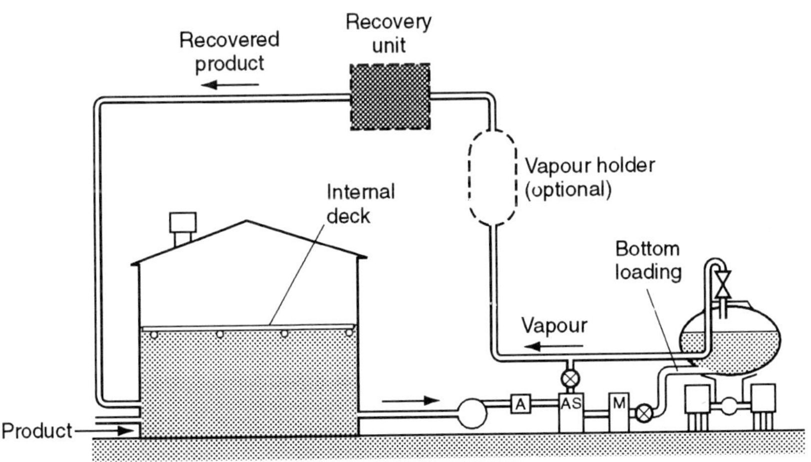

Figure 8.7 Vapour capture and recovery at terminal.

so-called air toxins such as benzene, which is a known human carcinogen and linked to adult leukaemia. Although the cause for further action at very low concentrations remains equivocal, specific legislation has been adopted/proposed on both sides of the Atlantic. The US EPA has requested actions at service stations and bulk gasoline plants (Fig. 8.7) to reduce benzene emissions by some 90%. There are also benzene rulings for waste water systems that may trigger expensive modifications. In the EC, particular regulations apply to streams containing 15% or more benzene. Loading and discharge of bulk gasoline/benzene/reformate streams need to be closed.

8.3.4 Odour control

Typically 80% of all public complaints regarding refineries are due to odours. Extremely malodorous compounds, such as mercaptans and hydrogen sulphide, are the source of many of these complaints.

An odour is often the first indicator of exposure to a chemical. However, the ability to smell the presence of a chemical does not of itself indicate toxicity or a health risk.

The common sulphur-containing industrial gases, hydrogen sulphide and methyl mercaptan, are among the most common odorants from a refinery. Although they have odour thresholds significantly lower than levels known to cause toxicity, they are nonetheless most often associated with annoyance at levels just exceeding their odour threshold.

The main reduction techniques available to refiners are:

- control of fugitive emissions;
- control of flares;
- control of fuel quality;
- scrubbing of odorous gases;
- incineration of odorous gases;
- bio-treatment.

8.3.5 Sulphur removal and recovery

Sulphur emissions are recognized as a major pollutant owing to the part they play in acid rain formation. SO_2 in the presence of catalyst gases such as NO_x forms SO_3, a very acidic species which is washed from the atmosphere during rain, and if the area on which it falls has little natural alkalinity, lakes and rivers can become acidic and not support higher life forms such as fish.

Sulphur dioxide emissions from refineries have decreased substantially in recent years owing to the greater availability of sweet refinery fuel gas resulting from increased upgrading capability (conversion of heavy oils to lighter) and authority restrictions. High-sulphur fuel oil can be a major source of SO_2 emission, as can be the catalytic cracker catalyst regenerator where some sulphur from the feedstock is burnt off. Methods of reducing these sulphur emissions are to use lower sulphur fuel oil and feedstocks in the catalytic cracker. Alternatively, both the fuel oils and feedstocks can be desulphurized in a hydrotreating unit. This option is expensive, requiring significant capital investment and incurring increased operating costs,

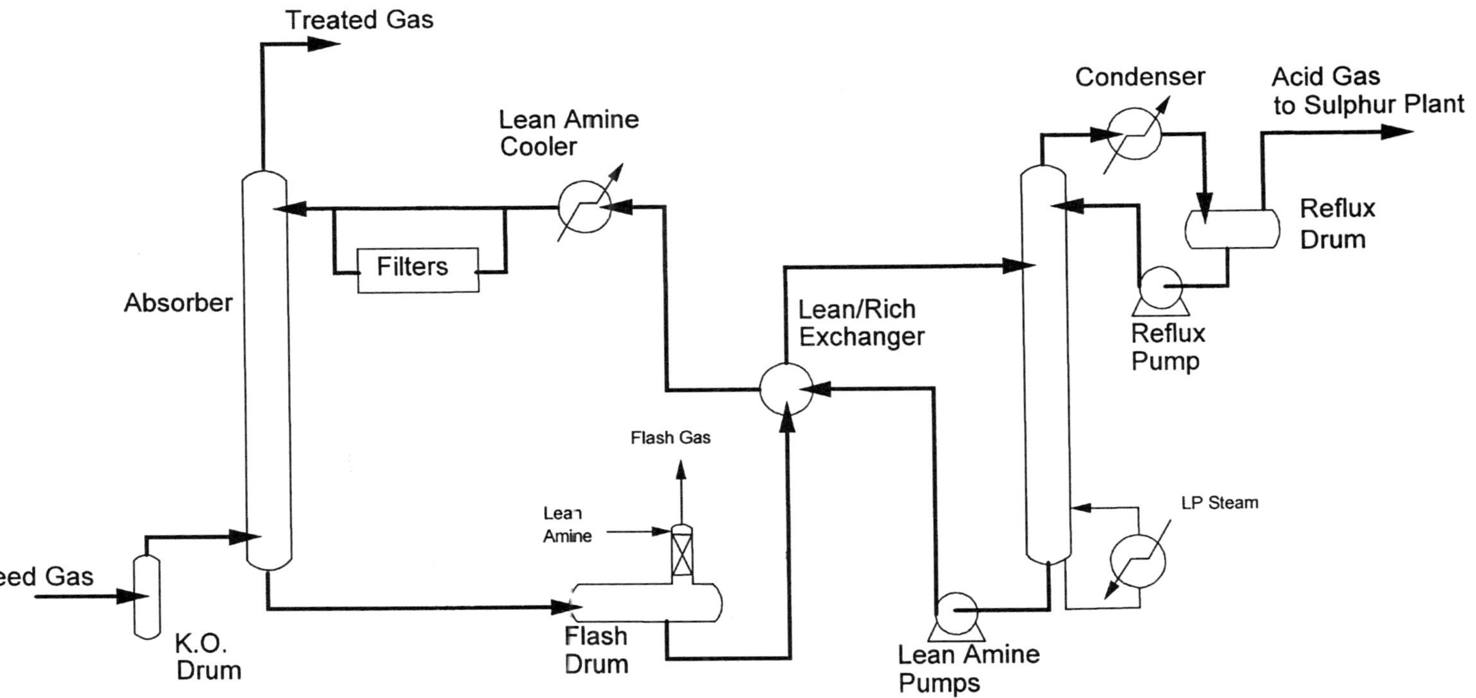

Figure 8.8 Typical flow diagram of amine unit.

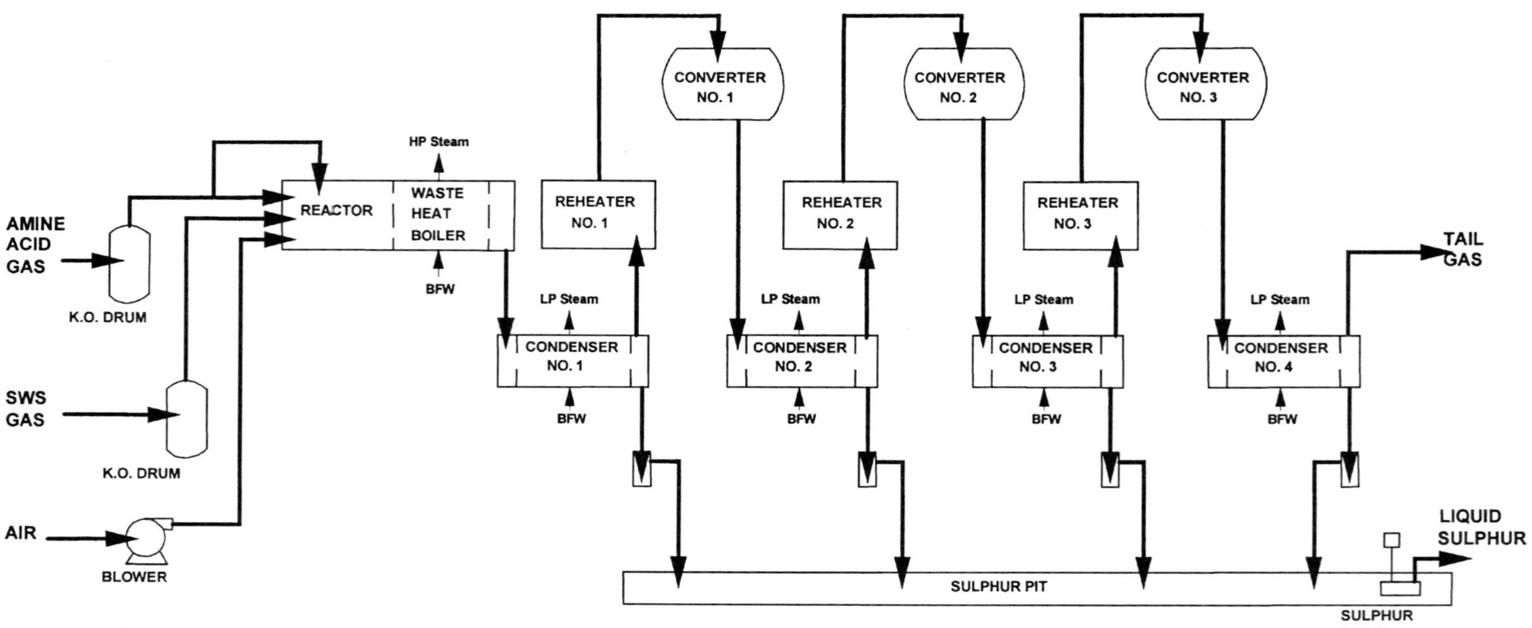

Figure 8.9 Typical three-stage Claus sulphur unit.

although coincidentally it also yields further light product and enhances feedstock quality.

H_2S formed in the gases through the processing of crude oil and other feedstocks is treated in an amine treater (Fig. 8.8) where H_2S is absorbed from refinery gases and the rich amine is regenerated to release an H_2S-rich stream. This stream forms the basic feedstock for refinery sulphur recovery plants (Fig. 8.9) where elemental sulphur is recovered in the Claus process. Depending on the complexity of these individual plants, sulphur recovery efficiency can vary from 92–93% up to 99.5%, although with substantially increased cost for each incremental improvement.

Other sour gases containing H_2S and ammonia are removed from collected refinery sour waters in a sour water stripper. The gas is removed and also forwarded to the sulphur plant, where a special stage destroys the ammonia.

8.4 Control of aqueous emissions

Water pollution is caused by the release of contaminants into water sources which are either damaging to aquatic life (either because of their toxicity or through their reduction of the normal oxygen level of the water) or aesthetically unpalatable.

Potential water contaminants in refinery effluent are:

- acids, alkalis (pH);
- oil (free and dissolved);
- sulphides;
- ammonia/nitrates;
- cyanides;
- heavy metals;
- heat;
- other organic materials;
- nutrients;
- settleable solids;
- colour;
- taste and odour producers;
- toxic compounds.

All water discharges from refineries should be treated at least to a minimum standard. Often composite measurements are used to monitor the quality of discharge. Empirical assessment of total contaminant levels include:

- total suspended solids; finely divided solid matter suspended in water;
- total dissolved salts; total inorganic salts dissolved in water;

- chemical oxygen demand (COD) — amount of oxygen consumed in the chemical oxidation;
- biological oxygen demand (BOD) — index representing content of biodegradable substances in the water;
- total organic carbon — determination of all organic carbon present;
- total nitrogen — determination of all nitrogen.

The major sources of refinery aqueous effluents are process water, ballast water, rainwater run-off and cooling water. The contaminants picked up by process waters and rainwater are eventually routed to sea or river, and if untreated can have a major impact on the aquatic environment. All countries have legislation to control the level of contaminants in refinery waste water. The use of advanced waste water treatment plants has led to continuing reductions in contaminant levels. This is reflected in the fact that '2018 tonnes of oil were discharged with the aqueous effluents from 95 European refineries in 1993 compared with 3340 tonnes from 95 refineries in 1990. This represents a reduction of 40%, despite the fact that the amount of oil processed in 1993 was 9% higher than in 1990. In the 95 European refineries surveyed there has been a 46% reduction in the ratio of oil discharged to oil processed since the 1990 survey, i.e. it fell from 6.7 tonnes per million tonnes of oil processed in 1990 to 3.6 tonnes per million tonnes in 1993. On a longer term view (Fig. 8.10), there has been a 95% reduction in oil discharged in European refinery effluents since 1969' [4].

8.4.1 Source control

Effluent treatment plants (ETPs) are often classified as end-of-pipe treatment, where all the waste water streams are collected and treated downstream of the process units and other sources. Effluent treatment is much easier and more effective if the contaminant loading can be controlled or limited at source so that the ETP is not overloaded but treats only the minimum quantity of contamination.

In addition to adequate equipment and hardware within the process plants, good source control requires both motivation and training of refinery staff, operators and maintenance personnel such that they are aware of environmental issues and able to avoid unnecessary release of contaminants.

Some examples of pollution control measures at source are given below.

Sour water strippers are used to remove hydrogen sulphide and ammonia selectively from a variety of process waters. In a typical sour water stripper the feed sour water is brought into counter-current contact with steam in a packed tower. H_2S and NH_3 are selectively stripped by the steam, which is

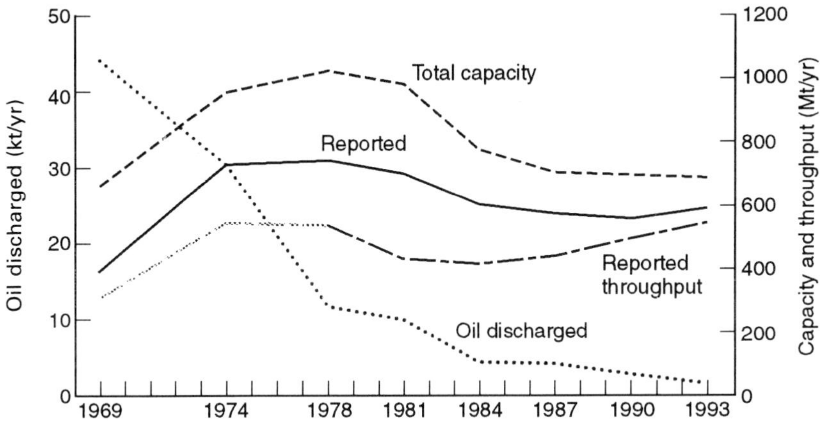

Figure 8.10 Oil discharged in the period 1969–93 [5].

then cooled and condensed. The incondensable off-gas containing the H_2S and NH_3 is routed to the sulphur plant. In this way H_2S is neither passed to the effluent treatment plant nor released to the atmosphere where it would cause a smell (section 8.3.5).

Cooling water systems are used to produce cold water to cool hot process streams. In once-through systems, natural cooling water is used to cool process streams and then returned to the environment. In addition to introducing heat pollution into seas and rivers, this system risks contamination by the process stream in the event of leaks occurring in the heat exchangers. An alternative system is to recirculate this water. The water can be re-cooled with more natural water, thus reducing the risk of contamination. Alternatively, a closed cooling water network can be employed where the water is cooled either in a cooling tower or cooling exchangers using refrigeration technology. In cooling towers water is cascaded over packing and intimately mixed with air. The heat is dissipated by sensible heat loss and evaporation. In cold climates this can mean the formation of a visible cloud of evaporated water above the towers. Although cooling exchangers using refrigeration technology avoid air emission problems, they do run the risk of tube leakage whereby refrigerant can leak into the water, requiring further treatment before final discharge. The advantages of closed cooling water networks include lower water consumption rates and minimal heat pollution at final discharge. However, they require the injection of chemical additives to hinder corrosion and bacterial growth and periodic purging to prevent build-up of dissolved solids.

Desalter operations produce oily waste water with a high salts content, which represents a high treatment load. The load on the effluent treatment plant can be minimized if this water is first directed to a break tank where, with the use of demulsifiers and sufficient residence time, the oil and water are separated through decantation. The oil is then recovered and reprocessed. Thus only the water drained from the break tank needs to be sent to the effluent treatment plant. The use of robust and reliable interface detectors on the water drains will help ensure that minimum oil is purged into the drain.

Clean water segregation is vital if the volume of effluent requiring final treatment before discharge is to be minimized. Where possible, rain- or storm water should be handled separately from refinery oily water streams so as to avoid contamination of these clean streams. One way to ensure minimal contamination of rainwater is to minimize the area of paving on site where oil and chemicals are likely to be spilt. This is usually possible by bunding in areas so as to limit any oil or chemical spills. The bunded areas can then be drained directly to the oily water sewers whilst the uncontaminated rain falling on clean paving can go to the clean water sewers. Similarly, well maintained tank roof drains are important if they are to direct clean rainwater to the clean water sewers. If the roof drains are blocked, rainwater can enter into the tank and become oily. It will then be drained from the tank as oily water and contribute to the effluent water treatment plant total load.

Leakage to sewers can occur when the sewer walls leak and ground water can enter the oily water sewers. This usually happens after rainfall when the water table can rise to the same level as the oily water sewers. It is important that sewers are inspected regularly to avoid in-leakage of ground water and the additional cost of its treatment.

Sampling systems should be closed where sample is returned to the process or collected where it is later recycled. Both of these techniques will help to reduce refinery loss and minimize the impact on the environment.

Tank draining should be routed to interceptors where oil can be separated out and recovered. Alternatively, interface detectors can be installed on the tanks to minimize oil loss to the drain.

Cross-contamination of clean water with dirty water can occur where the two lines run in close proximity to each other and when there is a risk of sewer damage. Collapse of the dividing walls between the two sewer systems can go undetected for some time, during which clean water can leak into the oily water system and create additional contaminated water.

Generation of solids in sewers can lead to operating problems in the effluent treatment plant, and should be avoided. Solids can be generated inadvertently in the sewers through chemical reactions which lead to precipitation or crystallization of solids compounds. Typical examples include the precipitation of calcium compounds when caustic solutions are mixed with water with a high calcium salt content, or salt precipitation when acid and alkali streams are mixed in the sewer system.

8.4.2 *Effluent treatment*

(a) Pretreatment. Aqueous waste which cannot be eliminated at source is usually treated using one of the following techniques before it is sent to the final effluent treatment plant.

Neutralization is a basic reaction for a number of wastewater treatment operations. It is the reaction between an acid and an alkali used to adjust the pH of the solution to within the desired range so the water is suitable for discharge. Neutralization may also be necessary to establish proper conditions before an oxidation–reduction chemical reaction, for precipitation of heavy metals as hydroxides, for proper clarification and for better adsorption.

Emulsion breaking is a pretreatment step used for certain oil–water mixtures. Emulsified oils are used in machine operations and as coolants. These emulsions must be broken so the oil and water can be separated. Emulsion breaking is usually accomplished as a batch process. The spent emulsions are collected in a holding tank which is equipped with agitator(s) and skimmers. The tank contents are held undisturbed for 2–8 h to give insoluble oils time to rise to the surface for removal. The tank contents are then agitated and emulsion-breaking chemicals such as coagulants, flocculants and wetting agents added. After the emulsion has been broken, the freed oil is separated using gravity separators or liquid–liquid cyclones. The pH of the water is then adjusted and the wastewater clarified by flotation or clarification.

(b) Primary treatment. Finally, all unavoidable contaminants or those most conveniently treated at end of pipe are passed to the effluent treatment plant (Fig. 8.11). The first step is primary gross oil removal, intended to take off free oil from the effluent. The effluent is passed through gravity separators (Fig. 8.12) or corrugated plate interceptors with sufficient residence time to allow free oil to rise to the surface, where it is skimmed off. This treatment has virtually no impact on soluble components. Substantial quantities of sludge can also collect in these separators.

(c) Secondary treatment. As a second-stage treatment, air flotation or filtration units are used for the removal of fine oil droplets from water. Flotation units (Fig. 8.13) work on the principle that oil droplets are carried to the surface by small gas bubbles. Air is introduced into the system, forming small bubbles which attach themselves to oil globules or suspended particles and float them to the surface, from where they are removed for further handling. Chemicals such as coagulants, acids and/or alkalis are often added ahead of the system to promote more complete removal. In filtration systems the oil is filtered from the aqueous stream using a filter medium such as sand or anthracite. This medium is then

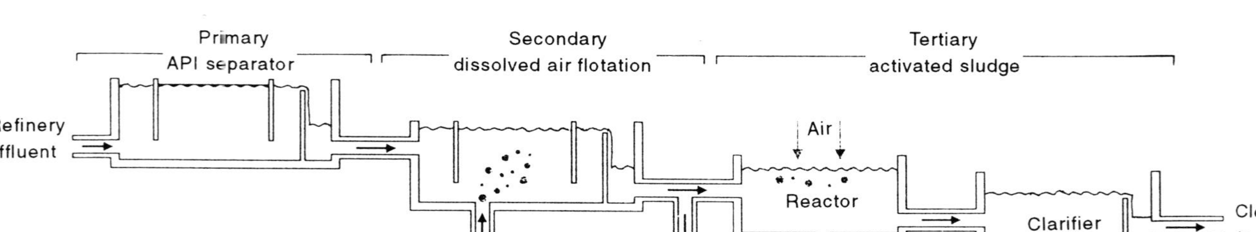

Figure 8.11 Typical three-stage effluent treatment process.

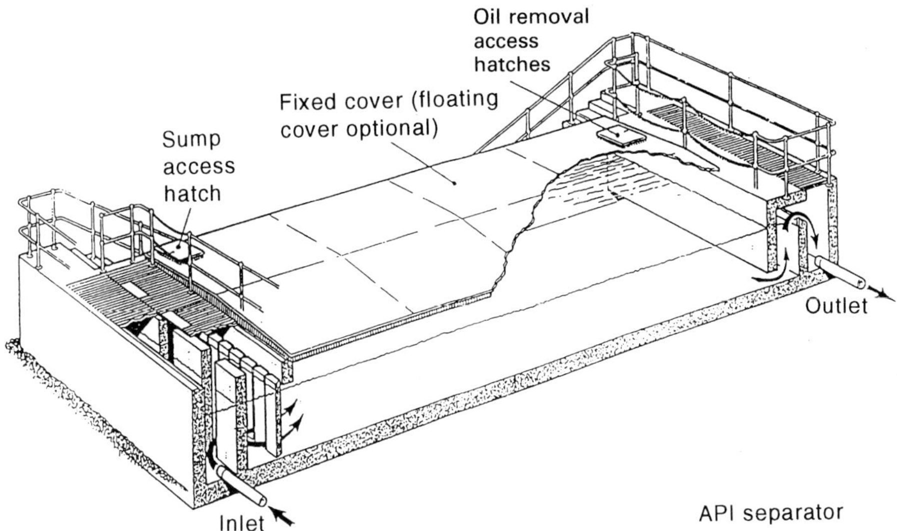

Figure 8.12 API separator.

backwashed to prevent excessive pressure drop and to remove the oil collected there. Again, there is no removal of soluble contaminants through the use of these techniques.

(d) Tertiary treatment. Once the majority of the free oil has been removed, biological treatment is used to remove the water-soluble constitutents in the effluent so as to reduce the biological oxygen demand (BOD) and specifically identifiable organic materials such as phenols. The process relies on the use of biological activity to degrade pollutants. When organic material is discharged into a receiving stream, a biological chain of events occurs, in which naturally present bacteria in the receiving stream metabolize and stabilize the organic material, consuming oxygen in the process.

For most waste waters, the destruction of organic material will take place under aerobic conditions in which a measurable oxygen residual is present. For certain concentrated organic wastes, anaerobic treatment (in the absence of oxygen) can be used, although its applicability is relatively limited, and it is usually followed by aerobic treatment for complete stabilization.

The most widely used biological process for industrial waste water is the activated sludge process. Incoming waste, with or without primary treatment for suspended and settleable solids removal, is mixed with return activated sludge and enters an aeration tank. This tank is aerated to

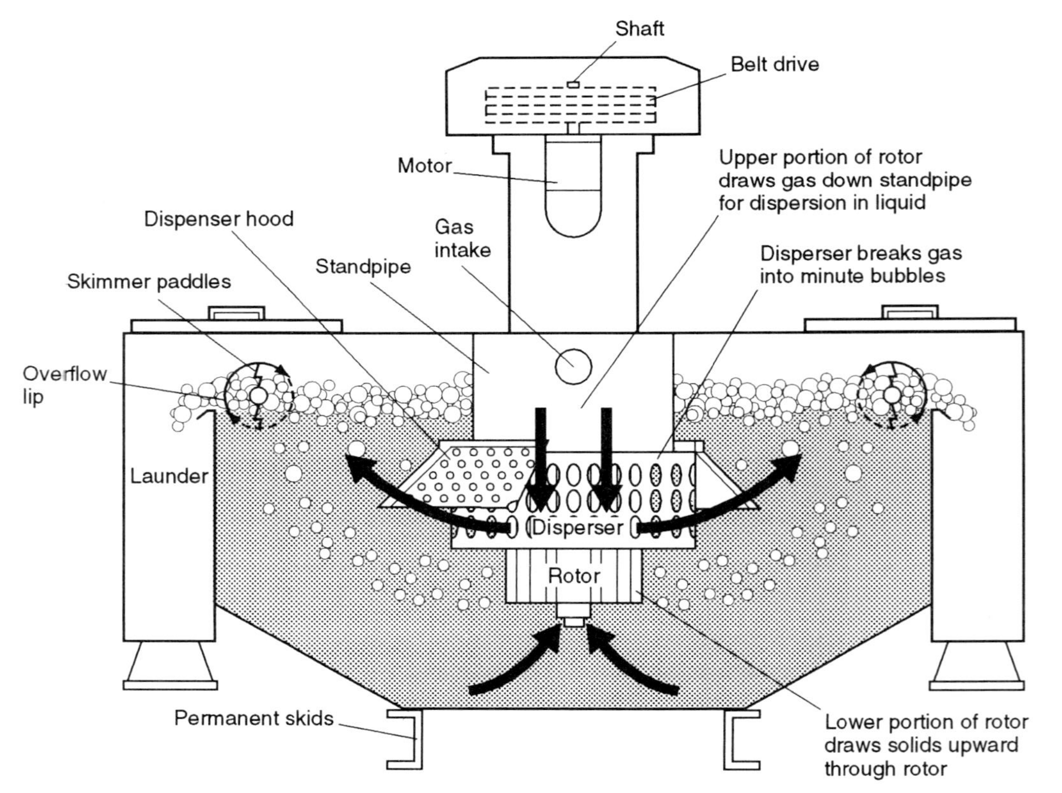

Figure 8.13 Depurator machine float cell (section view).

maintain residual dissolved oxygen so biological growth and activity occur. If necessary, nutrients can be fed in at this point. The wastes and the bacteria are held in contact long enough to stabilize the incoming organic material and accomplish the desired effluent quality. The mixed waste goes to a final settling tank where the bacteria settle from the water, the waste is discharged and the bacteria are recycled. Sludge will build up in the system in most cases to the point where some will have to be removed for ultimate disposal.

(e) Final polishing. Some refineries have a final polishing stage prior to final discharge of the effluents to remove any remaining bacteria, but this is by no means standard practice. Typically these consist of either a sand or a perforated drum filter or a final settling pond.

Sand filters utilize media such as anthrafilt or sand, although modern filters use mixed media, graded coarse to fine, in the direction of the water flow. The filters are regularly backwashed to remove the collected bacteria. In the perforated drum filter, water passes through the drum and the solids are retained on the drum. Solids are then removed from the drum and handled similarly to filter backwash water. Since slurry is produced at a more uniform rate, intermittent storage requirements may not be as critical as is the case with backwash water.

Settlement basins are artificial ponds or lakes used to hold water to effect the removal of suspended solids and insoluble oils. Lagoons are also used as retention ponds after chemical clarification to polish the effluent and to safeguard against upsets. The basin needs to have sufficient residence time for further biological activity.

A few refineries have used wetland technology to achieve final polishing. Reeds which thrive in wetlands can help the breakdown of effluent contaminants which are nutrients for certain microorganisms. The reeds' small roots host colonies of these microorganisms and the larger roots help to develop hydraulic transport pathways through the soil and supply oxygen to the bacteria.

8.5 Soil and groundwater protection

Pollution of air and surface waters can be readily identified. Pollution of soil and groundwater is more difficult to determine but is very important for the whole oil industry, including refineries. Wherever oil products have been handled for many years, it is highly likely that oil will have entered the ground and refineries are no exception. However, the significance of this varies with the hydrology of the ground under the site. In areas where the sub-soil is of very low permeability, then the problem remains localized within the site. In other cases, the contamination can spread outside the site, which may be particularly harmful where groundwater is used for drinking water supply, which, although important, is not the only cause of concern. Although the most obvious source of ground pollution from a refinery is oil, other pollutants can enter the ground. These are considered under section 8.6 on solid waste control.

8.5.1 Source control

Oil enters the ground in refineries in three main ways: through operational practices such as sampling or drawing water from tankage; through leaks from oily water sewers, underground pipe work, underground storage tanks and normal tanks; and from accidental spills of petroleum products into the ground.

Measures to control the first of these sources are basically the same as those already covered under the control of aqueous emissions (section 8.4), coupled with ensuring that areas where oil is regularly handled are covered with an impermeable surface and drained to the oily water sewer. Such areas should include maintenance areas such as heat exchanger cleaning areas.

Leakage of groundwater into sewers has already been mentioned in section 8.4. The reverse also occurs, leading to oil entering the ground. Again, regular inspection and repair, possibly by lining, are required.

Leakage from underground pipes is difficult to detect and can best be prevented by running oil-containing pipe work above ground wherever possible. Similarly, the use of underground storage tanks should be limited as far as possible.

A primary cause of land contamination is leakage from pipes and tanks. During operational situations it is often difficult to identify which tank is leaking. However, since continuing contamination gives rise to further waste and future liability for clean-up, refiners are making a major effort with leak detection for both tanks and underground piping.

8.5.2 Monitoring

Given that most refineries are likely to have at least some contamination of soil and groundwater, monitoring surveys are often necessary. These are required by national legislation in many European countries.

Before remediation of the soil and groundwater can be undertaken, it is important that a thorough assessment of the type and extent of contamination is completed. These assessments are usually phased. Many techniques have been developed for this purpose but they generally fall into two categories, non-invasive and invasive. The following are some of the more typical techniques used to assess the contamination of soil and water by the oil industry.

Non-invasive
- soil hydrocarbon vapour measurement;
- geophysical methods;
- resistivity methods;
- magnetic flux methods.

Invasive
- borehole drilling and sampling;
- cone penetrometer with soil conductivity or laser fluorescence sensors;
- vadose zone vapour probe; this is hydraulically advanced into the soil, a porous element exposed, a vacuum applied and a vapour sample taken, which is analysed immediately;
- groundwater wells.

A disadvantage of non-invasive methods is that they sometimes give a false impression of conditions underground and are therefore mainly used as screening techniques to allow focusing on the more expensive invasive techniques.

Once contamination has been detected and the source stopped, investigation of the hydrocarbon type and distribution will help to confirm the source and the migration mechanisms. This understanding allows steps to be taken to prevent the spread of contamination. Many methods are available but typically the installation of interceptors or cut-off trenches are used to collect the polluted water in the soil through natural drainage. The collected water is then removed for treatment.

8.5.3 Remediation

Once the ground and contamination conditions have been established, an assessment of the risk to human health and the environment is made. This helps to determine the remediation need and allows any clean-up standards to be set. Recent emphasis in treatment techniques for soil and groundwater in the oil industry have been on *in situ* treatment in contrast with previous methods which addressed *in situ* treatments such as land-farming/ biodegradation of sludges.

Perhaps one of the most applicable *in situ* techniques is that of biodegradation, where the controlled addition of oxygen and sometimes nutrients can accelerate the growth of microbes, which beneficially degrade contaminants. *In situ* biodegradation is a long-term passive treatment technique; an example is soil venting combined with air sparging, where air is bubbled into the groundwater and vapours from the soil are continuously drawn off. The vapours are analysed, and treated in a bio-system before discharge to the atmosphere. This has the effect of desorbing hydrocarbons from the soil and inducing more oxygen into the soil, which accelerates microbial growth. It works best where sandy ground has been contaminated with volatile hydrocarbons.

Methods used for treating contaminated groundwater include closed-loop systems, where groundwater is pumped to the surface, treated and then returned to the subsurface via wells. Surface treatment of the groundwater can include separation, air stripping, addition of oxidizing chemicals and bio-reactors.

8.5.4 Preventive techniques

A number of preventive techniques are available either to reduce the risks of groundwater pollution or to limit the spread of such pollution once it has arisen. Many of these methods are very costly, particularly when retrofitted. Their use should be considered after a rigorous assessment of the reduction in risk of the alternatives versus the cost in both financial and environmental terms.

Measures which can be considered include double bottoms in tanks, impermeable layers under tanks and impermeable surfaces for tank bunds. To prevent migration off-site, either mechanical or hydraulic measures can be applied. Mechanical measures include cut-off walls or ditches. Hydraulic methods include the pumping of groundwater from within the site so that the flow is inwards rather than outwards. This can be combined with remediation of contaminated groundwater. However, most of these methods in themselves constitute an alteration in the natural groundwater hydrology and hence themselves have environmental effects which need to be taken into account.

8.6 Control of solid wastes

Waste is defined as any material of no further primary use, and excludes aqueous and gaseous effluents from operating units. It is essentially any material that remains as an unwanted by-product of refining that needs to be disposed of. Whilst waste production in refineries and terminals is a smaller average percentage of total throughput compared with most other industries, it nonetheless represents a high cost in loss and potential environmental risk, and as such its minimization is a priority. However, generation of some waste is an inevitable consequence of refinery operation, and wastes generated fall into two categories, non-hazardous waste and potentially hazardous wastes, e.g. sludges with a high metals or hydrocarbon content. Generally refinery wastes fall into the first category, although regulatory authority definitions are changing to bring more and

Table 8.1 Typical wastes generated from refining operations

Waste	Source
Oily sludges	Tank bottoms, interceptor sludges, wastewater treatment sludges, contaminated soils, desalter sludges
Solid material	Oil spill debris, filter clay, acid tar, filter material, packing, lagging, activated carbon
Drums and containers	Metal, glass, plastic, paint
Non-oiled materials	Spent catalyst
Construction debris	Scrap metal, concrete, asphalt, soil, asbestos, mineral fibres, wood

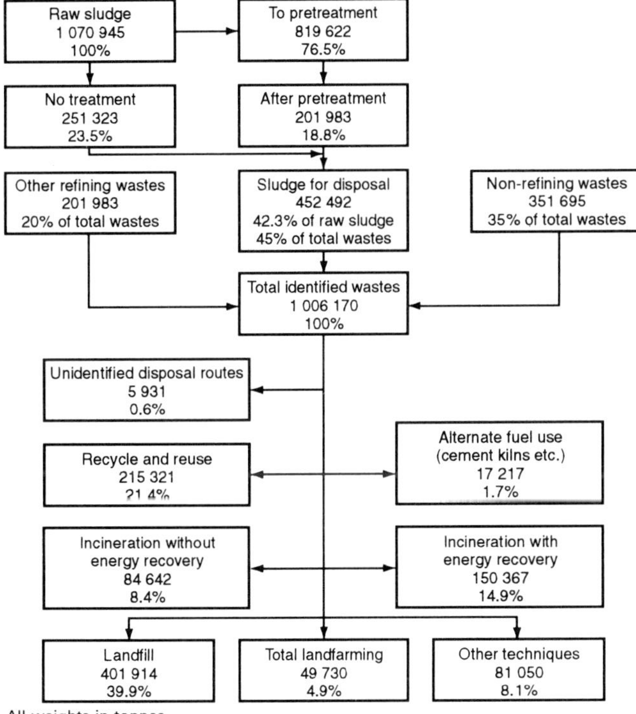

Figure 8.14 Summary of total waste generation and disposal routes (Western Europe) [6].

more waste types into the second category. Typical wastes generated from refining operations are shown in Table 8.1.

Segregation of different wastes is a first priority. Addition of a small quantity of hazardous waste may turn a large quantity of inert waste into hazardous waste. A number of routes are available for the disposal of refinery wastes. Figure 8.14 shows some of the routes used by the Western European refining industry.

8.6.1 Source control

It is most cost-effective to minimize the amount of waste at source, since waste represents a loss of either raw materials, intermediates or products which require both time and money to manage and recover. In addition, the generation of wastes and their subsequent recycling or disposal can present a range of regulatory, health and environmental risks or liabilities. Thus good source control is the most effective method of minimizing the impact of the refining operations on the environment. In a number of cases, relatively minor modifications can result in appreciable waste minimization. The following are some common and effective source control techniques used to minimize refinery waste:

- installing mixers on crude oil storage tanks to reduce sludge accumulation;
- closed-loop sampling systems on product tanks to reduce waste/slop oil production;
- use of antifoulants and corrosion inhibitors;
- dissolved air flotation (DAF) units; use of polyelectrolyte instead of inorganic flocculants to reduce the mass for final disposal;
- regenerative rather than once-through processes (e.g. Merox process instead of caustic treatment).

General good plant operation and economy in the use of chemicals will result in the minimization of wastes for disposal. Good housekeeping is essential to waste minimization. Seemingly unimportant procedural aspects in operations and maintenance may have a large impact on waste generation.

One way in which waste is generated is through spills and leaks in the plant. Thus proper material handling and storage will reduce waste generated by this method. Examples include:

- storage of drums off the floor to prevent corrosion through concrete 'sweating';
- bunding of storage/process area to contain spills;
- using larger containers instead of drums; larger containers are reusable when equipped for top and bottom discharge, whereas drums have to be recycled or disposed of as waste;
- equipping storage tanks with high-level alarms and automatic pump shut-offs;
- installing leak-proof valves;
- when there is a risk of leaks, the soil or floor should be rendered impermeable and a collection system provided.

Also cleaning, by its nature, generates waste. By choosing the right procedure and technique, this waste may be minimized or its nature altered so as to make it more easily disposable:

- drain equipment into closed systems where possible;
- use on-site pretreatment whenever possible, e.g. wash-steam filter material prior to dumping;
- minimize tank sludge prior to cleaning (through use of solvent and mixers).

Waste handling, when correctly done, optimizes the economics and minimizes the ecological impact of the final disposal.

8.6.2 *Waste treatment*

Since some waste generation is unavoidable, treatment of the waste is required to minimize its physical size before final disposal. The following are some of the more common techniques used by refineries to dewater or deoil the waste to decrease its quantity and to recover oil.

(a) Sludge dewatering. This is an intermediate process for the concentration of sludge for disposal. Sludge from a clarifier or a final biological sedimentation tank averages 1–3% solids by weight. Some thickeners handling particulate material yield solids running as high as 10% by weight. The first step is the use of a thickener, which is a holding tank for settling the produced solids more compactly through gravity. A thickener can increase solids concentration from 3% to 10–15%.

There are other intrinsic values in the use of thickeners. Inclusion of thickeners in the system enables the operator to control the clarifier for optimum clarity of effluent and still schedule further sludge handling. With biologically active sludge, a digester functions also as a thickener and sludge reducer through further biological activity. The compacted sludge is then further dewatered by filtration or centrifugation.

(b) Filtration. Filtration is a means of dewatering sludge suitable for treating sludges with a low oil concentration as oil can blind or smear the filter cloth. Filtration usually yields a rather oily cake, with little oil in the liquid phase. If the final disposal route of the sludge is incineration, leaving oil in the cake can reduce the fuel costs. As regards landfilling, filtration reduces the transport and disposal costs, because the sludge contains less water.

(c) Centrifuging. Centrifuging is particularly applicable when the oil content of the sludge is higher than 10%. The feed enters the machine and forms a concentric pool through which the solids settle to the outer wall. Solids are continuously removed by a scroll conveyor across a drying beach to discharge ports. The liquid flows counter-currently through a cylindrical section to an overflow dam of variable elevation, which affords pool-volume regulation.

(d) Drying of sludges. Drying of refinery sludges removes water and volatile organic compounds by heating using a steam coil. The vaporized materials are condensed and separated in a drum into an oil phase and a water phase. The solid phase is discharged.

The feed can be raw sludge or the solid phase from a filter press or centrifuge. Thermal treatment has proved to be effective in processing biological effluent treatment sludge, thereby converting the sludge into fertilizer or composting material. Alternatively, oil-containing sludges may be converted into low-grade fuel pellets which can be used in other industries.

(e) Solidification. Solidification is a process designed to improve waste handling and physical characteristics, decrease the surface area across which pollutants can leach, or limit the solubility of hazardous constituents, in which materials are added to the waste to produce a solid. It may involve a solidifying agent that physically surrounds the contaminant such as cement or lime, or it may utilize a chemical fixation process such as with sorbents. The resulting waste is usually an easily handled solid with low leachability.

(f) Stabilization. Stabilization is the conversion of a waste to a chemically stable form that resists leaching. This may be accomplished by a pH adjustment. Stabilization also generally results in a solidification of some sort.

Chemical stabilization is based on the reaction of lime with waste materials and water to form a chemically stable product. This technique is suitable to immobilize watery sludges to yield a powdery hydrophobic product which can be compacted. The immobilized product is water-repellent with a low risk of leaching. It hardens with time and has very good properties for civil engineering applications such as foundations, tank bases, bund wall and road making.

(g) Encapsulation. This involves complete coating or enclosure of a waste with a new, non-permeable substance. Micro-encapsulation techniques are based on the reduction of the surface-to-volume ratio of the waste by formation of a monolithic, hard mass with a very low permeability. Macro-encapsulation is the enclosing of a relatively large quantity of waste with a stiff, weight-supporting matrix and a seam-free jacket.

Encapsulation is suitable for on-site treatment of accumulated spent acid tars and oily sludges which are difficult to transport and to dispose of by other means. A disadvantage is that the treated product occupies a larger volume than the original sludge.

Because it can be applied on-site, the encapsulation process may be considered for single applications such as rehabilitating refinery sites after decommissioning or cleaning up an oil-polluted site after a spill. The decision to apply the process depends on the future use of the site and local legislation. The process is less attractive for the treatment of regularly produced sludges because of the increased mass generated for disposal.

8.6.3 *Waste disposal*

The following are some of the principal waste disposal and redemption techniques practised in the oil industry.

(a) Landfills. A landfill is a disposal route where waste is deposited in an artificial or natural excavation for an indefinite period of time. The deposition of wastes on land as a method of disposal will always be an activity which is controlled under legislation. In some countries it remains one of the cheaper methods of disposal, although the shortage of satisfactory sites and the difficulties in obtaining licences from the regulatory authorities is driving prices higher.

The key consideration in the operation of a landfill site is the protection of groundwater from contamination by the materials contained in the landfill. Wastes that contain water-soluble materials which can be leached by rainwater may ultimately contaminate nearby springs and streams, possibly rendering water sources unusable. Therefore, essential factors for such wastes are the following:

- The lining of the containment should be impermeable. Clay is the preferred material in many parts of Europe. In others, a lining of plastic sheeting is used. In some countries it is required to have multi-layer linings with integrated drainage systems for new landfills.
- Monitoring bore holes are used in order to inspect the effectiveness of the containment.
- The deposition of liquid wastes is not permitted except under rigorously controlled conditions. Whether or not liquid deposition is allowed, arrangements should be made for the collection and treatment of leachate.

A consideration for the disposer is that wastes deposited in landfill are not immediately destroyed but only stored. They must not be capable of reacting in a harmful way to generate heat or noxious gases. If flammable gases, e.g. methane, are generated they should be collected. Land is usually not usable until several years after a landfill operation is complete so that degradation of the material can take place. The overall economics of landfill or disposal methods are affected by several factors: the cost of transportation from the source to the site, the cost of the land and the cost of the landfill operation itself. The last factor is a minor part of the total.

With the cost of transportation and land increasing at alarming rates, this method may soon be less attractive than it has been previously.

Landfill is, however, the most practicable method of disposing of inert wastes such as building rubble.

(b) Incineration. Any process that uses combustion to convert a waste to a less bulky, less toxic material is called incineration. An incineration system must produce as complete a combustion as practical using an optimum selection of governing parameters such as time, temperature and turbulence, and provide air pollution control devices to minimize the emission of air pollutants. Many waste materials are readily combustible and the products of their combustion are harmless gases which are easily disposed of through vents or stacks to the atmosphere. In such cases, incineration is often the soundest method of waste disposal.

Some of the factors that characterize incinerators with good performance are:

- complete combustion;
- clear stack;
- low maintenance;
- minimum materials handling;
- minimum operating labour;
- adequate capacity;
- adequate availability;
- adequate flue gas treatment.

There are several types of incinerator designs to handle a variety of wastes, as shown in Table 8.2.

(c) Biodegradation. Many potentially hazardous chemicals present in refinery waste can be converted by microbiological methods into harmless compounds such as water and carbon dioxide. In general, the microbiological degradation of contaminants in soil is very slow in nature,

Table 8.2 Types of incinerators

Type	Feed	Comments
Fixed hearth incinerators	Solid, sludge and viscous oil	Low operating costs, small batch sizes
Multiple hearth incinerators	High water content sludge	High volumes, high operating costs
Fluidized bed incinerators	Partially dewatered sludge	Flexible wet sludge composition
Rotary kiln incinerators	Most wastes	Versatile and durable
Liquid fuel incinerators	Gasified or atomized liquids	–

because process conditions for such degradation are seldom favourable. To accelerate and optimize degradation the following conditions have to be fulfilled:

- sufficient number of microorganisms of the right strains;
- non-toxic concentrations of contaminants or other compounds;
- sufficient water;
- sufficient nutrients;
- sufficient oxygen for aerobic processes and a full depletion of oxygen for anaerobic processes;
- favourable temperature;
- sufficient availability of contaminants (preferably without high peak concentrations);
- pH of soil.

Several types of techniques are possible for the microbiological treatment of contaminated soil, as follows.

Land farming. Land farming systems have been used for the treatment of petroleum industry wastes for many years. The process involves the controlled application of waste on a soil surface in order to biodegrade the carbonaceous constituents by utilizing the micro-organisms that are naturally present in the soil. The conditions under which the degradation takes place are typically aerobic. The advantages of land farming are that it is a relatively cost-effective and simple technique, which is environmentally acceptable provided that it is properly designed, operated and monitored.

In most locations, permission from the authorities is required before a land farming facility can be started. In a number of countries the technique is not permitted at all.

Composting. Composting is a biological process where fresh organic wastes are transformed by decomposition into a stable humus-like substance. The processing is accomplished mechanically in a rotating cylinder. The waste is delivered to the cylinder in a moistened condition by the addition of water or sewage sludge. Air is added at low pressure and in controlled amounts throughout the length of the cylinder. In this manner an environment is created where the action of aerobic microorganisms ensures rapid decomposition of the wastes under inoffensive conditions. As with the backyard compost pile, the microorganisms which effect the decomposition are indigenous to the wastes themselves. The final process material is then screened and the compost is separated. The resulting compost is suitable for use as a fertilizer and soil conditioner.

Mechanized processes. The third category consists of wet and dry bioreactors and/or fermenters in which the soil is continuously mixed intensively. The biological composting/decontamination process can be accelerated if the necessary process conditions are closely controlled and monitored in pressure tight vessels. Typical hydrocarbons need a few hours to degrade, whereas PCBs require several days.

8.7 Recycling to minimize waste

Recycling waste materials for reuse may in many circumstances provide a cost-effective alternative to treatment and disposal. The success of recycling depends on both the ability to segregate recoverable and valuable materials from a waste and the ability to reuse waste materials as a substitute for an input material.

8.7.1 Reuse on-site

The optimum place to recover wastes is within the refinery itself. The following are some typical examples where waste has been recycled within the refinery.

(a) Use of cooling towers to close water circuits. Previously once-through cooling water discharged into the environment is now retained on-site, cooled and recycled to the process units. Thus the use of cooling towers allows an overall reduction in water intake and discharge.

(b) Reuse of crude distillation unit water in desalters. Water in the crude oil and recovered in the crude distillation unit can be used to substitute fresh water into the desalter, thus reducing total oily water production to be treated.

(c) Use of caustic cascades. Spent caustic from one process plant may still be sufficiently strong for use in another. For example, spent caustic can be further utilized in crude distillation units as a corrosion inhibitor or injected into bio-treaters for pH control.

(d) Centrifuge reprocessing of oil recovered. Recycling recovered oil eliminates the need for disposal and allows partial recovery of value in final product.

(e) Reprocessing of off-specification products. Reprocessing of off-specification product eliminates the need for disposal and allows the recovery of final product value.

(f) Reprocessing of oily emulsions. Reprocessing of oily emulsion (e.g. from an air flotation unit) in the distillation column eliminates treatment

with demulsifiers and disposal costs and allows conversion of the oil to product.

8.7.2 *Off-site recycling*

Wastes may be considered for use or reclamation off-site. Materials commonly reprocessed off-site by chemical and physical methods include oils, solvents and scrap metal. A strong commitment is required from the recycler not only to upgrade the waste materials for sale or exchange but also in finding suitable markets.

Examples of off-site recycling of refinery waste include:

(a) Recycling FCCU catalyst. Spent FCCU (fluid catalytic cracking unit) catalyst can be used as equilibrium catalyst in the start-up of new units.

(b) Cascading FCCU catalyst. Spent FCCU catalyst can be further utilized in other units operating at lower severity.

(c) Sale of FCCU catalyst. Spent FCCU catalyst may be used as an additive in cement manufacture. When the cement is used, the catalyst component forms insoluble hydrates with the chalk present in the cement mixture, which also gives beneficial fixation of the heavy metals present on the catalyst. It can also be used in brick manufacture.

(d) Disposal of spent catalysts. Industrial catalysts can contain heavy base metals and promoters/inhibitors such as phosphorus compounds. Sometimes these catalysts can be regenerated and reused. Alternatively, another disposal option is to return the spent catalysts to the manufacturers or to metals reclaimers for reprocessing. There is also a special group of precious metal catalysts used in oil refining, but they are not associated with disposal problems, and are always recycled because of their metal values.

(e) Sale of gypsum or sulphuric acid. SO_2 present in the flue gas desulphurization units can be converted through additional processing into gypsum or sulphuric acid, which can be sold.

(f) Drums/containers. Drums and other containers can often be recycled after suitable reconditioning.

8.8 Environmental management

Environmental management, like safety managment, is now an integral part of the management process in most oil refineries. Environment policies should be written to provide continuity and consistency in environmental protection programmes. The policy indicates the emphasis that the senior manager places on the programme, and outlines the major responsibilities and involvement of managers at each level and employees.

8.8.1 Environmental control

While a senior refinery manager will be responsible for the overall programme, there are many tasks and details to be looked after. A programme co-ordinator is often appointed in writing and given express authority necessary to administer the programme across functional lines of the organization. The co-ordinator should have adequate technical and administrative assistance to perform the planning, sampling and other functions of the environmental programme administration.

Many requirements for environmental protection measures are detailed in legislation and implementing standards, codes and regulations. Copies of all pertinent legislation should be clearly understood by the co-ordinator.

8.8.2 Environmental training

Environmental protection activities require specialized knowledge to organize and perform the managerial and technical tasks. A programme should be set up to include pertinent training for managers at the various levels and for employees whose work requires interfacing with environmental hazards. Each process, or each variation in process, can have differing impacts on the environment. Assessment of these impacts is critical to defining the degree of risk and selecting the appropriate controls.

Prior to construction, many processes require permits, licences or other written approvals. This may require research and an understanding of legislation to identify and secure these various requirements.

Following construction, modification or overhaul, facilities should have operational safety inspections to ensure that they are safe to operate. These should include the evaluation and contact of potential spills and other emissions that could harm the environment.

8.8.3 Environmental auditing

Auditing is often used to ensure that management of environmental procedures and control are assessed and opportunities for improvements and also weaknesses are identified and corrected. Figure 8.15 below shows the basic steps of an environmental audit programme. This process should be carried out at regular intervals and may involve both internal and external auditors.

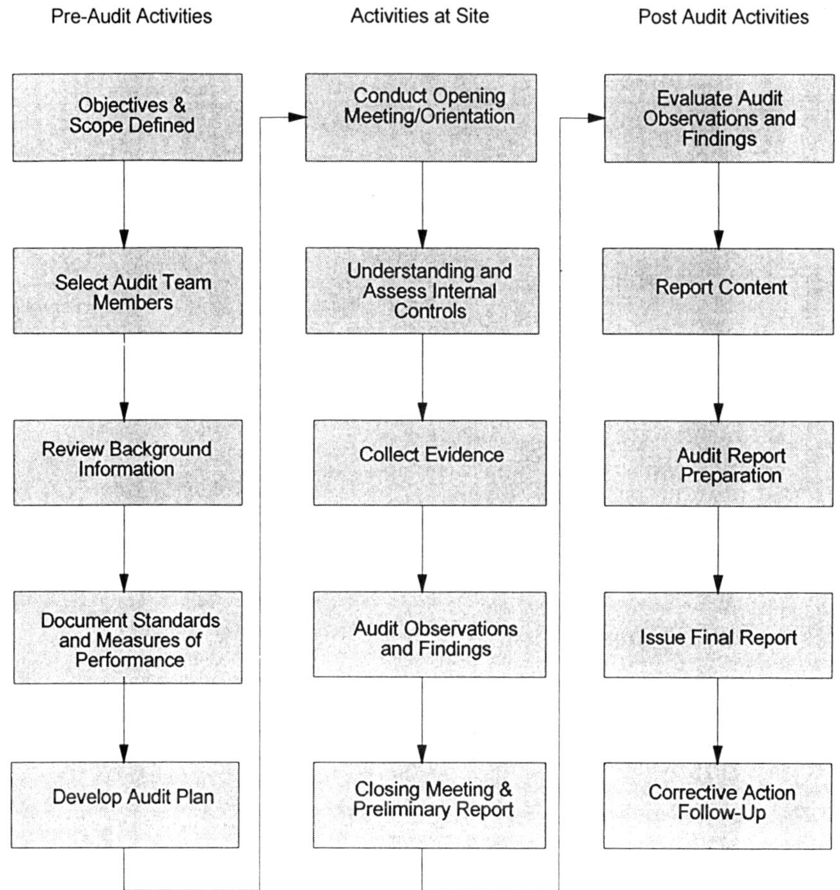

Figure 8.15 Basic steps of an environmental audit.

References

1. CONCAWE (1994) *CONCAWE Report*, 3(2), October.
2. CONCAWE (1993) *CONCAWE Report*, 2(1), April.
3. CONCAWE (1994) *CONCAWE Report*, 3(2), October.
4. CONCAWE (1994) *CONCAWE Report*, 3/94.
5. CONCAWE (1994) *CONCAWE Report*, 3(2), October.
6. CONCAWE (1992) *CONCAWE Report*, 2(2), October.

9 Distribution, marketing and use of petroleum fuels

T. COLEY

9.1 Introduction

A typical oil refinery operates continuously, manufacturing a wide range of products for a variety of end uses. The products pass from the processing units to refinery tankage but, as storage capacity at refineries is finite, their early transfer into the distribution network is essential for operation to continue at or near the designed throughput.

Refineries have the capability to manufacture many different types of material: gases, solvents, jet and burning kerosine, cracker and petrochemical feedstock, petrol, diesel fuel, domestic heating oil, residual fuel, lubricants, waxes and bitumen. Applications in which these products are used vary widely but a very high proportion of refinery output is burned as fuel.

9.2 Main refinery product types

The main fuel types, representing 80–85% of the production from a typical refinery, are gasolines, middle distillates and residual fuel oils. Gasolines are the grades of petrol used in piston-type engines for aircraft, passenger cars and equipment such as lawn mowers and small generator sets. Middle distillate fuels include aviation jet kerosine, burning kerosine, heating gas oil and diesel fuels for automotive, agricultural, industrial and marine engines.

Residual fuel oils, consisting largely of the heavy non-distilled residue from crude oil, are used in industrial heating systems and as diesel fuel in large, slow-speed engines for marine propulsion and industrial pumping, heating and power generator installations.

The balance is comprised of liquefied petroleum gases (LPGs), solvents, lubricants, waxes, petroleum coke and bitumen. Refinery fuel consumption accounts for about 2.5% of the total crude throughput at a simple hydroskimming refinery, consisting only of process units for distilling, catalytically reforming and hydrotreating. In a more complex refinery, with all the more severe processing needed to convert heavy products into lighter ones for gasoline and diesel manufacture, refinery fuel requirements are considerably higher, requiring up to 8% of the total throughput. This

Table 9.1 Regional fuel consumption demand patterns (1994)

Product group	W. Europe	Japan	Australasia	Canada	USA
Gasolines (%)	25.0	22.4	38.2	33.8	41.9
Middle distillates (%)	40.9	34.3	37.5	32.9	30.5
Fuel oil (%)	16.9	25.9	6.7	8.6	6.7
Others[a] (%)	17.2	17.4	17.6	24.7	20.9
Total (10^6 tonnes)	652.7	268.7	39.9	79.4	807.9

[a]'Others' consists of refinery gas, LPGs, solvents, petroleum coke, lubricants, bitumen, wax and refinery fuel and loss.

gives an indication of the real cost, in terms of both fuel requirement and increased emissions, of producing more 'environmentally friendly' fuel components.

Production patterns depend on demand in the particular markets supplied by the refineries and Table 9.1 shows the different percentage breakdown of oil product requirements for several regions of the world. The total consumption, in millions of tonnes, is also given for each of those regions, whose combined demand represents about two thirds of the world total consumption of petroleum fuels and other products.

Throughout the progression of a fuel from the refinery to the end-users' equipment, there are potential risks to the environment as a result of spillage, leakage and evaporation. Experience over many years has led to the development of regulations and Codes of Practice to avoid or minimize pollution of seas, rivers, canals, soil, groundwater and the atmosphere. At the present time, these controls are under close review in many regions of the world and, where appropriate, stricter legislation is being introduced.

The procedures adopted by oil companies to control pollution have evolved out of a combination of good operator training, properly designed containers, reliable connecting equipment and frequent inspection checks. These standard practices are complemented by technological developments in the field of automated metering and inspection devices, allowing periodic or continuous monitoring during product transportation, storage and transfer.

Risks to the environment are not over after delivery by the oil company into customers' fuel tanks. When the fuel is burned, the products of combustion will be discharged into the atmosphere and, as owners of motor vehicles are well aware, much new legislation has been introduced in recent years to minimize the amount of noxious emissions from vehicle exhausts.

These changes have posed tremendous challenges to both the motor and the petroleum industry and the very impressive developments in petrol and diesel engine technology will also be covered in this chapter.

9.3 Protection of the environment

A great deal of emphasis has been directed in recent years to the protection of the environment against pollution. Voluntary Codes of Practice and legally enforceable regulations have been introduced to provide means of controlling and minimizing both deliberate and accidental contamination of the environment from man-made sources of pollution.

Petroleum (literally, rock oil) is a natural material, formed over millions of years by the decomposition of organic matter under the very high pressure and temperature conditions prevailing deep inside the earth. Petroleum deposits are usually trapped between layers of impervious rock but movement of the earth's crust brought some of the deposits nearer to the surface from where, as a result of seepage through fissures in the rock, it was found and used by early civilizations. Crude oils vary widely but they are all made up of mixtures of hydrogen and carbon atoms, known as hydrocarbons, together with a variety of mineral impurities, depending on the geological strata with which the crude was in contact.

In the one and a half centuries since the first well deliberately seeking oil was drilled by 'Colonel' Drake in 1859, at Titusville, Pennsylvania, petroleum has become virtually essential for modern living, with a world annual consumption of around 3 billion tonnes. However, crude petroleum and its products are now considered by many organizations to pose a serious threat to the future, through contamination of the environment.

9.3.1 The atmosphere

Pollution of the atmosphere has become a high-level concern virtually throughout the developed world. A significant role is attributed to the use of petroleum products, with emissions from road transport vehicles being specifically highlighted.

Emissions from vehicle exhausts have been heavily targeted in recent years, but another concern which is currently receiving attention is that of evaporative losses from volatile petroleum products into the atmosphere. Initial controls were directed at minimizing the emission of fuel vapours from petrol-engined passenger cars, details of which will be discussed later. Steps are now being taken to reduce emissions of volatile organic compounds (VOCs) from storage and transport containers throughout the distribution system, both during transportation and whilst the product is being transferred from one container to another.

To put the situation in perspective, nation-wide surveys made in the UK during 1991 revealed petrol and diesel road vehicles as contributing an important proportion of the total emissions of carbon monoxide and dioxide, nitrogen oxides, black smoke and also VOCs. On a broader scale, the contribution of man-made hydrocarbon emissions into the atmosphere

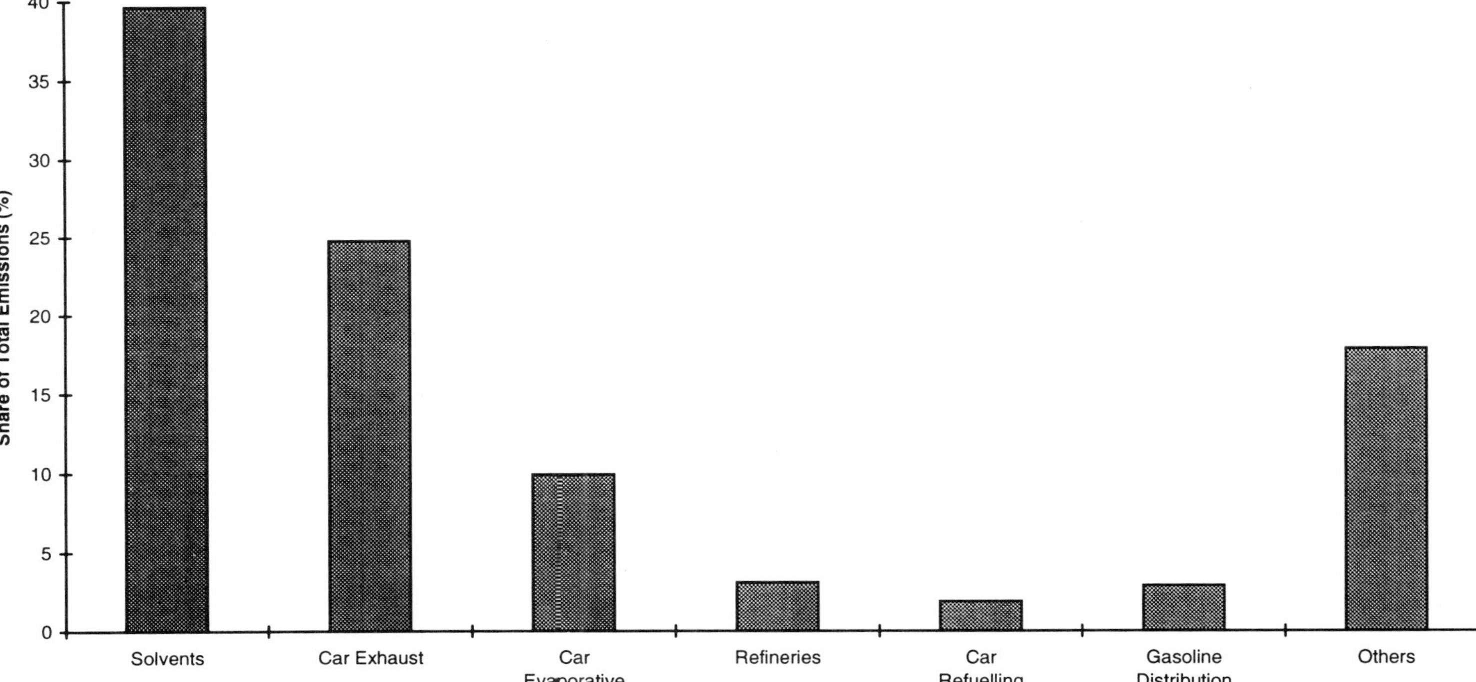

Figure 9.1 Man-made hydrocarbon emissions to the atmosphere (Western Europe): source contributions (%).

in Western Europe are presented in Fig. 9.1. This shows that the major contribution of 40% came from volatile industrial solvents, whilst 25% was identified as coming from the exhausts of petrol-engined cars. Evaporative losses from refineries, during distribution and refuelling and from the car itself, amounted to a further 18%. Man-made sources, in Western Europe alone, are estimated to emit annually about 10 million tonnes of VOCs, including hydrocarbons but excluding methane.

The same safeguards for equipment and personnel training are required to minimize uncontrolled VOC losses from oil company sites as those for avoidance of spillages when loading and unloading products from rail and road tankers.

9.3.2 *Sea waters: compliance with maritime regulations*

Strict controls are now in force to protect the seas from oil pollution. Collaboration on a world-wide basis by members of the International Marine Organization (IMO) resulted in the publication of the IMO/MARPOL Regulations (1991). These regulations and subsequent amendments (which are continuously being monitored and revised) define the procedures for disposal by oil tankers of oily waste after carrying out normal routine operations, such as emptying the bilges and cleaning and ballasting of crude and product tanks.

9.3.3 *Soil and groundwater*

In Member States of the European Union (EU), legislation for the protection of soil and groundwater quality has tended to be concentrated primarily on the protection of groundwater, which is a major source of drinking water. Legislation is directed at the control of any discharges to the environment which could result in contamination of groundwater.

(a) Potential sources of contamination

Pipelines. Leakages may occur as a result of failure of a pipeline due to corrosion or physical damage. Human errors at pumping stations and pipeline terminals can also result in oil spillages. Although pipelines tend to be mostly underground, they are not immune to leakage, so good operational procedures are necessary to avoid inadvertent pollution of the soil and groundwater.

In Western Europe, at the end of 1993, there were 215 separate cross-country pipelines with a total length of 21 600 km, for carrying both crude oil and finished products. Table 9.2 gives an analysis of pipeline incidents in Western Europe during 1993, involving spillages of more than 1.0 m^3. The data indicate not only the relatively modest number of occurrences

Table 9.2 Analysis of 1993 pipeline incidents in Western Europe

Main category	Number of incidents	Spillage (m^3)		
		Gross	Recovered	Net
Mechanical failure	2	251	233	18
Operational	0	–	–	–
Corrosion	3	2594	1581	1013
Natural hazard	1[a]	10	3	7
Third-party activity	4[b]	3110	1612	1498
Total	10[c]	5965	3429	2536

[a]Crack caused by earth movement after heavy rain.
[b]Includes one incident at a pumping station.
[c]Only one spillage affected potable water supplies.

but also the success of containment and recovery measures. More pertinently, it is worth noting that four of the ten incidents, which together were responsible for over 50% of the quantity spilled, were due to third-party action. This analysis suggests there is serious need for greater awareness of pipeline locations.

Couplings and connectors. Major spillages tend to occur during product transfer to or from ships, rail tankcars, road tankers or storage tanks, generally as a result of equipment failure compounded by human error. Regular inspection and maintenance of pumps, loading gantries, shut-off valves and other equipment in high-risk areas are an essential precautionary measure, together with effective supervision and training of operators.

As well as by checking that couplings are properly connected before delivery starts, the risk of spillage can be avoided by ensuring that the receiving tank is not over-filled. It is also necessary to ensure that the delivery line from the road or rail tanker has been emptied before disconnecting it from the receiving tank inlet connection.

9.4 Distributing the products

Finished products are delivered to the end-user customers through the distribution systems of oil companies and their agents.

Depending on the location of the refinery and its markets, the fuels will be transferred from refinery storage to the oil company's main terminal tankage for distribution in a variety of ways. These can include ocean-going and coastal tanker ships, barges on inland waterways, pipelines, railways and road tankers delivering to terminals, from where the products will be supplied by road to the network of depots within the local marketing area.

9.4.1 *Distribution systems*

(a) Tanker ships and barges. With crude oil exploration and production continuing in the established oil fields and in new areas all around the world, sea transportation in large crude oil carriers is the only practical way in which much of the crude can be delivered to the refineries and markets where it is to be processed and sold.

Water-borne transport can also be a logical option for the distribution of finished products. Several oil-producing countries, which formerly only exported crude oil have, for the best economic reasons, expanded their refining capacity and begun manufacturing finished fuel products meeting the specifications of foreign export markets. Some of the products may go by pipeline to nearby countries but other export grades are likely to be transported by sea to their distant markets.

Delivery by coastal tanker from indigenous refineries is common in many countries with lengthy coast lines. Where there is also a network of navigable inland waterways, as exists in Europe and the USA, the use of barge transportation provides a convenient, practical and economical way of moving large volumes of oil products.

(b) Pipelines. Pipelines are used to transport crude oil from the well to oil terminals, for loading at oil jetties into crude-carrying tanker ships and from tankage at receiving ports to the refinery. In many countries they also provide a practical, unobtrusive and economic way of distributing large volumes of the refined products around the marketing region.

An indication of how the pipeline distribution systems of Western Europe referred to above have developed is given in Fig. 9.2, showing the extensive network of product lines serving the main terminals in northern Italy, and Fig. 9.3, which shows the pipelines distributing products from refineries in northern France to Paris and to other terminals in their distribution areas.

There is, of course, a complex of pipelines for the distribution of imported crude oil from the ports to refineries in those areas, and a corresponding network of crude oil and finished product pipelines connecting oil fields, ports, refineries and terminals around the world, wherever petroleum and its products are handled.

(c) Rail tankcars. Most countries use railway networks for bulk transport of finished products to terminals and depots not served by pipeline. Railways provide an effective and economic distribution system, with a

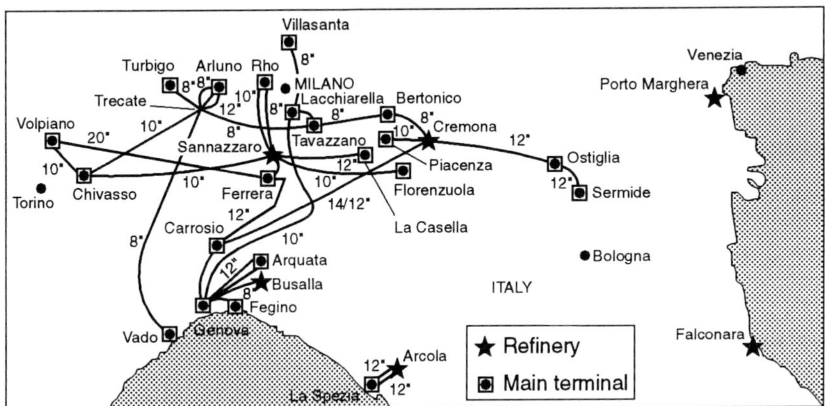

Figure 9.2 Product lines: northern Italy. (Reproduced with permission from *CONCAWE Report No. 5/94*, The performance of oil industry cross-country pipelines in western Europe – 1993 survey; published by CONCAWE, 1994.)

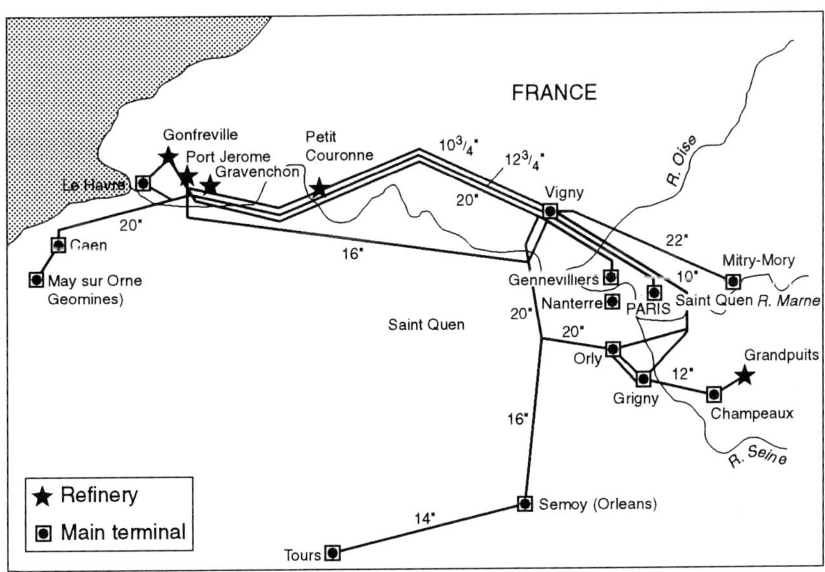

Figure 9.3 Product lines: Le Havre – Paris. (Reproduced with permission from *CONCAWE Report No. 5/94*, The performance of oil industry cross-country pipelines in western Europe – 1993 survey; published by CONCAWE, 1994.)

smaller likelihood of problems due to traffic congestion or accidents than on the roads.

Railcars are filled at the refinery or main terminal, generally from an overhead loading gantry discharging product through an inlet on the top of the tank. On arrival at its destination, the product will be unloaded by pumping from a bottom outlet of the railcar into the receiving tank.

Distillate fuels are normally carried in unheated tanks, except when climatic conditions during transit are likely to chill the fuel to below the temperature at which wax formation will occur and cause problems of incomplete tank emptying and non-homogeneity of the product. Residual fuels, which have much higher viscosities, are normally carried in insulated and heated tanks, to permit easy and complete unloading of the consignment.

(d) Road tankers. The situation with road tankers is similar in most respects to that with railcars. Gantry loading systems and screw-connected fittings for unloading are normally involved. As with railcars, insulated and heated tanks are required for residual fuels.

An additional factor, which could increase the chances of accidental spillages, is that road tanker drivers delivering to domestic, agricultural or small industrial consumers generally have to make several unloading operations to individual customers before all the tanker compartments are empty.

9.5 Anti-pollution controls

9.5.1 The atmosphere

Atmospheric pollution, by emissions from chimney stacks and vehicle exhausts and also as a result of uncontrolled evaporation of VOCs, is a growing concern, largely because of dire predictions about global warming and the hole in the ozone layer.

Limitation of the man-made emissions illustrated in Fig. 9.1 is being or will be enforced through legislation and regulations applying to the areas within the jurisdiction of the legislating or regulating authority.

The situation relating to atmospheric pollution controls will be described in the following sections. Although not covered in this chapter, corresponding control measures will need to be applied to processes at industrial plants where volatile solvents are used.

9.5.2 The high seas

Sophisticated modern navigational aids and weather forecasting techniques have helped to overcome or avoid many of the inherent hazards of sea travel but the risk of accidents cannot be entirely eliminated. In recent years there have been several highly publicized incidents with tanker ships, which resulted in massive spillages of crude oil. Although emergency actions have enabled significant amounts of spilled crude or heavy oil product to be recuperated, considerable damage has been caused to fish, sea birds and sea animals, and also to plant life, when the spillages have been close to the shore.

Sea-water pollution has also been caused when oil tankers have discharged oily waste water from tank ballasting and cleaning operations but, as was mentioned above, these practices are now being more tightly regulated and strict observance of the IMO regulations is now required of all oil tanker operators.

Contaminated water from tank washing has to be drained into a slop tank, from where it may be discharged, either into shore tanks or legally at sea, provided IMO regulations criteria are observed. Until 1992, it was possible to decant slops containing up to 100 ppm of oil into open water at sea, but the current permitted level has now been reduced dramatically to 15 ppm. Whilst the regulations generally allow disposal at sea of water from tank-washing operations, provided the maximum level of oil contamination is not exceeded, dumping is not permitted in some seas, such as the Baltic.

Similar tight constraints also apply when cleaning the bilges, an operation common to all ships. Contents of bilges have to be pumped into a separator tank, for recovery of the oil and to allow settling of the water phase before it can be diluted, as necessary, prior to discharging at sea. Coastal tankers tend to retain their contaminated waters from bilge emptying and also from tank cleaning and ballasting until they can be discharged into shore tanks and disposed of correctly.

Full and detailed records of these operations must be kept and be available for inspection at all times.

The absence of common international regulations for emission controls on tankers carrying gasoline on the high seas prompted the IMO to take action on vapour-collecting systems on tanker ships carrying cargoes of volatile products.

The IMO Standards have been drafted for vapour collection systems on sea-going tankers and for emission control systems at terminals. These take account of work carried out by the US Coast Guard and reported by the US Environmental Protection Agency (EPA). High-level or high-velocity vents are typically installed to provide means for vapour release during cargo loading or tank ballasting operations.

A recent study carried out in the European region measured the hydrocarbon levels in vapours emitted from the tank compartments of gasoline-carrying ships and barges during loading and these are illustrated in Fig. 9.4. Higher hydrocarbon levels are found with barges because of their relatively shallow tank compartments compared with those of sea-going ships, so the existing mixture of vapour from the previous cargo and air taken in during unloading tends to be richer in hydrocarbons.

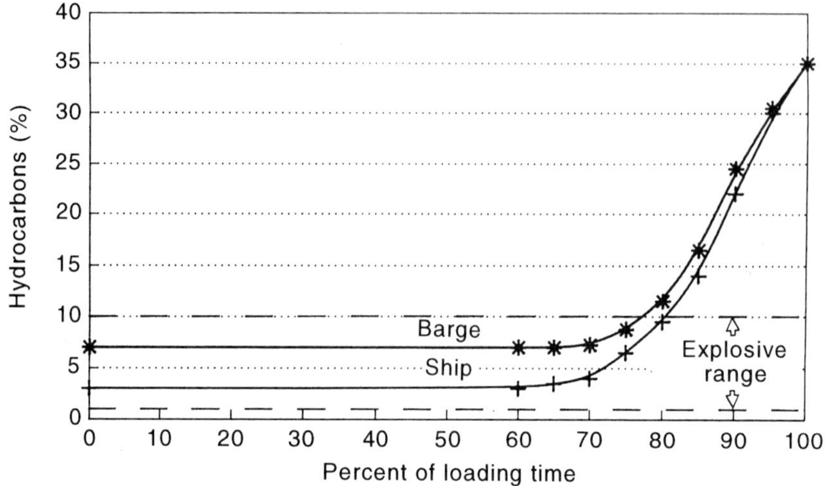

Figure 9.4 Hydrocarbon concentration profile during loading: waterborne transport of gasoline. Data based on weighted averages. (Reproduced with permission from *CONCAWE Report No. 92/52*, VOC emissions from the loading of gasoline into ships and barges in EC-12: control technology and cost-effectiveness; published by CONCAWE, 1992.)

It was found that the vapour concentrations from ship and barge tanks were in the flammable range for up to 80% of the loading time, a situation which necessitates extreme care being taken to control vapour emissions during loading operations. At sea, the high-level or high-velocity vents installed for safety reasons would have to be closed when vapour emission control systems are in use.

The actual hydrocarbon content of vapours emitted during loading is high, ranging from around 5% initially up to 35% and averaging around 10% throughout the loading period. With Europe's annual gasoline throughput of over 100 million tonnes, the total emissions of hydrocarbon vapour whilst vessels are being loaded represents a significant amount of environmental pollution.

9.5.3 Coastal and inland waterways

Procedural measures for coastal oil tankers and barges were introduced initially for reasons of safety and to reduce the exposure of crew members to hazardous vapours, but attention has now also been turned towards vapour recovery from gasoline or other flammable cargoes. The potential benefits from effective vapour recovery measures applied to sea-going ships and barges on inland waterways during loading operations are evident from the data presented above.

Standards for vapour emission control systems for barges carrying flammable liquids such as gasoline in the Rhine basin are set by the ADNR (Accord Européen Relatif en Transport International des Marchandises Dangereuses par Voie de Navigation Intérieure/Rhin). The requirements include a vapour-collection header with a high-velocity vent to atmosphere, means for connection to an onshore vapour recovery unit (VRU) and, at each tank outlet, a pressure/vacuum relief valve and detonation arrestor.

Although few, if any, ships or barges operating in and around Europe have onboard vapour-recovery systems, some have vapour-collecting systems which are suitable for, or adaptable to, connection with an onshore VRU, for emissions control during loading of volatile cargoes.

Technologies are available for vapour recovery during gasoline loading, drawing on experience associated with tanker truck loading facilities. They include the following options:

- absorption of the vapours in a low-volatility liquid;
- adsorption of the vapours on activated charcoal;
- condensation of the vapours in a heat exchanger set at a low temperature, such as −80 °C;
- separation by means of a hydrocarbon-specific membrane.

If a practical means of recovery is not possible, a further option would be thermal oxidation of the vapour in a flare. This will destroy the vapours and produce CO_2 together, possibly, with other undesirable emissions.

9.5.4 Soil and groundwater

In the early days of the oil industry, relatively little thought was given to contamination of the soil and groundwater. Later, possibly influenced by the commercial instinct to minimize the loss of valuable product, good housekeeping practices evolved and have contributed significantly to soil and groundwater protection.

In view of the probable high cost of cleaning up contaminated groundwater, the emphasis must be on preventing or minimizing the amount of leakage and spillage of oil products. The quality of material leaking or spilled is also of importance and several countries specify maximum concentrations of contaminants allowed in discharges into the air, surface water or soil. The Appendix to Swiss Ordinance for Waste Water Discharge 814.225.21, which came into effect on 1 January 1976, defines the permitted limits pertaining to surface water flows and impounded river water, to effluents discharged into surface waters and to effluents discharged into public sewers.

The constraints include limits on warming by cooling or waste waters; turbidity; colour; odour and taste; toxicity salt content; suspended and precipitated solids; pH (acidity or alkalinity); oxygen content; surface tension; and a number of specific contaminants, both inorganic and

Table 9.3 Parameters and maximum levels for organics in effluents discharged into surface waters in Switzerland

Parameter	Level permitted in effluents
Dissolved organic carbon (DOC)	15 mg/l maximum (24 h average)
Total organic carbon (TOC)	Not more than 7 mg/l above DOC level
Biochemical oxygen demand (BOD)	20 mg O_2/l (24 h average)
Aromatic amines	Each Canton sets conditions in agreement with Federal Office for Environmental Protection
Total hydrocarbons	10 mg/l
Chlorinated solvents (trichlorethylene, perchlorethylene, methylene chloride, etc.)	0.1 mg/l (measured as chlorine)
Phenols: volatile and not steam-volatile	0.05 mg/l

organic. A selection of parameters and maximum levels for organics in effluents discharged into surface waters is given in Table 9.3.

(a) Underground protection. Buried pipelines are potential sources of contamination in areas which are not protected by impermeable surfaces, so preventive measures have to be taken beforehand, as well as arranging for regular inspection procedures whilst the pipelines are in operational use.

Protection of underground pipelines is provided by the application of anti-corrosion coatings such as bituminized enamel or plastic tape. Cathodic protection, by means of sacrificial anodes or an impressed current, depending on soil characteristics, is normal practice. Leak detection equipment is also employed.

To segregate consecutive batches of product and also to clean deposits from the line, pipeline operators use solid plugs, known as 'pigs', which fit inside the pipeline and are pushed through by the fluid pressure. Care has to be taken to avoid spillages at the locations where the pigs are inserted and removed from the pipeline.

Technological developments of the pig concept are the metal loss detection intelligent pig, to search for pipeline corrosion, the ultrasound pig, for detecting small leaks, and the self-propelled pig, equipped with a video camera, for visual inspection of the inner surfaces of pipelines. These facilities make it possible for line owners to set up regular monitoring routines to minimize the likelihood of unexpected problems due to pipeline failure.

(b) Above-ground protection. Existing protection measures involve impermeable surfaces to prevent contamination due to leaks and spillages. These are normally installed in loading/unloading areas, underneath

overground storage tanks and in spillage basins and bunds, from where the spilled liquids can be recovered.

Some domestic heating oil tanks have a whistle fitted in the vent pipe to help prevent over-filling. More usually, it is avoided by first dipping the tank to verify what space is available and then setting the automatic cut-off on the pump control unit to deliver an appropriate quantity of fuel. A typical control unit printout sheet will show the driver's name, delivery date, customer name and identify the product grade, as well as giving the quantity delivered and the other details such as unit price, total cost and even the relevant value-added tax (VAT) rate.

(c) Treatment and recovery practices. Complete recovery of oil spillages is not always possible and, in such cases, disposal of the contaminated soil in landfill sites has often been employed. The imposition of more stringent regulations limiting the oil content of waste for land disposal is necessitating the use of more acceptable, albeit sometimes destructive, processes such as incineration.

Recovery of contaminated groundwater is usually achieved by means of oil–water separators. However, further treatment of the separated water may sometimes be needed to satisfy the purity standards required by local regulations.

Wastewater treatment procedures which might be used include:

- gravity separation in tanks or with plate interceptors;
- advanced treatment, such as filtration, sedimentation, flocculation or air flotation;
- biological treatment, such as with biofilters, activated sludge or aerated ponds.

The pipeline leakage in a mountainous region, which cracked due to an earth slip following heavy rain, reported in Table 9.2, caused some contamination of groundwater and soil pollution in the locality. Precautions were needed to protect the drinking water and 3 m^3 of the 10 m^3 spill were recovered by forming shallow channels and using water-washing to flush out the oil so it could be separated and recuperated. Some contaminated soil was removed for safe disposal elsewhere.

9.6 Marketing the products

The mode of delivery of fuels and other oil products to the customer will depend very much on the type of end-user equipment and the quantity to be delivered. In certain cases, the frequency at which fresh batches will be required can also have an influence.

9.6.1 Large industrial customer installations

Siting an oil refinery adjacent to or reasonably close to an oil-fired electricity generating station allows the fuel to be supplied on a continuous basis by pipeline. Another convenient arrangement is to locate industrial units on sites near an oil refinery, to facilitate pipeline delivery of fuel or feedstock. Provided normal precautions are taken to open and close the correct valves and to avoid overfilling the receiving tanks, there is minimal likelihood of spillage in such situations.

Where oil-fired power plants are remote from any refinery and if there is no pipeline supply available or possible, their large consumption requirements will be best served by railcar deliveries. Operational procedures to minimize the risk of spillages have been discussed above.

9.6.2 Small industrial and domestic customers

Regulations on oil storage facilities for small industrial plants and for domestic customers differ from country to country. In some European countries, however, strict standards have long been established for protection of the soil and ground water. Future trends suggest that these criteria will be adopted more widely.

Anticipated requirements are for above-ground storage tanks to have double bottoms, with a leak detection device in the space between the double bottoms. An impermeable surface or containment space beneath the tank may be specified as an alternative or even an additional requirement.

For buried steel tanks, a bunded containment with an impermeable surface lining will be required, possibly together with the use of an impermeable lining for the tank itself. Corrosion-resistant glass fibre-reinforced storage tanks are already mandatory for some buried installations. Buried pipework will also be subject to stringent controls, such as being laid in impervious channels and being pressure tested for leak tightness.

The use of technological developments such as high-level alarms to avoid over-filling and electronic devices for detection and prevention of leaks is likely to increase.

9.6.3 Service stations

Most service stations dispense the different grades and types of fuel from buried tanks. These installations must conform to the pertinent legislation and regulations of the licensing authority relating to environmental protection. These define the minimum standards for the location of underground storage tanks and will include such features as thickness of the concrete chamber for the tank, impermeability requirements, installation depth, layout of pipe runs and anchorage of the tank, to prevent it from floating if flooding should occur.

For safety reasons, vent pipes are run underground from each tank to a relatively remote location, from where they emerge to provide a high-level outlet for the vapours. Generally they are located conveniently close together. This will be advantageous as, before very long, it is likely that additional measures, such as carbon canisters for collection and recuperation of VOCs, will be required at petrol service stations.

Service stations are an important supplier–customer interface and, with many of them operating on a self-service basis, at least part of the responsibility for environmental protection is transferred from the oil company to the customer/end-user.

The most obvious expectation is that the customer will take care to avoid spilling fuel when topping-up the vehicle tank. This is helped by the automatic cut-off device in the pump nozzle, which stops the flow when the level reaches the nozzle. However, trying to fill the tank completely or inadvertently pressing the switch when the nozzle is out of the tank can result in spillage on to the ground. The further responsibility of the customer is to ensure that the vehicle's fuel system and engine are maintained in proper working order.

9.7 Environmental technologies related to product use

9.7.1 Fuels

The previous sections in this chapter have dealt with the working practices, regulations and legislation which have been introduced to minimize the risks of environmental contamination by oil fuels. Once the fuel is transferred into the customer's system, the potential pollutants other than the fuel itself, as a result of leakage, are the products of combustion.

Fuel specifications initially came into being to provide the end-user with products conforming to a realistically defined standard of purity and performance requirements. Other criteria which also needed to be satisfied included legality, safety and handling, reliability and, more recently, environmental considerations. As an example, Table 9.4 shows the make-up of an automotive diesel specification, with justifications for the particular fuel characteristics.

Interestingly, the first fuel property to come under scrutiny as harmful to the environment was sulphur, which is still regarded as one of the major pollutants in virtually all fuels. The influence that environmental protection legislation is having on sulphur levels and the trend with other fuel properties will be discussed in the following sections.

Table 9.4 Automotive diesel fuel specification criteria

Justification	Parameter
Legalistic	Distillation
	Sulphur
Purity	Water
	Sediment
Safety/handling	Flash point
	Cold properties
Performance/reliability	Cetane number
	Density
	Viscosity
	Cloud point
	Cold filterability
	Distillation
	Ash
	Acidity
	Carbon residue
	Stability
Environmental protection	Sulphur

9.7.2 *Marine diesel engines and fuels*

Large marine diesels are slow-speed engines that can operate satisfactorily on the relatively poor quality, low-cost residual fuels for marine use, which are generally referred to as bunker fuel oils. Auxiliary engines for pumping and other functions normally run on a distillate diesel fuel, which is also used for smaller, faster revving marine propulsion engines.

(a) Shipboard pretreatment. Onboard pretreatment of marine bunker fuels is required for several reasons. Heaters will be installed to enable the fuel to be pumped from the ship tanks and further heating will be necessary to melt waxy solids and lower the fuel's viscosity to the level required to ensure good atomization by the fuel injector. However, before reaching the injection pump, the fuel has to be cleaned by centrifuging and filtration, to remove sediment and other solid impurities which could damage the fuel system components.

(b) Ignition quality. The cetane number indicates the readiness of a fuel to ignite spontaneously when injected into the hot, compressed air in the combustion chamber; the higher the cetane number, the shorter is the ignition delay. Residual fuels are low in cetane number but the slow running speeds of the larger marine engines allow sufficient time to accommodate the delay between injection and ignition.

Residual fuels are available in several grades, relating to their viscosity. The highest viscosity grades, which also have the highest sulphur contents, are the heavy residues from the refinery distillation units.

(c) Sulphur levels. Recent surveys of marine fuels supplied from bunkering ports in the UK, France, Belgium and The Netherlands to ships operating in the busy English Channel area showed maximum sulphur levels no higher than 3.5% m/m (i.e. by mass), although up to 5.0% m/m is permitted by the ISO 8217 specification. The lower viscosity (and more costly) grades are obtained by blending the heavy residue with different proportions of a distillate stream. The lighter residual fuels have lower sulphur contents, with maximum levels up to 3.0% m/m for the medium grades and around 2.5% m/m for the low-viscosity grades.

With the current emphasis on environmental protection, the high sulphur levels of bunker fuel oils have resulted in ships usually being required to be operated on a distillate fuel when in port. Distillate marine fuel grades have maximum sulphur levels, as specified by ISO 8217, which range between 1.0% and 2.0% m/m and will be cleaner burning than the residual fuels. The survey referred to above estimated a total of 100 kilotonnes/year of sulphur emissions in the study area, of which 26% were in port. A breakdown of the estimated emissions is given in Table 9.5.

One of the combustion by-products of any fuel containing sulphur is sulphur dioxide (SO_2), which readily reacts with other oxygen atoms to become sulphur trioxide (SO_3). This, in turn, will combine with the water of combustion (H_2O), which condenses as the exhaust gases cool, forming sulphuric acid (H_2SO_4), a very corrosive pollutant.

Reducing sulphur emissions from marine engines. Although high sulphur levels are still predominant in marine bunker fuels, environmental concerns are targeted at reducing the sulphur content. This can be achieved to some extent by selecting lower sulphur crude oils or altering the proportions of high- and low-sulphur components used in blending marine bunker fuels. However, if neither option is possible on a long-term basis, the only possible answer may be to treat the residue to reduce its sulphur content by means of hydrogen treatment.

The desulphurization or hydrodesulphurization (HDS) process treats products containing high molecular weight and other sulphur compounds with hydrogen, at elevated temperatures and pressures, in the presence of

Table 9.5 Marine sulphur emissions (kilotonnes/year) in the English Channel area

Location	Bunker fuel oil				Distillate fuel
	High viscosity	Medium viscosity	Low viscosity	Total	
At sea	46.4	18.6	5.2	70.2	5.8
In port	14.6	5.5	2.3	22.4	4.9
Total	61.0	24.1	7.5	92.6	10.7

a catalyst. In the reactions which take place, sulphur from the compounds forms hydrogen sulphide (H_2S). The H_2S from the process normally goes to a sulphur recovery unit, where it is converted into elemental sulphur.

HDS is used routinely to reduce the sulphur levels of diesel fuel, kerosine and some lighter distillate streams. A process for residue desulphurization (RDS), which requires higher temperatures and pressures, has been developed from HDS to produce low-sulphur fuel oil components. RDS units are generally designed for 80–85% removal of sulphur from the feedstock. Consumption of hydrogen is high and residue desulphurization is a fairly costly process.

A study was carried out by CONCAWE, the oil companies' European organization for environment, health and safety (and the major source of information given in this chapter), to look at ways to reduce sulphur oxide emissions from ships. Two considerations studied were reduction of the fuel sulphur content and removal of sulphur oxides from ship exhaust gases.

The first option, described above, is carried out at the refinery but the second is the responsibility of the ship owner/end-user. The procedure for removing sulphur oxides (SO_x) from the exhaust gases is by sea-water scrubbing, which makes use of the ability of sea water to neutralize SO_x.

Sea water is sprayed into a flash chamber/particle scrubber, where evaporation of the water cools the gases, which then pass into a packed column washing tower, where a counter-current flow of sea water removes the SO_x. A heat-exchange system reheats the cleaned gases before they leave the funnel, to avoid a white plume of water vapour. Further work on this technique is needed to overcome some operating problems but the broad findings have been environmentally favourable. Although the discharged wash water is markedly more acidic than normal sea water, when discharged at sea it will be diluted rapidly in the ship's wake and will not have any effect outside the discharge mixing zone.

9.7.3 Fuels for large industrial power plants

High-viscosity and often also high-sulphur fuels are used in many of the large stationary engines used for electric power generation plants, industrial pumping installations and steam-raising plants. Such fuels were also commonly used as boiler fuel by steam ships but these have largely been replaced by diesel-engined ships, which were discussed above.

Being located on land, the environmental constraints on industrial plants are generally more stringent than those currently in force for marine engines, which emit their more sulphurous pollutants when on the high seas.

(a) Reducing sulphur emissions from large industrial plants. In addition to being required to run on lower sulphur fuels, other clean-up operations may be necessary to reduce the level of sulphur compounds and other pollutants in the flue gases.

A study was carried out a decade ago on ways of reducing sulphur emissions from residual fuels in the ten-member European Economic Community (EEC). Other countries have joined since then and the European Union (EU) now has 15 members, contributing to the total emissions of sulphur and other atmospheric pollutants, who are all bound by EU law to conform to the lower emissions levels. At the time of the study, it was estimated that by the year 2000, power stations would be responsible for 25–35% of the total sulphur emissions from oil-burning equipment.

RDS was one of the ways considered for reducing sulphur emissions, although it was felt that alternative low-sulphur fuels, such as low sulphur coal or natural gas, could make RDS economically unattractive. The trend projected in 1986 for sulphur emissions from inland fuels up to the year 2000 is shown in Fig. 9.5. This indicates the influence of high and low oil demands and the impact which investment in RDS would have, but RDS still remains unattractive.

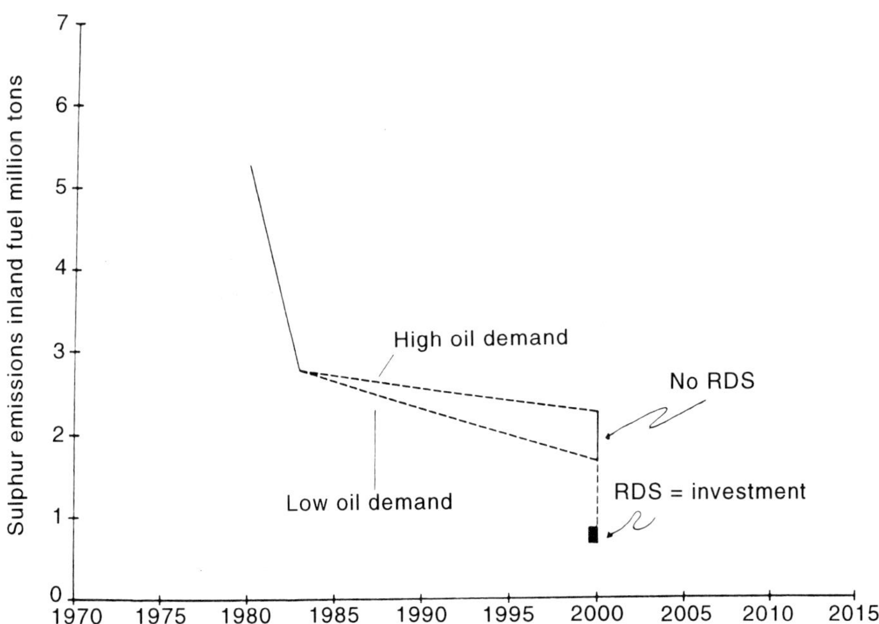

Figure 9.5 Sulphur emissions, inland fuel. (Reproduced with permission from *CONCAWE Report No. 5/86*, Sulphur emissions from combustion of residual fuel based on EEC energy demand and supply, 1980–2000; published by CONCAWE, 1986.)

An option for the plant operator, which was calculated to be more cost effective than RDS, was flue gas desulphurization (FGD). Although costly, it was estimated that FGD applied at power stations would replace the need for between five and ten refinery RDS units achieving somewhat lower levels of desulphurization.

9.7.4 *Fuels for small industrial and domestic installations*

(a) Heating gas oil quality constraints. The most commonly used fuel for small industrial and domestic heating systems is a distillate grade, commonly referred to in the oil industry as gas oil, as it was formerly used to supplement the heat content of town gas manufactured from coal. Heating gas oil, or domestic heating oil, is similar to automotive diesel fuel but with a slightly less stringent specification.

From the viewpoint of environmental pollution, one of the principal contributors is, once again, sulphur. In Western Europe a maximum of 0.3% m/m sulphur is fairly typical, with a few countries specifying 0.2% m/m. Future pan-European legislation is likely to impose a maximum sulphur content of 0.2% m/m.

The distillation characteristics of a gas oil can have an influence on its smoking tendency. If the gas oil contains some high-boiling components, they may not burn completely inside the combustion chamber of the furnace and will be emitted from the chimney as particles of soot. In Europe, the Common Customs Tariff agreement legally defines a gas oil as a petroleum oil having a maximum of 65% recovered at 250 °C and a minimum of 85% recovered at 350 °C, in the ISO 3405 distillation procedure.

As there is no limit on the distillation end-point of the gas oil, some heavy fractions, which may not burn completely, could be present. The responsibility for ensuring an acceptably low level of smoke rests with the owner and the engineer who services the burner. The permitted viscosity range of 1.5–5.5 cSt (mm^2/s) is suitable for typical pressure-jet burners, providing good mixing of the atomized fuel with the combustion air.

New house building developments now tend to be designed around heating systems operating on either gas or electricity, usually decided by the developer or determined by the availability of the utility suppliers' services in the particular locality. Although the purchaser has virtually no say in the choice of system, there is also no need for a separate fuel storage tank, with its potential for costly environmental control measures at some future date.

(b) Quality controls on domestic kerosine. In the UK, kerosine has been a popular fuel for domestic use in small, portable, wick-fed heaters,

vaporizing burners and even for some pressure-jet burners/boiler units, where its clean burning characteristics result in low maintenance and service needs.

Heating kerosines are manufactured to specifications which define, amongst other properties, their maximum sulphur content. The current British Standard for kerosine (BS 2869) allows a maximum of 0.2% by mass of sulphur for the regular grade kerosine used in burners with a flue or chimney to vent the exhaust gases outside the building. A much lower level of 0.04% by mass is specified for wick-fed appliances, which are not normally connected to a flue.

In addition to the need for a low sulphur content, other important quality criteria for wick burners are a high smoke point and a low char value. The smoke point, measured in millimetres, is the maximum wick height at which fuel vaporized from the wick will burn without smoking. The higher the smoke point, the greater is the amount of heat (or light) which can be produced. The char value indicates burning quality and is measured by the charred material which remains on the wick after a specific amount of kerosine has been burned. The greater the ratio of char to kerosine, the poorer is the burning quality of the fuel. These properties will be influenced by the distillation end-point, by hydrotreating to reduce the sulphur content and by refinery 'sweetening' processes, such as caustic washing and Merox treating. The Merox processes were developed by Universal Oil Products (UOP) to oxidize extremely unpleasant-smelling mercaptans (thiols) to non-odorous disulphides.

In other countries where domestic kerosine is used, permitted sulphur levels vary widely, e.g. from as high as 0.25% m/m in India down to 0.015% m/m in Japan.

(c) General guidelines for efficient burner operation. For reasons of economy and environmental protection, the burner should be cleaned and serviced regularly by the owner or a specialist service engineer, to ensure it is operating at or close to optimum efficiency and that there is little or no visible smoke.

Diurnal temperature variations will cause a fuel tank to breathe, drawing in air as the temperature falls at night. This can result in moisture from the air condensing and settling to the bottom of the tank. Drain taps on the storage tanks should be opened from time to time, to drain out any settled water and the tank itself checked for rusting which could, in time, cause leaks.

9.7.5 *Aircraft engines and fuels*

Aviation fuels are, understandably, subject to strictly controlled specifications, to minimize the likelihood of engine failure whilst in flight. There are some differences between military and civil grades but the main safety considerations apply to both applications.

(a) Gasoline for piston-engined aircraft. Most large passenger and freight aircraft are powered by jet engines but piston-engines tend to be predominant in the light aircraft used for private aviation and also for commercial operations such as flying schools, air taxis and crop spraying. There are also some piston-engined military aircraft.

The number of grades of aviation gasoline is decreasing (and the volume in relation to jet fuel consumption is modest), but currently four grades are available. Aircraft piston engines tend to require a high-octane fuel and those octane levels, particularly the higher ones, have to be attained by the addition of an octane booster, tetraethyl lead (TEL). This and a similar additive, tetramethyl lead (TML), were widely used in automotive gasolines but lead additives are in the process of being eliminated from the automotive grades.

The 80 octane grade is either low in lead or, for the UK market, completely lead-free. The other grades are 100 LL (low lead), 100 leaded and a leaded high-performance 115 octane blend. The maximum levels of TEL are given in Table 9.6.

The 115 octane grade is now almost non-existent and there is limited demand for the 80 octane grade. This means that only 100 octane aviation gasoline (Avgas) is being produced in significant quantities, to universally accepted UK/US specifications.

(b) Aviation jet kerosine. Measures taken to avoid fuel-related problems which could cause pollution of the atmosphere include the use of antioxidant and metal deactivator additives. These improve the fuel stability and suppress the formation of gums and sediment which could restrict normal operation of the fuel system components. Other additive types used in the aviation fuels are corrosion inhibitors and lubricity additives to protect the fuel system from chemical and mechanical damage, anti-icing additives to dissolve any free water in the fuel, which might freeze at altitude and block fuel lines, and biocides to prevent fungal and

Table 9.6 Lead contents of aviation gasolines

Octane grade	TEL (ml/US)[a]
80	0.5
100 LL	2.0
100	3.0
115	4.6

[a] 1 US gal = 3.785 dm^3.

bacterial growths which could block or restrict flow in the fuel system or cause corrosion.

The cleanliness and high chemical purity of aviation fuels mean that they have a very low conductivity. To avoid the risk of spark-induced explosion and fire, anti-static additives are routinely used to dissipate charges of static electricity which could build up as a result of the high pumping rates used when transferring fuel into storage tanks and during aircraft refuelling.

Modern jet aircraft engines consume fuel at a very fast rate and the steady expansion of air travel for business and holidays is drawing attention to their contribution to the total environmental pollution. A large proportion of the jet engine exhaust is dispersed at high altitudes, but travellers and residents near airports are keenly aware of pollution caused by exhaust fumes and droplets of unburned fuel. A reduction in sulphur content below the maximum level of 0.3% m/m permitted in current jet kerosine specifications seems the most likely first course of action.

9.7.6 Engines for rail transport

The number of coal-fired steam engines operating in Europe has been dramatically reduced during the past 50 years but, while many have been replaced by electric locomotives drawing power from overhead cables, there are large numbers of diesel engines in operation, in diesel/electric locomotives, shunters and railcar units. Railroad diesel engines generally run on automotive diesel fuel or a fairly similar grade and, as with autodiesel, the fuel property regarded as the principal contributor to atmospheric pollution is sulphur.

In 1987, the Council of Environment Ministers in the European Commission issued a directive reducing the maximum sulphur content of all gas oils, except those used by shipping, to 0.3% m/m by 1989 and allowing a stricter limit of 0.2% m/m to be set in heavily polluted areas. At the same time, a proposal for a single limit to apply throughout the Community was put forward and the European diesel fuel specification, issued in 1993 by the European Standards Organization, CEN (Comité Européen de Normalisation), imposed a maximum limit of 0.2% m/m throughout the EU.

9.7.7 Automotive engines

A significant proportion of legislation to protect the environment from oil-associated pollution has been directed specifically at the growing road transport sector. This has necessitated vast investment, of money and resources, by the motor and petroleum industries into research programmes to meet the increasingly tight limits on exhaust emissions.

The results of their endeavours to meet the required levels of emissions from vehicle exhausts, within the prescribed time-scale, will be discussed in the following sections.

(a) Spark ignition automotive engines. Design trends in spark ignition engines over recent years have resulted in lower rates of fuel consumption, often coupled with enhanced performance.

A significant factor has been the increase in the size and/or number of valves per cylinder. Initially two valves, one inlet and the other exhaust, were normal for small automotive engines. Then a second inlet valve was introduced and, at the present time, engines with four valves per cylinder, two inlet and two exhaust, are becoming more and more common. Larger or multiple valves per cylinder, providing less restriction to the flow, have allowed better filling of the cylinder with the air–fuel mixture during the intake stroke and more complete exhausting of the burned gases, resulting in improved engine efficiency.

Another trend has been the progressive replacement of the carburettor by the fuel injection system, almost throughout the size range of passenger car engines. Effectively, the carburettor is no longer fitted, as it cannot provide the close control of air/fuel ratio (AFR) needed for three-way catalyst emission control systems. Although in its early stages of development fuel injection into the inlet manifold gave better throttle pedal response and acceleration, it was achieved at the cost of higher fuel consumption. However, developments in injector design and fuel control systems brought marked improvements in fuel economy.

Current engine models largely use multi-port injection and sophisticated electronic engine management systems. These use sensors to monitor a variety of engine operating conditions and are able to ensure that the optimum of air/fuel ratio, spark advance and other key factors are always attained.

Although the immediate advantage of more fuel-efficient engines is felt by the motorist who buys the fuel, there are also environmental benefits, because of the overall reduction in specific fuel consumption per vehicle. This at least partly compensates for the growing number of vehicles on the road!

After combustion, the gases are emitted into the atmosphere through the exhaust pipe. The original purpose of the vehicle exhaust system was reduction of noise, which, for at least a decade, has been regarded as a pollutant. Recent legislation has imposed the need to modify the gaseous components into less noxious forms. Modification of the gases is achieved by three-way catalytic converters, in which the platinum, palladium and/or rhodium catalyst converts carbon monoxide, unburned hydrocarbons and nitrogen oxides in the exhaust to less harmful carbon dioxide, nitrogen and water.

In the USA and Japan, three-way converters have been mandatory in the exhaust systems of new cars since 1975. They became mandatory in Europe in 1993 and their use is increasing elsewhere in the world. They were the primary reason for the change from leaded to unleaded gasoline because lead will poison the catalyst and nullify its effectiveness.

Discontinuing the use of lead antiknock additives necessitated other changes, as deposits from combustion of the additive had protected exhaust valves and seats from damage by the hot exhaust gases. Most engines required a different metallurgy for the valve seats to avoid the problem of burning and recession when running on lead-free gasoline.

The technique to control evaporative losses of VOCs from cars is to use onboard carbon canisters to absorb the vapours. Figure 9.6 gives a typical layout for the vapour control system and shows the function of the carbon canister. The canister contains activated charcoal which absorbs the gasoline vapours. The vapours are trapped in the canister whilst the engine is stopped, but when the engine is running the gasoline vapours are drawn into the inlet manifold and the in-coming air purges the canister.

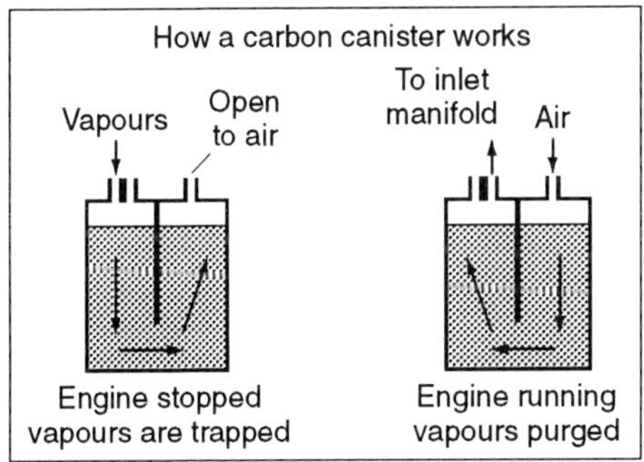

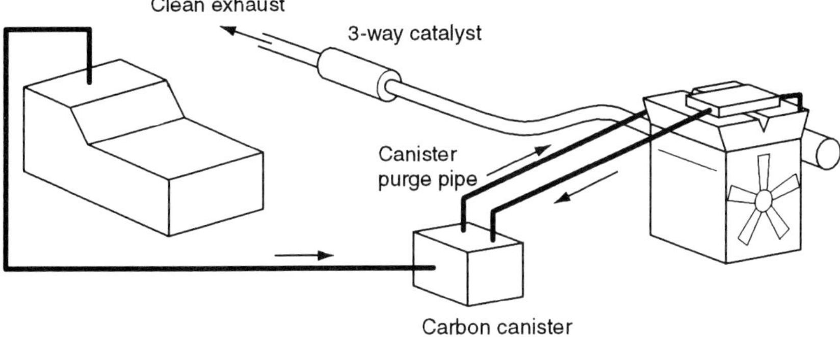

Figure 9.6 The closed fuel system. (Reproduced with permission from *Gasoline Vapour Emissions – a European Concern*; published by CONCAWE, 1990.)

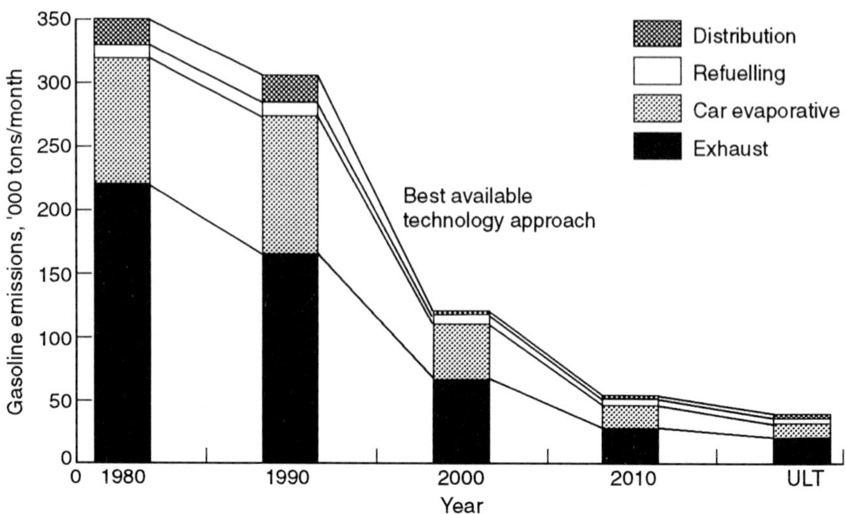

Figure 9.7 Impact of closed system on gasoline emissions for May–September in EC-12. (Reproduced with permission from *CONCAWE Report No. 3/90*, Closing the gasoline system – control of gasoline emissions from the distribution system and vehicles; published by CONCAWE, 1990.)

The impact of a closed system for gasoline, making use of the 'best available technology' (BAT) approach to control emissions from distribution, refuelling, car evaporation and the exhaust, is illustrated in Fig. 9.7.

The emissions, in kilotonnes/month, are based on figures collected during the summer period, when ozone formation is more likely, owing to higher ambient temperatures. The 'ultimate' (ULT) level of control would eventually be achieved when three-way catalysts and enlarged carbon canisters are fitted to all gasoline cars.

Dialogue between the European Commission and the relevant motor and oil industry associations (ACEA and EUROPIA) resulted in the European Programme on Emissions, Fuels and Engine Technologies (EPEFE). The European Auto/Oil programme had the objectives of providing test data on the influence of exhaust clean-up technologies and fuel properties on gasoline and diesel vehicle emissions.

Gasoline quality developments. A common feature of most gasolines, for over 50 years until fairly recently, was the presence of an additive, TEL. The need for better quality gasoline became evident in the 1920s, when attempts to improve engine efficiency by the use of higher compression ratios revealed the phenomenon of pinking or knocking, which can lead to catastrophic engine damage.

Extensive research studies during the early 1920s identified TEL as having the greatest effectiveness as an antiknock agent and the best potential for commercial development. For several decades TEL was virtually the only antiknock additive but later a closely related product, TML, was also commercialized. The use of lead-based compounds to increase the octane rating of gasolines continued until, in the 1970s, concern about the harmful effects of lead on humans resulted in many countries legislating against its use. The progressive reduction in the maximum lead content of the top gasoline grade in Europe and the USA is shown in Fig. 9.8. As lead is often present in crude oils, a maximum lead content of 0.013 g/l is permitted in the European unleaded gasoline specification EN 228.

Maintaining the volume of gasoline produced, in addition to its octane quality, without the use of lead has required significant restructuring of gasoline manufacturing and blending operations. Although other metallic compounds were found to give antiknock benefits, they have been rejected

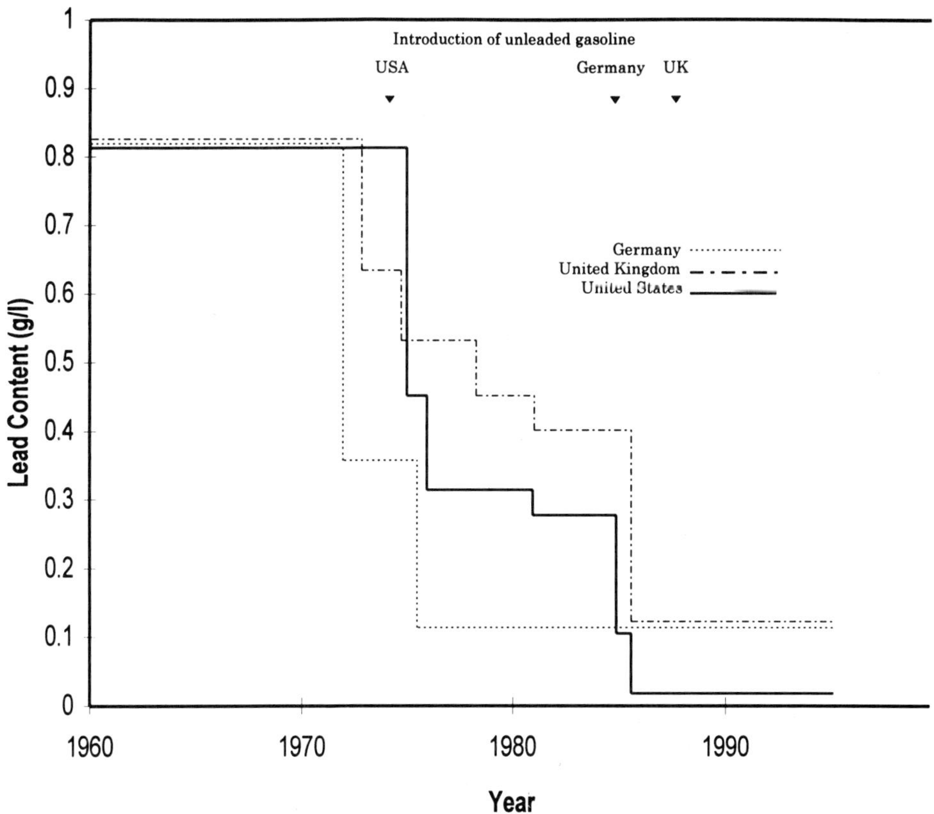

Figure 9.8 Trends in premium gasoline lead content.

for various reasons, which include deposit formation, wear, toxicity and cost. Some organic antiknock compounds were also rejected as being less cost-effective than lead alkyls or further refinery processing.

Studies of other ashless compounds identified some promising candidates with antiknock capabilities and, consequently, oxygenated compounds such as alcohols and ethers have become widely used as blend components in lead-free gasolines. Many countries around the world now have 'clean' or reformulated gasolines, which are designed to give reduced levels of vehicle exhaust and evaporative emissions.

Reformulated gasolines reduce emissions in a number of ways: their lower vapour pressure reduces evaporative emissions; smog-forming benzene and other hydrocarbons in exhaust emissions will be reduced because of the lower content of aromatics and olefins; oxygenates in the fuel will result in lower levels of carbon monoxide and hydrocarbons in the exhaust, whilst the lower sulphur content will also be beneficial. The total effect on exhaust emissions resulting from switching to the ARCO (Atlantic Richfield Co.) Emission Control gasoline, EC-1, introduced in Southern California in 1989, are evident from the data given in Table 9.7.

The stringent controls imposed in certain areas of the USA have initiated directionally similar actions elsewhere, particularly in Japan and Sweden, whilst other European countries are following.

In the USA, reformulated gasoline has been defined as a result of the regulation–negotiation (reg–neg) between the Environmental Protection Agency (EPA) and the oil industry, which is summarized in Table 9.8.

Lead antiknock additives are still being used in many parts of the world, but they have been eliminated or are being phased out from those countries taking a strong line on environmental protection. However, other types of additive are being used to help reduce pollution from gasoline engine exhausts, by maintaining the condition of the engine and fuel system. Amongst those currently employed are additives to reduce the increase in octane requirement (anti-ORI) that occurs as vehicles accumulate mileage and lay down deposits in the combustion chamber and

Table 9.7 Reductions in vehicle emissions with EC-1 fuel

Type	Approximate reduction (%)
Exhaust emissions:	
Hydrocarbons	5
Carbon monoxide	10
Nitrogen oxides (NO_x)	6
Benzene	43
Evaporative emissions	22

Table 9.8 USA 'reg–neg' agreement on reformulated gasoline

Specification item	Target value (psi/kPa)	Refinery average limit (psi/kPa)	Absolute limit (psi/kPa)
RVP Class A[a]	7.2/49.7	7.1/49.0	7.4/51.1 max.
RVP Class B	8.1/55.9	8.0/55.2	8.3/57.3 max.
Oxygen	2.0% by mass	2.1% by mass	1.5% by mass min.
Benzene	1.0% by vol.	0.95% by vol.	1.3% by vol. max.
Heavy metals	None without EPA waiver		
T90E	Average no greater than refiner's 1990 average		
Sulphur	As above		
Olefins	As above		
Detergent additives	Compulsory, but not yet defined by the EPA		

[a]For 1995–96 only; 1997–99 VOC emissions must be reduced by at least 16.5% relative to 1990 average gasoline.

valve ports; antioxidants to improve oxidation stability and prevent the formation of sediment and gums; detergents to prevent deposit formation which could impede the function of the fuel injector or carburettor; and additives to prevent 'soft' valve seats from recessing when operating on low-lead or unleaded gasoline.

Other additives which may be included are anti-icers to prevent ice forming in the carburettor during warm-up in near-freezing conditions and high humidity, demulsifiers to prevent problems due to water pick-up from tank bottoms, corrosion inhibitors to protect metal surfaces and spark-aider additives to improve the spark when operating at borderline air/fuel ratios and improve emissions by helping to prevent misfiring.

The EPEFE programme referred to above was completed during 1995 and conclusions drawn from the accumulated data will provide the basis for new EU directives on vehicle emissions and fuel quality, up to the year 2010.

The evolution in gasoline and diesel passenger car exhaust emissions standards in Europe, as set by directives issuing from the European Commission, is illustrated in Fig. 9.9.

(b) Automotive diesel engines. Recent years have seen a tremendous increase in the numbers of diesel-powered passenger cars in the developed countries, with the exception of the USA and Canada, where the gasoline engine is still the most popular for personal transport. The diesel share of new passenger car registrations in Europe is approaching 20% and close to 100% for commercial vehicles in most of the world.

One of the main attractions of the diesel car over the gasoline model is its lower fuel consumption. Disadvantages in earlier years were the poorer performance, higher noise levels and higher cost of the diesel version. However, the modern passenger car diesel can provide almost the same

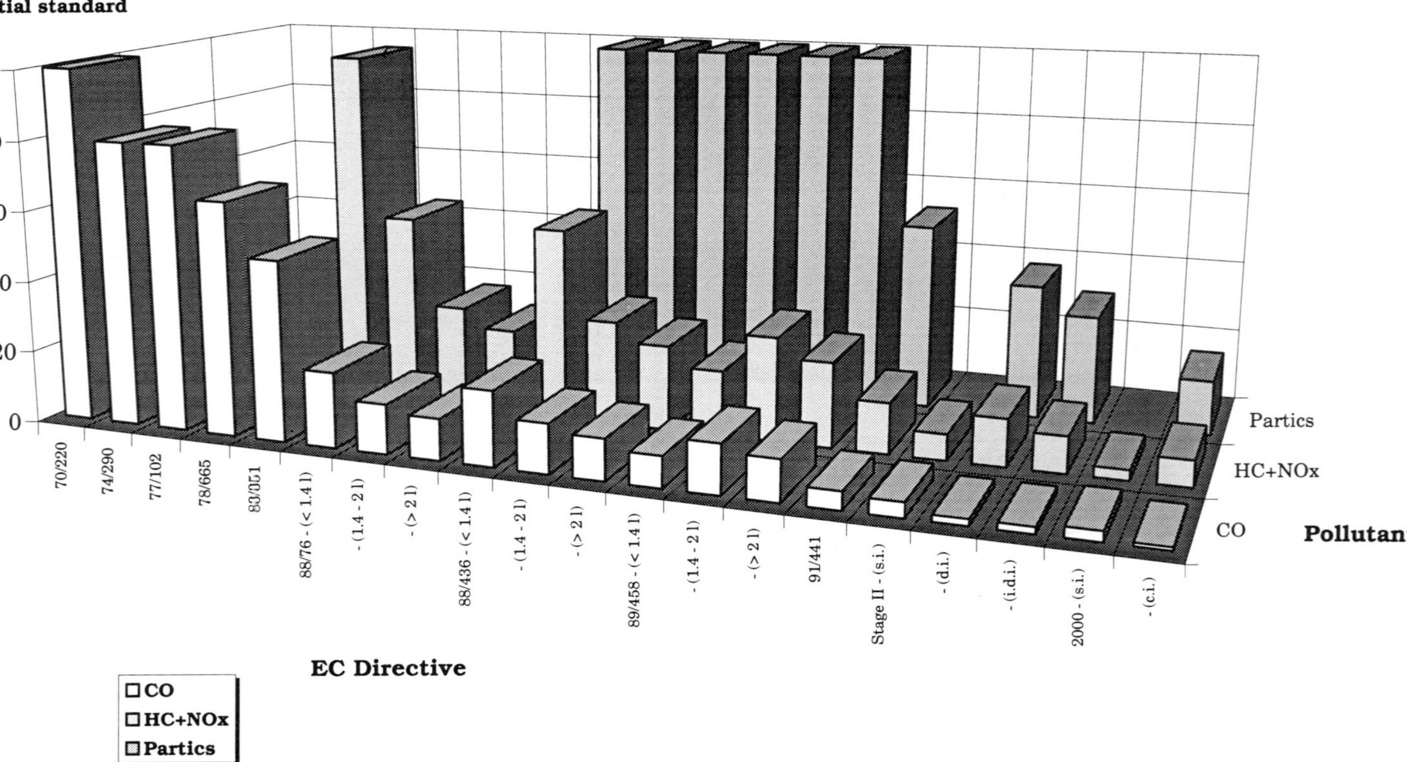

Figure 9.9 EC passenger car emission standards 1970–.

performance as its gasoline counterpart, with better fuel economy and an acceptably low noise level, and with little or no price difference. These impressive improvements have been achieved as a result of extensive (and expensive) research programmes into the diesel combustion process.

Included in studies on engines for passenger cars and commercial vehicles were the benefits of turbocharging, aftercooling and exhaust gas recycling, together with changes to fuel injectors, injection timing, higher injection pressures, exhaust catalysts and particulate traps. Sophisticated engine management systems, to monitor engine conditions and ensure that the right amount of fuel is injected at the optimum moment in the engine's operational cycle, have also had a significant effect. Turbochargers, intercoolers and also electronic management systems are becoming the standard for a significant segment of commercial truck and bus fleet operators.

The EPEFE programme to collate data on the influence of fuels and engine technologies on emissions showed that, in the light-duty diesel test cycle, increasing cetane number and decreasing density had the greatest influence in lowering hydrocarbon emissions. The back-end volatility of the fuel, T95 (95% recovery temperature), and the polyaromatic content also had significant effects on emissions, with NO_x emissions levels going down as polyaromatics decreased but increasing as the T95 was reduced. However, heavy-duty (HD) engines (which consume about 70% of the diesel fuel sold in Europe) showed opposite trends.

Criticism of diesel vehicles by the general public is largely associated with exhaust smoke, which is usually the only emission (from either gasoline or diesel vehicles), other than steam and water during warm-up, that is visible. Black smoke consists mainly of particles of carbon (soot) associated with small quantities of partially burnt hydrocarbons from the fuel and the small amount of engine lubricant which finds its way into the cylinder. Sulphates, which are formed during combustion of sulphur in the fuel, will also contribute to particulate formation.

Lowering the fuel sulphur content will decrease particulate emissions but technological developments on the hardware side include particulate traps and catalytic converters. Catalytic converters are used in small passenger car systems and the heavy hydrocarbon content of the exhaust gas, which normally contributes to the total mass of particulates, is reduced by the oxidation catalyst.

Particulate traps tend to be too bulky for passenger cars and are more suitable for large commercial vehicles. They are installed in the engine exhaust system and trap the soot particles in wire mesh or ceramic fibres or in porous ceramic foam. Particle retention will vary according to the type of trap and also with time, as some types are blockable. They all require frequent regeneration to burn off accumulated soot deposits by means of electrical or fuel-fed heaters.

A further development is a continuously regenerating particulate trap, which oxidizes the particles into carbon dioxide and water. This is achieved by using a platinum catalyst to convert some of the NO_x in the exhaust to nitrogen dioxide, which will stimulate oxidation of the carbon particles. It works as a continuous process and operates at normal diesel exhaust temperatures, as low as 275 °C. Particulate traps are still under development, as both cost and durability remain a barrier to series production.

Diesel fuel quality developments. Automotive diesel fuels are normally made by blending straight-run petroleum fractions from the atmospheric distillation unit with 'cracked' gas oils from one or more of the cracking processes, in proportions depending on the specification criteria to be met.

The different sources of gas oils which could be used as blend components for automotive diesel fuel are shown in Fig. 9.10. However, not all refineries are likely to have the full selection available. The graph indicates the approximate ranges in density and cetane quality of typical gas oils from each refinery process, including that of the major blend component, straight-run gas oil from the atmospheric distillation unit. The shaded area shows that the European auto diesel specification imposes a very tight constraint on the refiners, who have to produce fuels of more than 49 cetane number which fall within the limited tolerance band for density.

Demand for domestic heating oil is shrinking but, as heating oils usually contain a high proportion of low-cetane cracked gas oil, they are not suitable as direct replacements for diesel fuel. However, a cetane improver additive may be used to restore the ignition quality if some cracked gas oil is incorporated to increase diesel fuel yield.

Consumption of automotive diesel fuel is increasing in most parts of the world and cold flow improvers are widely used to enable refiners to meet the demand. Diesel fuels are complex mixtures of hydrocarbons, with a typical boiling range of 160–370 °C. As much as 20% of the fuel can be relatively heavy paraffinic hydrocarbons which have a limited solubility in the fuel and, if cooled sufficiently, will come out of solution as wax. The polymeric flow improver additives work by modifying the way in which the wax crystals form, making them smaller and less likely to restrict the fuel from flowing through the vehicle fuel system.

Unlike gasolines, diesel fuels do not contain any non-petroleum refinery components such as alcohols or other oxygenated products. For that reason, clean, 'reformulated' diesel fuels tend to result from further or selective processing of conventional petroleum-based components to produce fuels having extremely low levels of undesirable, pollution-forming characteristics.

Product enhancement additives are becoming widely used to improve fuel characteristics which are not (yet) specified but which can bring

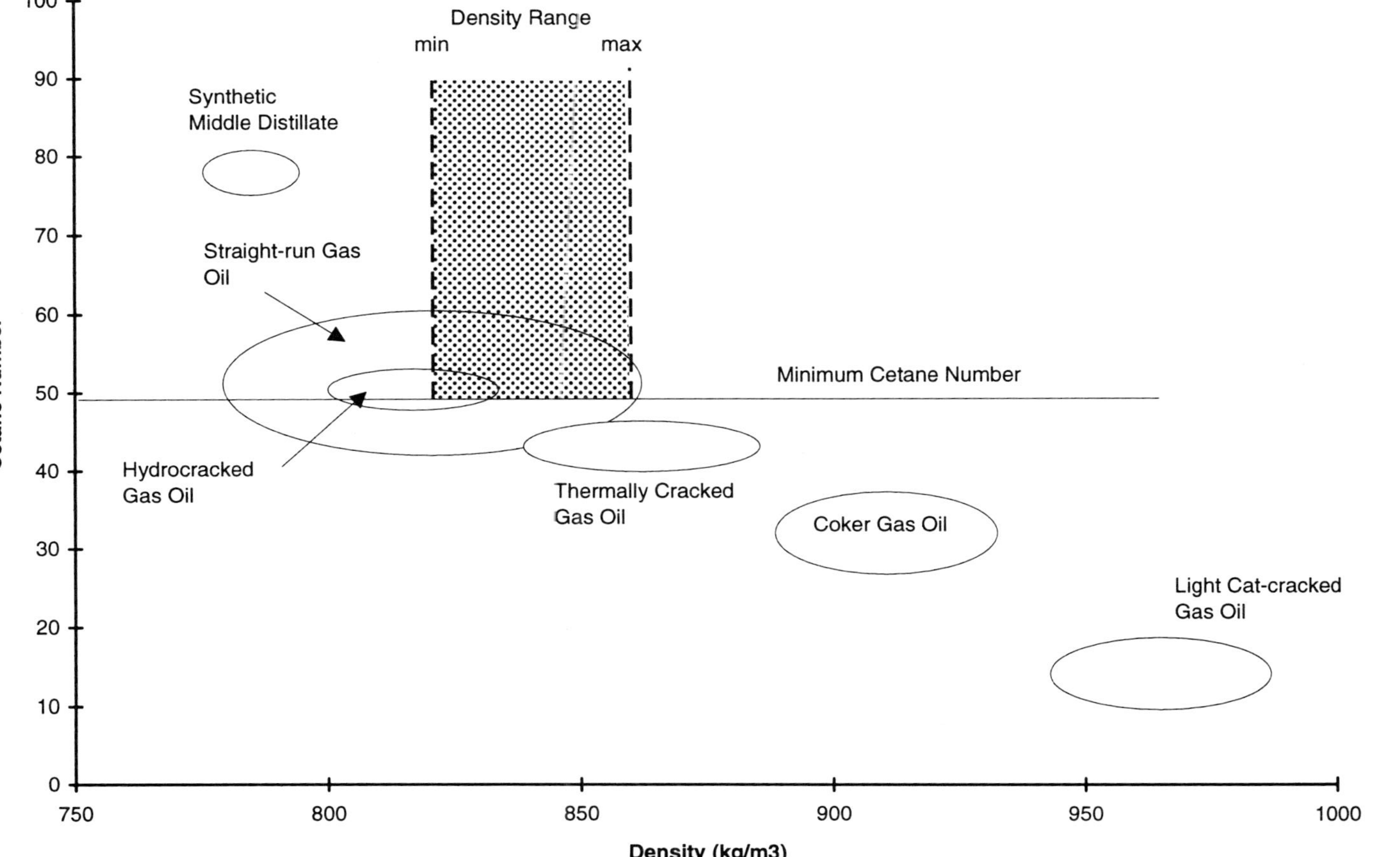

Figure 9.10 Density and cetane number characteristics of gas oil types. (Reproduced with permission of Shell International Oil Products.)

environmental benefits. These include detergents, antiwear agents and lubricity additives to protect fuel injection systems from harmful deposit build-up and wear, antifoamants to minimize the risk of spillages during refuelling and corrosion inhibitors to protect storage tanks and vehicle fuel systems.

A technique for producing a very clean diesel fuel has been developed by a major oil company. The Shell Middle Distillate Synthesis (SMDS) process converts natural gas into a liquid, from which a light gasoline fraction, kerosine and gas oil can be obtained by distillation. These products are virtually free of sulphur and nitrogen and, being wholly paraffinic, the gas oil has a very high cetane number. At present SMDS production capacity is limited but it has good potential.

Sulphur reduction in automotive diesel fuel has, once again, been the primary environmental objective. Additionally, specifications for cetane number, final boiling point and aromatics content have also been tightened because of their influence on carbon monoxide, nitrogen oxides, hydrocarbons and particulates (soot) emissions in the exhaust.

In 1993, CEN, the European Standards Organization, issued the first pan-European specification for automotive diesel fuel, with which the 18 Member States are bound to comply. To take account of the wide range of temperature conditions across Europe, a series of climatically related requirements were drawn up, from which each member could select appropriate summer and winter grades. These cover cold properties, density, cetane quality and distillation. All members must comply with the generally applicable requirements, consisting of flash point, carbon residue, ash, water and particulate matter (dirt), copper strip corrosion, oxidation stability and sulphur content.

Sweden recently introduced two special ‘city diesel’ grades, with tax incentives to encourage their use. Characteristics of these new urban environmental grades and the standard diesel grades are given in Table 9.9.

These new urban grades are proving popular, not just because of their lower price and not only in Sweden. The environmental benefit of using such fuels has stimulated at least one British company, with a chain of fuel service stations, to market imported urban diesel fuel. The low sulphur levels will have some beneficial effect on the amount of acid-forming materials in the exhaust and also help reduce particulate formation.

Large volume manufacture of these fuels is technically impossible. Maximum production capability has been estimated at about 25% of European diesel fuel consumption. Also, as the tax incentives indicate, manufacturing costs are extremely high.

A significant proportion of conventional diesel fuels are now treated with a variety of additive types, usually in the form of a ‘package’ prepared to suit the specific requirements of individual oil companies. Most contain

Table 9.9 Swedish diesel fuel classifications

Fuel characteristic	Urban diesel 1	Urban diesel 2	Standard grades
Sulphur (% m/m) (max.)	0.001	0.005	0.20
Aromatics (% v/v) (max.)	5	20	–
Polyaromatic hydrocarbons (PAH) (% v/v) (max.)	0.002	0.01	–
Distillation:			
IBP (°C) (min.)	180	180	–
10% (°C) (min.)	–	–	180
95% (°C) (max.)	285	295	370 summer (S) 340 winter (W)
Density (kg/m^3)	800–820	800–820	820–860 (S) 800–845 (W)
Cetane number	50	47	49 (S) 45–58 (W)
Tax rate (\$/m^3)	126	165	199

a cold flow improver additive to avoid problems during normal winter conditions.

Other additives which may be present include cetane improvers for easy starting, stabilizers to prevent sediment and gum formation, detergents to control deposits, anti-corrosion additives for metal protection, anti-wear additives to protect fuel-lubricated surfaces of injection pumps and anti-foamants to permit fast refuelling rates and reduce the likelihood of spillages. As the odour of diesel fuel can be very persistent and not particularly pleasant, some fuels may be treated with a re-odorant to make it more acceptable. A manganese-based additive, methylcyclopentadienyl manganese tricarbonyl (MMT), has been found effective in reducing particulate formation and diesel fuel treated with this product is being marketed in the UK by the Mobil Oil Company.

9.7.8 *Into the next millennium*

Despite the increasingly severe legislation against environmental pollution by the road transport sector, the motor and petroleum industries, aided by the additive companies, have been able to meet the targets. European specifications for gasoline and diesel fuel will not change markedly until the year 2000, but requirements in the USA (and California in particular) will give the lead for stricter constraints on emissions elsewhere in the world.

Considerable work is in progress on alternative fuels but gasoline and diesel fuel will continue as the principal fuels for the vast majority of road transport. Three-way catalysts for gasoline vehicles are now mandatory in many countries but there is room for improvement of the catalyst, to maintain its emission control effectiveness for longer periods.

At the present time, the diesel engine is under heavy criticism because of the potentially carcinogenic hazard of the microscopic PM_{10} (particulates less than 10 μm in size), which are claimed to be retained in the lungs after inhalation. Banning diesel-powered vehicles, on which the infrastructure of most developed countries depends, has been suggested but that is as impracticable as total elimination of particulates. What appears more likely is the imposition of even higher excise duty on diesel fuel (and also probably on gasoline) to restrict its use by private motorists.

Continuing research towards cleaner burning automotive engines and fuels will certainly be necessary to reduce exhaust emissions further. The successes of the past three decades provide hope that the interests of future generations will be protected, but enhanced inspection and maintenance procedures will be necessary to ensure close adherence to the emissions criteria.

Further reading

CONCAWE (1986) Sulphur emissions from combustion of residual fuel oil based on EEC energy demand and supply, 1980–2000. *CONCAWE Report No. 5/86.*

CONCAWE (1988) Capital and operating cost estimating aspects of environmental control technology – residue hydrodesulphurization as a case example. *CONCAWE Report No. 88/51.*

CONCAWE (1988) The control of vehicle evaporative and refuelling emissions – the 'on-board' system. *CONCAWE Report No. 88/62.*

CONCAWE (1990) European soil and groundwater legislation: implications for the oil refining industry. *CONCAWE Report No. 4/90.*

CONCAWE (1992) The environmental benefits and cost of reducing sulphur in gas oils. *CONCAWE Report No. 3/92.*

CONCAWE (1992) VOC emissions from the loading of gasoline into ships and barges in EC-12: control technology and cost-effectiveness. *CONCAWE Report No. 92/52.*

CONCAWE (1992) *CONCAWE Product Dossier No. 92/103: Gasolines.*

CONCAWE (1992) *CONCAWE Review*, **1**(2), October.

CONCAWE (1993) The European environmental and refining implications of reducing the sulphur content of marine bunker fuels. *CONCAWE Report No. 1/93.*

CONCAWE (1993) *CONCAWE Review*, **2**(2), October.

CONCAWE (1994) The contribution of sulphur dioxide emissions from ships to coastal deposition and air quality in the Channel and southern North Sea area. *CONCAWE Report No. 2/94.*

CONCAWE (1994) Trends in oil discharged with aqueous effluents from oil refineries in western Europe – 1993 survey. *CONCAWE Report No. 3/94.*

CONCAWE (1994) Motor vehicle emission regulations and fuel specifications: 1994 update. *CONCAWE Report No. 4/94.*

CONCAWE (1994) The performance of oil industry cross-country pipelines in western Europe. *CONCAWE Report No. 5/94.*

CONCAWE (1994) *CONCAWE Review*, **3**(2), October.

CONCAWE (1995) *CONCAWE Review*, **4**(1), April.

Department of the Environment (1993) *Diesel Vehicle Emissions and Urban Air Quality, Second Report of the Quality of Urban Air Review Group*, H.M. Stationery Office, London, December.

Owen, K. and Coley, T. (1995) *Automotive Fuels Reference Book*, 2nd edn, Society of Automotive Engineers, Warrendale, PA.

10 Lubricants

C.I. BETTON

10.1 Introduction

The activities of the oil industry are aimed primarily at the production of fuel. The proportion of crude oil that is refined into a lubricant base oil is only 1% of the total (Institute of Petroleum, 1992). It could be argued that the base oils produced by a refinery are a by-product of the refining process and that the integrated oil company regards lubricant production accordingly. Lubricants, however, represent a high technology, specialist, high added value group of products with high potential environmental impact. The environmental aspects of lubricants extend beyond the obvious direct impacts to secondary impacts such as energy savings due to improved performance.

The perfect lubricant from an environmental point of view would consist of a material that:

- was obtained from a renewable resource;
- did not require a large amount of energy to produce;
- was a perfect lubricant in that it reduced friction to very low levels;
- was unaffected by heat and pressure;
- did not contain any potentially toxic or harmful components;
- was not 'used up' during the process of lubrication;
- was not dependent on temperature in order to function;
- was readily degradable to non-harmful components if spilled;
- worked for the lifetime of the device being lubricated;
- was recoverable and reusable.

Unfortunately, such a material does not exist nor, based on the current state of knowledge of how lubricants work (Mortier and Orszulik, 1992), is it ever likely to exist.

What is clear from the above list is that the basic ways in which a lubricant can be said to be environmentally beneficial can be grouped into:

- performance;
- components;
- effect on the environment if spilled (toxicity);
- rate of removal from the environment if spilled (degradability).

The most important of these aspects is performance. Performance must never be sacrificed for short-term environmental gains; environmental advantages are in addition to, not instead of, the maximum level of performance that can be attained. To take an example of a motor car engine oil, modern engines are not capable of existing on the lubricants based on castor oil that were adequate for the engines of the 1920s, even though such a lubricant would meet many of the 'environmental' criteria listed above. What it is easy to ignore are the substantial benefits in fuel efficiency and emissions reduction that are achievable from a modern engine, compared with an early 'gas guzzler'. If an old-fashioned oil were used in a modern engine, the engine would be destroyed in a few thousand miles and the environmental costs of replacing that engine in terms of the energy and foundry emissions, etc., from the production plant would far outweigh any benefits from the use of the 'old-style' lubricant. Performance is and must always be paramount when assessing the environmental aspects of any lubricant; that is not to say, however, that it should be performance at any cost. Recent developments in metalworking oil technology with the elimination of materials such as chlorinated paraffins from some formulations (Howell, Lucke and Steigerwald, 1996) demonstrate that as knowledge of environmental and health effects increases, formulations can change accordingly. Such changes are not without cost and are not easily achieved, as in all cases performance needs to be maintained if not improved.

10.2 Performance

Performance of lubricants is measured in different ways depending upon the use to which the lubricant will be put. There are standard tests for some types of lubricant such as engine oils that convey to the customer some information concerning the quality of the oil that is purchased (SAE, 1996). It is not my intention to describe or review these methods here. There are, however, some points regarding the standard industry tests that are worth making, particularly when considering the most up-to-date lubricant technology and the environmental benefits that can be accrued.

Performance tests for lubricants are controlled by national (API) and international (ACEA) organizations, and some individual motor manufacturers also have their own test requirements (SAE, 1996). The tests are based upon the principles of ready availability and reproducibility. A test that is run by one institution must be, and be seen to be, the same as that by a different institution; the results must be comparable. As can be imagined, huge investment is put into establishing test methods and setting up facilities to run the tests; companies are therefore reluctant to chop and change tests without compelling reasons. The result of these restrictions is that the tests used to classify lubricants according to the internationally recognized standards are often lagging behind contemporary technology. The major lubricant companies and the motor manufacturers, being aware of the deficiencies in the standard classification tests, have developed their own methods of assessing performance that are in addition to the standard requirements. A series of motor car engine oils, for example, may all meet a particular standard but the performance of oils may be very different in actual driving conditions, with those of the major lubricant producers being far more effective than some others that are not designed to the same high standards. With lubricants you really do get what you pay for.

In conclusion, when considering the environmental benefits of any lubricant, the most important factor to be considered is the performance. A quality lubricant formulated according to recognized standard performance criteria from a major manufacturer will confer the highest level of environmental benefit possible by conferring long life due to reduced wear, lowest possible emissions due to reduced friction and maximum protection from corrosive attack due to correct formulation and lowest possible losses due to low consumption in use and compatibility with seal materials, reducing leaks.

10.3 Components

As mentioned earlier, a performance lubricant requires the presence of additive chemicals in order to enable it to work effectively. A typical lubricant therefore consists of a base fluid in which are dissolved a number of different chemicals, each performing a unique function. The additive chemicals are generally more expensive than the base fluid, so from a business point of view it is important to formulate the lubricant with sufficient additives to achieve the desired performance – but no more. Environmentally, the principle 'the least is best' applies. In a perfect world, no lubricant would ever reach the environment except in those instances where the design makes such losses inevitable, e.g. greases on railway points systems, chain bar lubricants on chain saws and any two-stroke engine oil such as those used in outboard motors, motor cycles etc.

10.4 Base fluids

The choice of a base fluid for a lubricant is dependent upon the desired characteristics of the final product. I shall not consider this aspect of performance in this section. If further information is required, several excellent reference works are available that will assist the reader in understanding the performance aspects of base fluids (Prince, 1992; Randles *et al.*, 1992).

From an environmental point of view, base fluids can be split into three categories: mineral oils, synthetic oils such as poly-α-olefins and esters, and highly refined and hydrocracked mineral oils, which some people regard as synthetic whilst others consider to be of natural origin depending perhaps on whether one is a buyer or a seller!).

10.5 Mineral oils

Mineral oils are produced to a performance specification from the refining of crude oil. They vary in viscosity and composition, the major constituents being *n*-alkanes, isoalkanes, cycloalkanes (also known as naphthenics) and aromatics. The molecular weight distribution of the various components largely determines the performance characteristics (Prince, 1992). Analytical determination of the individual hydrocarbon components in a base oil is not a simple procedure (Bennet, Girling and Bounds, 1990; Gough and Rowland, 1990; Betton, 1994). Gas chromatography in conjunction with mass spectrometry (GC–MS) is used to demonstrate the components of an oil. Typical CG–MS traces for two formulated oil products are shown in Fig. 10.1 (Bennet, Girling and Bounds, 1990). The oils themselves are characterized by an unresolved hump (unresolved complex mixture or UCM) starting at 20 min retention time (Gough and Rowland, 1990). This reveals that a typical base oil consists in part of monoalkyl and T-branched alkanes; there are 536 possible acyclic T-branched alkane structures with carbon numbers between C_{20} and C_{30} (Gough and Rowland, 1990). Of this complex mixture of hydrocarbons, only a small proportion are water soluble. The presence of ring components (naphthenics and aromatics) further increases the complexity of the base oil mix. Comparison of the traces of the whole product shown in Fig. 10.1 with those for the water-soluble fraction of the same oil shows that the identities of the majority of the water-soluble components are attributable to the oil additives used to enhance the base oil properties (i.e. imido-succinates, sulphur compounds, methaacrylates). Base oil hydrocarbon components in the aqueous phase are present only in minor quantities – the additive components seen in the aqueous phase are not even noticeable when looking at the traces for whole product (Bennet, Girling and Bounds 1990).

The environmental effects of the base oil depend on two factors, the toxicity and the degradability. Toxicity is dependent on the availability of the material and, as we have seen, only a small proportion of the total mix is water soluble. Soluble components tend to be the lower molecular weight hydrocarbons; the higher the molecular weight, however, the higher is the acute toxicity (Coleman *et al.*, 1984). The outcome of this paradox is that mineral base oils have low acute toxicity to aquatic

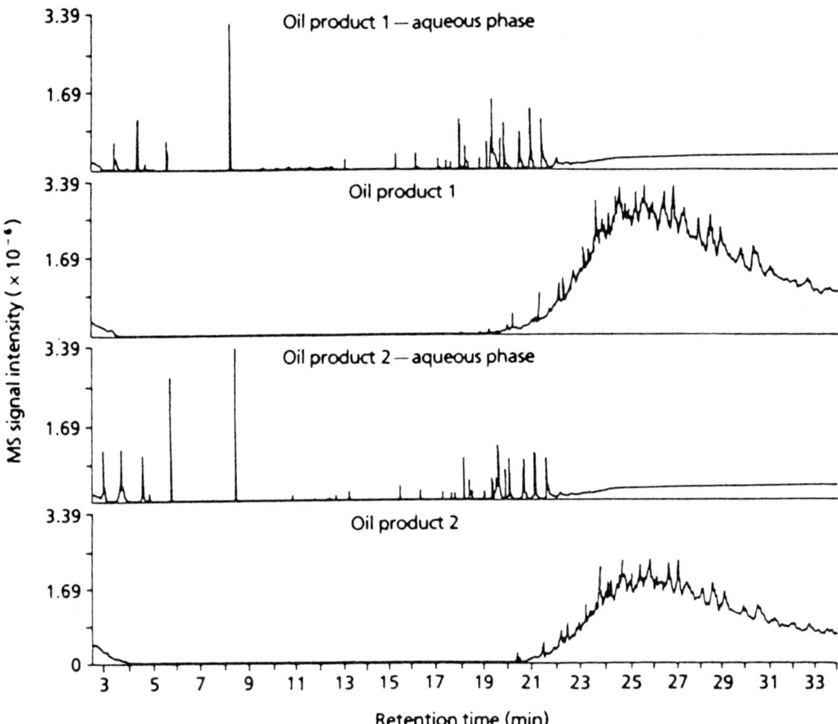

Figure 10.1 GC–MS trace for two formulated oil products and their aqueous phases at equilibrium. (Reproduced from Bennet, D., Girling, A.E. and Bounds, A., *Chemosphere*, 1990, **21**, 659–69).

organisms (BenKinney *et al.*, 1991; Barbieri, BenKinney and Naro, 1993; Betton, 1994). However, the rate of degradability has a capacity to affect the environmental impact of the base oil.

Bacteria can utilize various molecules as an energy source or by incorporating them into new bacteria – biomass. This process is known as biodegradation. In general, the more linear a molecule is then the easier it is for bacteria to make use of it. As we have seen earlier, the various molecular structures present in mineral base oils are branched and ring structures with only a small proportion of linear hydrocarbons. For this reason, mineral base oils are regarded as poorly degradable. Such materials do ultimately degrade, however, as has been dramatically shown by the various catastrophic spills of crude oil that have occurred in Alaska and the coasts around Europe and the Middle East (Benyon and Cowell, 1974; Clark, 1982; Green and Trett, 1989; Pritchard and Costa, 1991).

If, however, it is desired to improve the environmental performance of the base fluid, then alternatives to mineral oil are required. Fortunately, bearing in mind our prime concern regarding product performance, the alternatives to mineral oil are also 'better' performers technologically.

10.6 Synthetic base oils

There are two principle types of synthetic base fluids in use, polyol esters and poly-α-olefins (PAOs). A full description of the technology of these materials is given elsewhere (Mortier and Orszulik, 1992). The advantages, environmentally of these materials are described below.

10.6.1 Polyol esters

There are three main types of polyol ester used in lubricants (Randles *et al.*, 1992):

- pentaerythritol esters — $C(CH_2OCOR)_4$
- trimethylolpropane esters — $CH_3CH_2C(CH_2OCOR)_3$
- neopentyl glycol esters — $(CH_3)_2C(CH_2OCOR)_2$

These materials are structurally very similar to the naturally occurring glycerides (fatty acid esters of glycerol) found in living systems (Edwards and Hassall, 1971; Cain, 1990). As bacteria have evolved systems capable of metabolizing glycerides, they are readily able to make any small biochemical adjustments necessary to utilize the polyol esters as sources of energy or as anabolic feedstocks. In consequence, the polyol esters are generally readily biodegradable (Cain, 1990; Battersby, Pack and Watkinson, 1992).

10.6.2 Poly-α-olefins

Poly-α-olefins (PAOs) are generally hydrogenated oligomers of an α-olefin, usually α-decene. Full details of production techniques can be found elsewhere (Randles *et al.*, 1992). For the purposes of the current discussion, it is important to recognize that PAOs are produced to meet viscosity requirements, and are classed according to the viscosity at 100°C, i.e. PAO 2, PAO 4, PAO 6, etc. The larger and more complex the structure of the molecule, the higher is the viscosity and the lower the degree of biodegradability.

10.7 Hydrocracked mineral oils

As mentioned earlier, mineral oils consist of a mixture of different classes of hydrocarbon. Further treatment of the 'standard' base oil by a combination of high pressure, hydrogen and passing over a catalyst (for details see Prince, 1992) causes the following changes in composition:

- hydrogenation of aromatics and other unsaturated molecules;
- ring opening, especially of multi-ring molecules;
- cracking to lower molecular weight products;
- isomerization of alkanes and alkyl side-chains;
- desulphurization;
- denitrogenation;
- reorganization of reactive intermediates, e.g. to form traces of stable polycyclic aromatics.

The environmental consequence of these changes is to increase the proportion of molecules present that are able to be utilized by bacteria, hence these products tend to exhibit excellent biodegradability (Cain, 1990).

10.8 Additives

The purpose of an additive in a lubricant is to impart those properties to the lubricant that are essential to its function but are not present in the base fluid. In consequence, the number and type of additives required to make any lubricant effective is dependent on the base fluid and end use of the lubricant. For details of the function of additives, see Mortier and Orszulik (1992).

Additive chemistry is a combination of high technology and alchemy (to the outsider!) and the additive companies spend much time, effort and money developing additives to meet the requirements of the equipment manufacturers and the lubricant companies. Exact formulations of both individual components and additive packages are jealously guarded commercial secrets. The additive manufacturers under the guise of the Additives Technical Committee (ATC) have, however, published some background information on the effects of additives and their potential environmental impact (Linnett *et al.*, 1996). They have also commissioned research into the toxicity and biodegradability of various component additives in order to generate data required for the classification and labelling of dangerous substances. These data have shown that the component additives have generally low aquatic toxicity when studied using standard test methodology in fish, daphnia and single-celled algae. The most toxic of the components are the zinc-based antiwear/antioxidant additives. These are normally present in a formulated engine oil at approximately 1–2%.

As the virgin oil and neat additive rarely enter the environment, other than as a result of an accidental spill, and bearing in mind the low toxicity of the base oil component, it can be legitimately argued that the composition of an oil does not present a hazard. The benefits that are gained from a quality lubricant in terms of extended engine life, reduced emissions and fuel savings clearly outweigh any minimal detrimental effects that may occur due to the components.

10.9 Actual environmental effects

It is not virgin oil that enters the environment. Used oil either leaking from cars via faulty seals and joints, or via the exhaust, and do-it-yourself oil changes, the proceeds of which are simply dumped on the soil or down the drain, are the major environmental inputs of lubricants. Estimates of the fate of lubricants sold in the European Union were made by the European oil companies' organization CONCAWE (CONCAWE, 1985) and are shown in Table 10.1.

The result of these estimates can be visualized in the centre of the lanes of any motorway as a black coating of oil or as the stains in any car park bay under the area where the engine comes to rest. The volumes of used oil involved are considerable and have been estimated as representing the equivalent of one *Exxon Valdez* per month over the area of the European Union countries (Betton, 1992). What are the consequences of such apparently large-scale inputs?

Research was carried out at the University of Sheffield in the UK on the environmental impact of roadway run-off from the M1 motorway at four separate sites (Maltby *et al.*, 1995a,b). The effects of the lubricants lost from the traffic using the motorway were studied by comparing the biology and chemistry of the receiving water downstream of the run-off entry point with the situation upstream. In this way the only factor affecting the streams was judged to be the run-off from the motorway.

The most striking feature of the studies was the minimal effect on the environment of the run-off. Of the seven streams initially surveyed, only one of those, that with the smallest natural flow of water, showed any

Table 10.1 Estimates of fate of lubricants sold in the EU (CONCAWE, 1985)

	Tonnes per year ($\times 10^3$)	%
Total EU lubricant sales	4500	100
Consumed	2350	50–55
Recycled	700	15
Burnt as fuel	750	17
Unaccounted for	600	13
Poured down drain deliberately	100	2

effect on the biology of the system. This was measured by comparing the number and diversity of animals and plants in the area downstream of the motorway drainage input with the area immediately upstream. In the one affected site, a decrease in diversity was charactized by fewer sensitive species and an increase in those species typically resistant to the effects of pollution. Hydrocarbons characteristic of used engine oil were found in significant quantities in sediment taken from downstream sites, but not from upstream samples. Water samples were not found to contain any significant contamination. Laboratory investigations in which samples of contaminated sediment were extracted and separated into water-soluble (containing metals), aliphatic, naphthenic and aromatic fractions showed that the principal cause of toxicity was in the aromatic fraction. The other hydrocarbon components and the metal-containing portions did not appear to have any significant toxic effect. This is in line with the low toxicity of the components as discussed earlier.

Identification of the actual toxic components has revealed that the polycyclic aromatic hydrocarbons (PAHs) phenanthrene, pyrene and fluoranthene account for up to 76% of the observed toxicity. These particular PAHs have been found not to possess any carcinogenic potential (IARC, 1983).

10.10 Biodegradability

Oils are biodegradable. Accidents such as the *Torrey Canyon*, *Amoco Cadiz*, *Exxon Valdez* and *Braer*, in addition to deliberate pollution such as occurred during the 1991 Gulf War, have led to enormous sums of money being spent on clean-up and scientific investigation not only at the time of the incident but also for prolonged periods where recovery of ecosystems has been followed. Biological activity is primarily responsible for the recovery on both a macro- and micro-scale. Biodegradation by microbes is an essential part of the regenerative process (Cain, 1990; Battersby, Pack and Watkinson, 1992; Betton, 1992; Painter, 1992). It should be remembered that even if it were possible to eliminate all lubricant inputs, the environment would still be subjected to large volumes of oils and hydrocarbon materials from natural seepage. We are after all considering a natural product that has leached into the biosphere for many millennia and species have evolved to deal with long-term, low-level exposure to such chemicals.

When tested in standard OECD tests for ready biodegradability (OECD, 1981), oils do not perform well (Cain, 1990; APAVE, 1992). The standard OECD protocols require either a knowledge of the chemical structure to calculate theoretical values of oxygen uptake or CO_2 evolution or a determination of experimental values for these parameters.

Information on the purity or the relative proportions of major components of the test material is required to interpret the results obtained, especially when the result lies close to the 'pass' level.

Of the five test methods currently recommended by OECD for assessing ready biodegradation, the Sturm test is the one that has gained the most widespread acceptance for examining the biodegradability of oil products. A modified version of the MITI test has also been successfully applied. In addition, the Co-ordinating European Council for the Development of Performance Tests for Lubricants and Engine Fuels (CEC) has published a test method, Biodegradability of Two-Stroke Cycle Outboard Engine Oils in Water (CEC, 1993), which has been widely used in Europe by both industry and contract test houses for all types of oil products and poorly soluble hydrocarbons (Cain, 1990; Betton, 1992). This method, however, relies on the use of Freon, a substance whose manufacture is no longer permitted under the terms of the Montreal Protocol on ozone-depeleting substances. The life span of the CEC test is therefore severely limited and new methods are under development within the oil industry (CONCAWE, 1993). For a detailed discussion of the biodegradation of oils, see Cain (1990) and Betton (1992, 1994).

Although it is not appropriate to concentrate here on the mechanism of biodegradation and its relationship to lubricants, it is important to consider the fundamental question of whether the environmentally desirable characteristic discussed in the Introduction need or indeed should be applied to lubricants.

The question of whether biodegradability is a desirable characteristic in a lubricant has been the subject of much, often heated, debate among product developers for many years. In the following paragraphs an attempt is made to highlight some of the areas of concern that have been raised and to give reasons why on balance biodegradability is desirable, always providing, of course, that performance is not compromised.

10.10.1 Biodegradation is not necessary in a lubricant

As shown earlier, a large proportion of the lubricant that is sold is 'lost' and unaccounted for. Lubricant *is* deliberately dumped into the environment. This total environmental burden does degrade, albeit at a relatively slow rate. Lubricants specifically designed to be more readily degradable will be less likely to foul the environment via leaks, spills or deliberate dumping.

10.10.2 A biodegradable lubricant will encourage dumping at the expense of collection and disposal

It is fundamental that environmental benefits should be in addition to performance, as was discussed earlier. If that is the case in a product it is probable that a biodegradable lubricant will be at the upper end of the price range. Individuals who specify such lubricants and who are prepared to pay for them are not the type of people who will deliberately dump oil. It is uninformed and socially unconcerned people and those who buy the cheapest product in a chain store who are likely to be involved in dumping.

10.10.3 A biodegradable lubricant will degrade in the engine

Biodegradability depends on bacteria to do the degrading. The environment of a motor car engine, with its extremes of temperature and pressure, is not conducive to the maintenance of bacterial life. In addition, bacteria tend to live in water and not oil; it is only when dealing with emulsions and water contamination that conditions conducive to bacterial growth and degradation of lubricants can occur.

10.10.4 A biodegradable lubricant will result in high concentrations of toxic residues that are detrimental to the environment

As we have seen, the additive components of oils are not particularly toxic, and degradation of the base oil will not leave a toxic residue. The work on road run-off has shown that it is PAHs, formed during combustion and deposited in the lubricant, that are responsible for the small degree of toxicity found. These materials are also those with the simplest structure and possess some degradative potential, in addition to which they are subject to degradation by ultraviolet light. It should also be remembered that PAHs are naturally occurring products of combustion and have been present in the environment for as long as there have been fires. Systems have evolved to cope with these materials.

10.10.5 Biodegradation is not necessary, as motor manufacturers are now producing sealed lubricant systems

A motor car may well not leak oil for the first few years of its life, although as examination of the Chief Executive's parking space will eloquently demonstrate, this is not always the case! Motor cars are now lasting for much longer periods and it is apparent that there are and will always be a large proportion of the cars on the road that leak oil to the environment. A biodegradable lubricant would be effective in minimizing the effects of those losses.

10.11 Collection and recycling of used oils

The recycling of used lubricants has been practised to various degrees since the 1930s and particularly during the Second World War when the scarcity of adequate supplies of crude oil during the conflict encouraged the reuse of all types of materials, including lubricants. Environmental considerations regarding the conservation of resources and sundry 'oil crises' have maintained interest in the concept of recycling up to the present day. A recent review (CONCAWE, 1996) has examined the environmental costs and potential benefits of the whole issue of collection and disposal of used oil in great detail.

It is essential to recognize that all used oils should be collected for controlled disposal. Some products, such as transformer oils and hydraulic oils, can be readily collected from large industrial concerns, regenerated to a recognized standard and returned to the original source.

Oils from automotive sources will include mono- and multi-grade crankcase oils from petrol and diesel engines, together with gear oils and transmission fluids. Used industrial lubricants that have been inadequately segregated may also be included. Apart from any degradation products from the in-service use of the oil, a wide range of contamination is possible, including the following:

- water – combustion by-product, rainwater/salt water ingress;
- fuels – residual components of gasoline and diesel fuel;
- solids – soot, additive and wear metals together with rust, dirt, etc.;
- chemicals – used oil can be used as an unauthorized means of hazardous waste disposal;
- industrial oils – inadequate segregation of oil types can allow contamination by fatty or naphthenic products.

Provided that efficient management systems are in place, many industrial oils should be largely contained and not escape into the environment. There are many potential sources of used industrial products; however, reprocessing is not an option for a large number of these products which are synthetic and fatty-oil based. Some specific types of industrial oils are suitable for relatively simple reprocessing before being returned to their original service. Typical processing methods involve filtration and removal of water and volatile decomposition products under vacuum, and can sometimes be carried out at the plant using mobile equipment.

Legislation around the globe is increasingly controlling the collection and disposal of all waste materials, including lubricants. Large-scale (greater than 3 m^3) waste oil collection vessels at service stations must now be licensed in the UK, as must any company that transports or treats waste lubricants.

The method of disposal that is utilized will be dependent on many different factors, however. Availability of appropriate treatment facilities, raw materials, type of product being collected, levels of contamination and so on will all affect which is the most appropriate disposal route. A full life-cycle analysis of each situation would be required before a definitive choice of disposal option could be made. However, such analyses are complex, time consuming and necessarily subjective. The following used oil disposal routes are considered to offer the 'best environmental option':

- re-refining to base oil using modern technology to reduce PAH concentrations to acceptable levels (e.g. using severe hydrotreatment or solvent extraction);
- reprocessing to industrial fuel using modern technology (e.g. Trail-blazer process);
- recycling through a refinery as a low-sulphur fuel oil blendstock;
- direct burning as fuel in cement kilns;
- burning after mild processing in road stone coating plants (care must be taken to ensure that emissions of chlorine containing components do not exceed acceptable limits);
- gasification to produce fuel gas or petrochemical feedstock.

The following disposal routes are considered to involve unacceptable levels of pollution due to emissions to soil, air or watercourses:

- direct burning in space heaters (emissions of heavy metals and other products of low-level combustion causing localized pollution);
- re-refining using acid/clay and other old technologies (the majority of plants currently in operation) producing acid tars and oiled clay requiring specialist disposal;
- road oiling (high risk of groundwater contamination).

The economics of environmentally acceptable used oil disposal will be dependent upon the availability of local facilities. None of the environmentally acceptable methods listed above are considered to be financially self-sufficient without the application of some form of subsidy. This is largely related to costs of collection and transport. For a full discussion of the economic arguments, see CONCAWE (1996).

10.12 Conclusion

Environmental technology as applied to lubricants is related first and foremost to performance. The benefits to be gained from reduced wear and friction are substantial and in general outweigh all other aspects. There are environmental aspects of a lubricants performance, however, that if addressed could reduce what is after all a surprisingly small impact due to the inevitable losses that occur during use.

The perfect 'environmental' lubricant that was outlined in the Introduction does not, nor will it probably ever, exist. It is hoped that it is now apparent, however, that some of the ideal characteristics are not needed, and that those that are desirable are so for reasons of aesthetics and a desire to keep the environment clean, rather than for a compelling need to reduce toxicity and impact on ecosystems.

References

APAVE (1992) NF-Environment. Huiles et lubrificants moteurs. Projet de reglement technique. Rapporteur Michel Genesco. APAVE, Paris.

Barbieri, J.F., BenKinney, M.T. and Naro, P.A. (1993) Acute and chronic toxicity of petroleum base stocks to aquatic organisms. *Proceedings of Society of Environmental Toxicology and Chemistry (SETAC) Ecological Risk Assesment: Lessons Learned*, Houston, TX.

Battersby, N.S., Pack, S.E. and Watkinson, R.J. (1992) A correlation between the biodegradability of oil products in the CEC L-33-T-82 and Modified Sturm tests. *Chemosphere*, **24**, 1989–2000.

BenKinney, M.T., Novick, N.J., Cross, J.S. and Naro, P.A. (1991) Environmental characteristics of petroleum products and base stocks: an ecology assessment. *Proceedings of Society of Environmental Toxicology and Chemistry (SETAC) Conference on Environmental Interfaces: Scientific and Socioeconomic*, Seattle, WA.

Bennet, D., Girling, A.E. and Bounds, A. (1990) Ecotoxicology of oil products: Preparation and characterisation of aqueous test media. *Chemosphere*, **21**, 659–69.

Benyon, L.R. and Cowell, E.B. (eds) (1974) *Ecological Aspects of Toxicity Testing of Oils and Dispersants* Applied Science, Barking.

Betton, C.I., (1992) Lubricants and their environmental impact, in *Chemistry and Technology of Lubricants* (eds R.M. Mortier and S.T. Orszulik), Blackie, Glasgow, pp. 282–98.

Betton, C.I., (1994) Oils and hydrocarbons, in *Handbook of Ecotoxicology* (ed. P. Calow), Blackwell, Oxford, pp. 244–63.

Cain, R.B. (1990) Biodegradation of lubricants, in *Proceedings of 8th International Biodegradation and Biodeterioration Symposium, Windsor, Ontario, Canada, 25–31 August 1990* (ed. H.W. Rossmoore), pp. 249–75.

CEC (1993) Biodegradability of two-stroke outboard engine oils in water. *Report CEC L-33-A-93*, Co-ordinating European Council for the Development of Performance Tests for Lubricants and Engine Oils, London.

Clark, R.B. (ed.) (1982) *The Long Term Effects of Oil Pollution on Marine Populations, Communities and Ecosystems*, Royal Society, London.

Coleman, W.E., Munch, J.W., Streicher, R.P., Ringhand, H.P. and Koffer, F.C. (1984). The identification and measurement of components in gasoline, kerosine and no. 2 fuel oil that partition into the aqueous phase after mixing. *Archives of Environmental Contamination and Toxicology*, **13**, 171–8.

CONCAWE (1985) The collection, disposal and regeneration of waste oils and related materials. *CONCAWE Report No. 85/53*. CONCAWE, Brussels.

CONCAWE (1993) *Biodegradation Task Force on Method Development. Committee Research Programme*, CONCAWE, Brussels.

CONCAWE (1996) Collection and disposal of used oil, *Special Task Force WQ/STF 26 Report*, CONCAWE Brussels.

Edwards, N.A. and Hassall, K.A. (1971) *Cellular Biochemistry and Physiology*, McGraw-Hill, London.

Gough, M.A., and Rowland, S.J. (1990) Characteristation of unresolved complex mixtures of hydrocarbons in petroleum. *Nature (London)*, **344**, 648–50.

Green, J. and Trett, M.W. (eds) (1989) *The Fate and Effects of Oil in Fresh Water*, Elsevier Applied Science, Barking.

Howell, J.K., Lucke, W.E. and Steigerwald, J.C. (1996) Metalworking fluids: composition and use, in *The Industrial Metalworking Environment – Assessment and Control.*

Proceedings of the Symposium at Dearborn. Michigan, USA, November 13–16, 1995, American Automobile Manufacturers Association, Washington, DC, 13–22.

Institute of Petroleum (1992). *UK Petroleum Industry Statistics: Consumption and Refinery Production 1990 and 1991*. Institute of Petroleum, London

IARC (1983). *IARC Monographs on the Evaluation of the Carcinogenic Risk of Chemicals to Humans. Volume 32: Polynuclear Aromatic Hydrocarbons*, International Agency for Research on Cancer, Lyon.

Linnett, S.L., Barth, M.L., Blackmon, J.P. *et al.* (1996) Aquatic toxicity of lubricant additive components. *Proceedings of the Esslingen Tribology Colloquium*, 9–11 January, Vol. 2, 871–82.

Maltby, L., Forrow, D.M., Boxall, A.B.A. *et al.* (1995a) The effects of motorway runoff on freshwater ecosystems: 1. Field study. *Environmental Toxicology and Chemistry*, **14**, 1079–92.

Maltby, L., Boxall, A.B.A., Forrow, D.M. *et al.* (1995b) The effects of motorway runoff on freshwater ecosystems: 2. Identifying major toxicants. *Environmental Toxicology and Chemistry*, **14**, 1093–102.

Mortier, R.M. and Orszulik, S.T. (eds) (1992) *Chemistry and Technology of Lubricants*, Blackie, Glasgow.

OECD (1981) *Guidelines for Testing of Chemicals. Section 2: Effects of Biotic Systems*, OECD, Paris.

Painter (1992) *Detailed Review of Literature on Biodegradability Testing*, OECD, Paris.

Prince, R.J. (1992) Base oils from petroleum, in *Chemistry and Technology of Lubricants* (eds R.M. Mortier and S.T. Orszulik) Blackie, Glasgow, pp. 1–31.

Pritchard, P.H. and Costa, C.F. (1991) EPA's Alaska oil spill bioremediation project. *Environmental Science and Technology*, **25**, 372–9.

Randles, S.J., Stroud, P.M., Mortier, R.M. *et al.* (1992). Synthetic base fluids, in *Chemistry and Technology of Lubricants* (eds R.M. Mortier and S.T. Orszulik), Blackie, Glasgow, pp. 32–61.

SAE (1996) SAE Information Report: International Tests and Specifications for Automotive Engine Oils – SAE J2227, August 1995, *SAE Handbook 1996*, Society of Automotive Engineers, Warrendale, PA, pp. 1210–5.

11 Long-term environmental concerns and their implications for the oil industry

P. CALOW

11.1 Some obvious forecasts

The oil industry and its products impinge on all aspects of the environment: the seas from its extraction and transportation; the soils from the impacts of extraction, pipelines, processing and use; and the atmosphere from its gaseous products. It is implicated in 'high profile' pollution: accidents and ecosabotage at wells and refineries; decommissioning of marine installations; major accidents and spills at sea; the contribution of its gaseous products to the greenhouse effect and to respiratory diseases. So no crystal ball is needed to make the first forecast: that pressure on the industry from environmental issues will not diminish over either the short or long term. Indeed, the oil industry is likely to come under increasing pressure for increasing attention to be given to increasingly effective environmental protection technology.

At the same time, as with the chemical industry, there is a social dilemma: an ever increasing emphasis on preventing the environmental ills of the industry despite an ever increasing social appetite for its output in terms of fuel, derivative chemicals and a myriad of other benefits that we obtain from it. There are also no less important benefits to employment and the economy, not only in the producer states but also in others that have products and services that rely upon it. The second obvious forecast, therefore, is that restrictions on production, transportation, storage, marketing, use and disposal are all going to have to balance the benefits of measures taken for environmental protection against the costs to the economy, if not explicitly certainly implicitly.

It is easy to hope that we can have industrial output without having to pay environmental costs, or alternatively to impose restrictions on industrial output without social costs. Exploring these hypotheses is at the heart of the sustainable development debate. To do it effectively will require that the cost–benefit analyses referred to above are in fact made increasingly explicit and transparent, and this is likely to lead to two further trends: on the one hand increasing attempts to internalize environmental costs and on the other to make plain, in a way that is obvious to the consumer, the source of the products from which benefits are being enjoyed.

There is one final general forecast, namely that the oil industry and all that depend upon it are fundamentally unsustainable. The resource is finite and, except on a time-scale that is so long as to be irrelevant, effectively non-recyclable. So, despite upward revision of estimates of world stocks (see below), sooner or later they will run out. For the sake of a continuing civilization there has to be pressure on technology to delay this, through increasing efficiency, but there will be more and more urgent pressure to come to terms with alternative energy sources.

This final chapter aims to use these general forecasts as a basis for making more specific ones about challenges from environmental issues and opportunities for technology. Much of this will be driven by developing legislation, in particular from the principles of policy on which it is based and from how it is implemented, and these are addressed in the following two sections. Finally, the crystal ball clouds in addressing the issue that will have most impact on the industry and the environmental consequences that emanate from it: demand based on projections of global population size and its distribution between developed, developing and transitional economies. All these issues have global, national and local perspectives and another challenge will be in applying appropriate measures at appropriate levels, and also in ensuring effective coordination between them. It will be obvious that all major players – industry, government and regulators, scientists and technologists and consumers – have a role to play, but increasingly consumer pressure will be galvanized to encourage more environmentally friendly approaches in industry. Hence paying attention to the perceptions of the public – consumer concerns – will become increasingly important.

11.2 Principles of policy that are likely to shape the immediate future

Environmental issues are therefore likely to present continuing and increasing pressures in the development of the oil industry and its technology at all points in the chain from exploration, to extraction, transport, treatment, use and disposal. The issues are in fact crystallizing around a small number of major policy principles that will affect thinking and legislation at all levels from global to local in the forseeable future. Chief amongst these are sustainability, precaution, prevention and polluter pays. Although there is much cross-referencing, and sometimes even contradiction, between them, it is convenient to address each separately.

11.2.1 Sustainable development

Brought into focus by Brundtland (World Commission on Environment and Development, 1987), this holds centre stage following the UN Earth

Summit in Rio de Janeiro in 1992 and because of its assimilation into both international and national policy and legal instruments, is likely to continue to do so. Its appeal is that it implies the possibility of finding development 'that meets the needs of the present generation without compromising the ability of future generations to meet their own needs'. That this should be so is not self-evident but, as already discussed above, it encourages making explicit the complex interactions between industrial development, wealth creation, social welfare and environment.

Industrial development certainly has negative effects on environmental resources, but this ought to be less of a concern for renewable ones. At the same time, industrial development brings social, welfare benefits that can have positive effects on the environment by facilitating increasing social awareness and providing a spur to technological progress that is environmentally friendlier. There are various philosophical positions that can be taken with regard to if and how these interactions can be optimized, and that can be characterized in various shades of green (O'Riordan, 1991). However, the pragmatic consequence is that at all scales industries and technologies will be favoured that are based on cyclical rather than linear processes, renewable rather than non-renewable inputs and that maximize efficiency and minimize waste, especially that likely to lead to environmental harm.

The point has already been made that in these terms the oil industry is fundamentally unsustainable. Estimates of proven recoverable reserves have undergone several upward revisions in the past 20 years but the stocks are still finite and the second law of thermodynamics ensures irreversibility and hence linearity of most of the processes into which the reserves are directed. The pressures on technology, therefore, will be to (a) enable exploitation of more difficult and non-conventional reserves such as oil shales and tar sands, (b) improve the efficiency of all processes in the chain from extraction to use, but especially in the latter, and (c) enable use of alternative and especially renewable energy sources.

There is also the question of the extent to which the release of waste from the use of oil products is sustainable. The material itself and its breakdown products are clearly natural and will enter into the natural cycles of carbon, so that in principle this is sustainable. In practice, of course, the natural cycles are being overloaded with excessive release of oil and derivatives. There will, therefore, be a challenge for technology to counteract these tendencies, and also to provide remedies for the harm caused. One caution here, though, is that given the natural tendency of oil and derivatives to be subject to degradation by normal environmental processes, it is important that remediation should not be allowed to lead to more problems than it causes, as might be the case with the use of detergents or bioengineered microbes in cleaning up oil spills.

Of course, some of the reaction products from oil raw materials created by and used in the chemicals industry are novel and will not be easily subject to natural processing. Here risk assessment procedures are becoming increasingly important in determining the need for restrictions on production, marketing, use and disposal of existing products (e.g. Calow, 1995) and even in the design of new products. Risk assessment is likely to become increasingly important as a tool in environmental protection (see below). However, equally, the public perception of risk is likely to be as important as the science behind the assessment, as exemplified by the *Brent Spar* incident (Briefing, 1996) and the bovine spongiform encephalopathy (BSE)/Creutzfeldt–Jakob disease (CJD) issue in the UK (Editorial, 1996a).

Finally, the point must not be lost that there is a positive link between industrial output and social welfare that can create the conditions needed for societies taking environmental protection increasingly seriously and, through research and development, stimulating technological innovation that itself can be environmentally friendlier. The relationship between economic growth and energy consumption is, however, complex, with national variations in life style and climatic conditions having an important influence. Thus, GNP per capita is currently greater in Japan than the USA, but energy consumption per capita in Japan is approximately half that in the USA (Tolba and El-Kholy, 1992). As we come to recognize and understand these international differences there will be a tendency to emulate economies, social systems and technologies that reduce energy intensity (energy consumption/GDP).

11.2.2 Precautionary principle

Deriving from the German *Vorsorgerprinzip*, the precautionary principle advocates controlling action even when the links between an adverse environmental effect and its cause are not clear (O'Riordan and Cameron, 1994). There is therefore an obvious tension between this approach and that based upon a scientific assessment of risk of a process, or material, causing problems, and basing management procedures on attempts to reduce, not necessarily remove, risks. The tension is perpetuated in the legislation when, for example, in the EC Treaty following Maastricht there is under article 130 a statement that Community environmental policy '*shall* be based on the precautionary principle' (author's italics) and only '*take into account*' available scientific data (Krämer, 1995).

Risk assessment can, of course, involve caution. One example of this is in incorporating large safety margins into the procedure (Calow, 1995). Another actually involves technology. Thus, risk assessments can be used to establish minimum or maximum levels that have to be achieved by industry; but over and above this, as a precautionary measure, there may be a requirement to use the best available technology and techniques

(BAT). For example, this is an important part of integrated pollution control legislation required in the UK Environmental Protection Act 1990 and is likely to be an important feature of integrated, pollution prevention and control legislation in the pipeline in the European Union.

Encouragement of BAT, under the precautionary umbrella, will undoubtedly be a continuing theme intended as a ratchet for continuous improvement, but the law of diminishing returns will also inevitably impose itself and, through the drive for cost effectiveness, is likely to provide some kind of brake on the ratchet. 'Not entailing excessive cost' (NEEC) is included as a counterpoint to BAT in the UK legislation. There are also arguments that can be made in the name of sustainable development for this (see above). If we have limited resources for environmental protection, which seems likely but is not accepted by all, excessive use in one area will lead to less for others, overall leading to less effective environmental protection. The message, therefore, for technological development will be that innovation will inevitably have to be tempered by cost effectiveness. Yet the precautionary principle and BAT will remain as important slogans for the pressure groups, and in certain circumstances will become all-important as the public perception of risk becomes prominent. Another message, therefore, is that in planning action within the industry, at whatever level, whether in terms of new development, decommissioning or remediation, whether by technological or other means, effective methods will have to be developed for sounding out public opinion and taking it into account.

11.2.3 Prevention

Prevention rather than cure was one of the first principles to be listed in the European Commission Environmental Action Programmes (Haigh, 1995), and now, as the US Pollution Prevention Act 1990, is a framework instrument in the USA that requires the Environmental Protection Agency to consider prevention as a guiding principle in all that it does. Prevention is often seen as the bedfellow of precaution, but it need not be. Prevention can equally be risk driven. The point is that to prevent does not necessarily mean to prevent release and contamination, but to prevent effect. It will become increasingly important for risk assessment to address this issue and to demonstrate effectively, to the satisfaction of science and society, that there are levels of contamination that can be assimilated by ecological systems without appreciably increasing the chance of ecological harm. This, in turn, will require increasing attention to be given to defining what is meant by ecological harm (Calow, 1994).

Another effect of the prevention philosophy, though, is that it will encourage going beyond end-of-pipe solutions to action at source, taking into account the complex nature of industrial processes and systems and the social systems that interact with them. Thus there will undoubtedly be pressure to develop technologies that minimize the release of contaminants such as CO_2 and SO_2 likely to harm the environment, but beyond that there will be efforts to find alternative ways of carrying out the activities that create them and to encourage life-style changes that reduce demand for these activities.

There is an attempt in the USA, in the EU through the Fifth Environmental Action Programme (EC, 1992a) and globally through Agenda 21 (UNCED, 1992), to encourage 'shared responsibility' (a concept brought into focus in the EU Action Programme) between all major players – industry, scientists and technologists, and consumers – in achieving increased prevention. Those involved at all points in the oil industry will need to participate actively in this process; and recognize the opportunities as well as the constraints that are likely to emerge from it.

11.2.4 Polluter pays

Again, this is a policy principle that has been influential for some time. First made explicit by the OECD in a statement of principle in the mid-1970s (OECD, 1975), it has subsequently been confirmed in a number of international and national policy documents. Of course, it leads ultimately through cost transfers to the effects that consumers and users pay. However, this is not only anticipated but encouraged, providing price signals in the market place that can influence consumer choice. We shall return to this below.

An important challenge for full implementation of the principle is that all costs should be internalized – including not only the charges for administering pollution prevention, fines and clean-up costs associated with breaches of regulation, but also the values attached by society to environmental resources that are used up in production either as raw materials or in consequence of using the environment as a repository for wastes. Much attention is being given to this valuation either by reference to the trading of ecological entities in real market places or the interpretation of consumer preference by game playing in hypothetical market places (e.g. Pearce, 1993). In that it is an attempt to make explicit – internalize – environmental costs, it is compatible with the development of criteria for testing the sustainable development hypothesis (see above). above).

Apart from the way in which these explicit valuations are going to influence the industry, by influencing consumer choice (see below), there are other consequences for technological developments within the industry. Increasingly, and for reasons already made clear, taking costs and benefits into account will become prominent in developing and designing systems. The costs can be computed relatively easily for research, development, construction, operation and decommissioning, but these technological costs will have to be set alongside similarly quantified effects from the system on the environment. The results of this cost–benefit analysis will depend as much on the way the environmental effects are quantified and valued as on the way the technological costs are calculated, and so it will be important for all interested parties, not just the economists, to keep a critical overview of all the elements in this exercise.

11.3 Influences of instruments of control

Most of the environmental protection measures of the past have been effected through command and control instruments: prohibitions and restrictions of various kinds. Recently, more and more attention has been given to the use of market instruments that get signals into the market place either by labels and registers or prices to influence consumer choice and to use that in turn, by promoting action at source, to favour more environmentally friendly products and processes (RECIEL, 1995). In the EU, there are instruments available for both negative labelling (identifying hazards, through Directive 67/548 on packaging and labelling of dangerous substances) and positive labelling (through the ecolabel Regulation 880/92). There are also challenges both for carrying out effective hazard identification on complex oil products (Betton, 1994) as a basis for negative labelling and for developing effective life-cycle assessment procedures that can be used as a basis for positive labelling and product design (see above).

Economic instruments have also been applied widely to oil materials and their breakdown products. Thus differential taxes have been applied effectively to control the use of leaded versus unleaded petroleum in the UK and EU; a carbon/energy tax on fuels is being contemplated in the EU and has already been implemented in some Member States (Denmark, Finland, Sweden); the trading of permits to emit polluting gases within predetermined limits has been used in North America. Escrow accounts involving bonds against possible accidents and the subsequent costs of clean-up have been contemplated (Costanza and Perrings, 1990), and could become of interest to cover the consequences of spills during the transportation of oil, leaks from pipelines, major accidents at refineries and decommissioning of installations whether for extracting or processing.

Voluntary programmes are also becoming more widely considered and used, e.g. in taking responsibility for covering the costs of clean-up and remediation in the event of major accidents (Ong, Chaper 2). There is also the possibility of a voluntary agreement with automobile manufacturers to improve fuel efficiency that is being explored within the EU (ENDS, 1996). The advantage of voluntary programmes is that having established required targets, the flexibility given in achieving them encourages innovation and enables industry to find solutions that are most compatible with its own needs. The burdens of establishing, implementing and monitoring are removed from the regulators. Of course, the difficulty is with credibility and the possibility of cheating. Therefore, hand-in-hand with the development of these voluntary programmes will go a need for rigorous and credible surveillance systems that can be operated routinely, cost effectively and possibly remotely. This again raises opportunities and challenges for technology.

11.4 Where the emphases will be

The oil industry consists of an interrelated chain of processes and materials from well and platform to distribution, processing and use. Managing one part of the chain with respect to its environmental effects is bound to have consequences for others so that it will be important to deal with it holistically, i.e. with integrated assessment, management and control procedures, a point made repeatedly through the preceding chapters.

Yet certain parts of the chain have particular prominence either because of public perception and/or their quantitative involvement in the use of and outputs from the oil industry. Thus, irrespective of their importance in terms of a long-term and global perspective, major accidents at sea will continue to command public attention and will retain prominence. Therefore, there will be continuing pressure to develop appropriate risk assessment procedures that can capture the likelihood of unlikely happenings in uncertain conditions and circumstances at sea (Chapters 2, 5 and 9). Similarly, it will be important to develop appropriate, cost-effective management responses, but these need not always involve technological innovation. Thus, the *Sea Empress* disaster off the coast of Wales in February 1996 prompted calls for a requirement that all tankers should have double hulls. However, double hulls may not be the solution. In fact, they probably would have made little difference to the *Sea Empress* disaster or indeed to the *Braer* wrecked off Shetland in 1993. It would be more cost effective instead to tackle root causes – poor maintenance, inadequate navigational skills and language difficulties (Editorial, 1996b). These are issues for governments at both national and international levels. However, again, the public perception of risk and of how the risk should be managed will be important (see above), so finding ways of gauging this and taking it into account in decision making will be crucial.

However, in terms of the shear quantities of consumption and outputs, transport is where the emphasis has to be in the oil industry chain. It accounts for ca. 30% of commercial energy consumption and 60% of the total global consumption of liquid petroleum. Moreover, emissions from transport represent a very high share of overall emissions. In Europe, they account for ca. 90% of all lead emissions, 50% of all NO_x emissions, 30% of all VOC emissions and 20% of all CO_2 emissions. This is where the focus is and will eventually be, with specific international initiatives under the EU Fifth Environmental Action Programme (EC, 1992a) and Agenda 21 arising out of the UN Conference (UNCED, 1992).

The EU proposes a 'sustainable mobility' strategy (EC, 1992b, 1995) involving a combination of improved development planning at all levels that allow for alternatives to road transport; improved co-ordination in the planning of and investment in transport infrastructure networks and facilities including incorporation of real costs; improvements in the competitive position of environmentally friendly modes, such as railways, inland and sea navigation and combined transport; encouragement of collective transport in the urban setting, continued technical improvement of vehicles and fuels; and the promotion of more environmentally rational use of the private car. The impact of EU legislation is still importantly related to technical improvements of vehicles to enhance efficiency and reduce emissions and noise, although other developments such as in telematics and electronic traffic control are also receiving attention. A recent analysis of the European Environment Agency (EEA, 1995) concludes that 'the contribution of [these] initiatives, particularly their short-term impact, is difficult to observe at present'.

11.5 A cloudy crystal ball

In one sense, many of the issues raised and discussed above are but details in circumstance that are largely driven by the changing demand of an increasing world population that is increasingly developing. To make forecasts about the consequences of these trends for the demand for oil and its environmental effects is fraught with uncertainties about the level of population growth, economic growth, energy intensity and institutional arrangements. Recent estimates of world energy demand to 2020, taking these uncertainties into account, range from a ca. 100% increase to slightly less than a 50% increase over 1990 levels (= 8.7 gigatonnes of oil equivalent) with similar, proportional effects on polluting emissions (World Energy Council, 1993). An important element of these projections, however, will be the extent to which technology evolves to reduce reliance on non-renewable sources and improves efficiency, and the extent to which this can be transferred into developing countries. Ultimately this will depend upon the extent to which governments can create conditions for technological development, transfers and international co-operation. As far as the first is concerned, important elements are likely to be elimination of disincentives for technological development, an argument for ensuring full costs, including those associated with environmental harm, are attached appropriately to energy use, encouragement of research and development by ensuring appropriate support from both government and private sources and encouragement of involvement of appropriate personnel by support of educational and recruitment programmes. Interestingly an imponderable in all this is the extent to which energy-use research is going to benefit or suffer from a reducing investment in military programmes. Ultimately, technology transfer will depend upon the extent to which national governments are going to be open-minded enough to see the virtues for their own benefits of encouraging best practice in developing states, and the extent to which technologies of the developed world are appropriate for developing states and those in transition. Much will depend on the extent to which the momentum of the UN Rio Conference can be maintained and enhanced.

References

Betton, C.I. (1994) Oils and hydrocarbons, in *Handbook of Ecotoxicology*, (ed. P. Calow) Blackwell, Oxford, Vol. 2, pp. 244–63.

Briefing (1996) Risk: a suitable case for analysis? *Nature (London)*, **380**, 10–4.

Calow, P. (1994) Ecotoxicology: what are we trying to protect? *Environmental Toxicology and Chemistry*, **13**, 1549.

Calow, P. (1995) Risk assessment: principles and practice in Europe. *Australian Journal of Ecotoxicology*, **1**, 11–3.

Costanza, R. and Perrings, C. (1990) A flexible assurance bond system for improved environmental management. *Ecological Economics*, **2**, 57–76.

EC (1992a) *Towards Sustainability*. A European Community Programme of Action in Relation to the Environment and Sustainable Development, COM(92)23, Vol. II, European Commission, Brussels.

EC (1992b) *White Paper on the Future Development of the Common Transport Policy: A Global Approach to the Construction of a Community Framework for Sustainable Mobility*, COM(92)494, European Commission, Brussels.

EC (1995) *Common Transport Policy Action Programme, 1995–2000*, COM(95)302 Final, European Commission, Brussels.

Editorial (1996a) Give us the beef about beef. *New Scientist*, **149**(30 March), 3.

Editorial (1996b) On the rocks, *Financial Times*, 21 February.

EEA (1995) *Environment in the European Union. Report for the Review of the Fifth Environmental Action Programme*, EEA, Copenhagen.

ENDS (1996) Lengthy EC negotiations ahead on CO_2 strategy for cars. *ENDS Report*, **252**, 37–8.

Haigh, N. (1995) *Manual of Environmental Policy: the EC and Britain*, Catermill Publishing, London.

Krämer, L. (1995) *EC Treaty and Environmental Law*, 2nd edn, Sweet & Maxwell, London.

O'Riordan, T. (1991) The new environmentalism and sustainable development. *The Science of the Total Environment*, **108**, 5–15.

O'Riordan, T. and Cameron, J. (eds) (1994), *Interpreting the Precautionary Principle*, Cameron & May, London.

OECD (1975) *The Polluter Pays Principle: Definition, Analysis, Implementation*, OECD, Paris.

Pearce, D. (1993) *Economic Values and the Natural World*, Earthscan Publications, London.

RECIEL (1995) Issue on 'Economic Instruments and the Environment'. *Review of European Community and International Environmental Law*, **4**, 287–357.

Tolba, M.T. and El-Kholy, O.A. (eds) (1992) *The World Environment 1972–1992*, Chapman & Hall, London.
UNCED (1992) *Agenda 21*, United Nation Conference on Environment and Development, Conches, Switzerland.
World Commision on Environment and Development (1987) *Our Common Future*, Oxford University Press, Oxford.
World Energy Council (1993) *Energy for Tomorrow's World*, St Martin's Press, New York.

Index

Page numbers appearing in **bold** refer to figures and page numbers appearing in *italic* refer to tables.

enviormental effects of the oil idustry.